Astrophysics Update

Springer
Berlin
Heidelberg
New York
Hong Kong
London
Milan
Paris
Tokyo

John W. Mason (Editor)

Astrophysics Update

With 104 Figures, 8 Tables and 1 Colour Plate

Editor
Dr John W. Mason
Olympus Mons
51 Orchard Way
Barnham
West Sussex, PO22 0HX
UK

The background to the cover design features the first detailed, all-sky picture of the very early universe obtained by the Wilkinson Microwave Anisotropy Probe (WMAP). The WMAP image reveals 13 billion+ year old temperature fluctuations (shown as colour differences) that correspond to the 'seeds that grew to become galaxies'. The centre image on the front cover is a design depicting cosmic history from a tiny fraction of a second after the Big Bang through to the development of the large-scale structures of galaxies we see today, 13.7 billion years later. Both images reproduced courtesy of NASA/WMAP Science Team.

SPRINGER–PRAXIS BOOKS IN ASTROPHYSICS AND ASTRONOMY
SUBJECT *ADVISORY EDITOR*: John W. Mason B.Sc., M.Sc., Ph.D.

ISBN 3-540-40642-5 Springer-Verlag Berlin Heidelberg New York

Library of Congress Cataloging-in-Publication Data

Astrophysics update/ editor, John Mason.
p.cm. – (Springer-Praxis books in astrophysics and astronomy)
Includes bibliographical references.
ISBN 3-540-40642-5 (acid-free paper)
1. Astrophysics   I. Mason, John (John W.)   II. Series

QB461.5.A885 2004
523.01–dc22

Bibliographic information published by Die Deutsche Bibliothek

Die Deutsche Bibliothek lists this publication in the Deutsche Nationalbibliografie; detailed bibliographic data are available from the Internet at http://dnb.ddb.de

Apart from any fair dealing for the purposes of research or private study, or criticism or review, as permitted under the Copyright, Designs and Patents Act 1988, this publication may only be reproduced, stored or transmitted, in any form or by any means, with the prior permission in writing of the publishers, or in the case of reprographic reproduction in accordance with the terms of licences issued by the Copyright Licensing Agency. Enquiries concerning reproduction outside those terms should be sent to the publishers.

© Praxis Publishing Ltd, Chichester, UK, 2004
Printed in Germany

The use of general descriptive names, registered names, trademarks, etc. in this publication does not imply, even in the absence of a specific statement, that such names are exempt from the relevant protective laws and regulations and therefore free for general use.

Typeset in LaTex by the authors and Springer-Verlag
Cover design: Jim Wilkie

Printed on acid-free paper

Foreword

Astrophysics is advancing faster than ever before, and encompassing a broader range of subjects: even those of us who are professional researchers find it hard to keep up. This compact, but comprehensive, collection of up-to-date survey articles – each written by an expert, and in an accessible style – is therefore especially welcome.

When the history of modern science is written, recent developments in astronomy and cosmology will provide some of the most fascinating and momentous chapters. One can get a feel for the rate of recent progress by contrasting the contributions in this volume with what might have been written on the same topics as little as five years ago. Indeed, the themes of some of the articles – for instance the comparative study of extrasolar planetary systems – then barely existed as scientific topics in their own right. Today, we have a far clearer understanding of galaxies, and of the huge black holes that lurk at their centres. Surveys of very faint objects at high redshifts now offer a far clearer picture of what the Universe was like when galaxies were young. We are also starting to understand the 'extreme physics' of gamma-ray bursts and high energy cosmic rays. And cosmology, once an arena only for speculation, has been transformed into a quantitative subject, rich in observational data. The size and shape of our Universe have been pinned down by a combination of studies conducted both from the ground and from space.

Relatively little of this progress is attributable to armchair theorists. The real heroes and heroines are those who design and build ever more powerful and sensitive instrumentation. Modern computer technology permits fast, efficient analysis and widespread distribution of data. Astrophysicists can also compensate for their inability to conduct real experiments by using computers to simulate the behaviour of stars and galaxies in virtual universes. It is good that these styles of research are well covered in this book.

As always in science, each advance may raise new questions and bring existing puzzles into sharper focus. How did planets, stars and galaxies form? Why does our Universe consist of roughly 5 percent ordinary matter, 25 percent dark matter (of unknown nature) and 70 percent even more mysterious dark energy? Astrophysics presents many challenges like these, and this book lays out some of the groundwork for tackling them. All astrophysicists are indebted to John Mason, and to the authors of the individual contributions, for this timely and valuable book.

Cambridge, UK *Martin Rees*
July 2003

Preface

Progress in astrophysics and cosmology has been extremely rapid in recent years, and it is becoming increasingly difficult to keep up with the pace of developments in these fields. *Astrophysics Update* is a collection of topical and timely reviews in key areas of astrophysics and cosmology, each written by leading experts in their fields, but in a style which will be accessible to a broad readership. We hope that the format will appeal to professional astronomers and postgraduate students who are interested in reading about topics outside their own specific areas of research. In this volume, we have attempted to highlight those subjects that are currently rich and active research spheres - areas where major advances or discoveries are being made, and where there is rapid progress in observational techniques and theoretical understanding.

In Chap. 1, I. Neill Reid and Suzanne Hawley review recent developments in our understanding of star and planet formation. They adopt an empirical approach, concentrating on observational results for intermediate- and low-mass stars, brown dwarfs and planets. They present the latest derivations of the initial mass function, for both field stars and star clusters, including searches for extremely low-mass brown dwarfs in young stellar associations. In the growing discipline of extrasolar planetary formation, the chapter summarizes the statistics of currently known exoplanet systems, considers recent observations of circumstellar dust disks and discusses the distribution of properties of the parent stars.

More than 100 planets are known outside our Solar System, and Hugh Jones discusses exoplanets and their properties in Chap. 2. We have a good idea of the planetary mass function down to about one-tenth of a Jupiter mass around primary stars ranging from spectral type F to M though mainly around solar type stars. Most Jupiter-like mass planets are to be found at Jupiter-like distances though there is an important minority at much smaller radii. These strongly favour a planetary migration scenario with some mechanism to stop migration to periods shorter than three days. The majority of exoplanets have been discovered around primary stars with particularly high metal content. One intriguing result has been the high eccentricities of exoplanets in comparison with the Solar System and the similarity of the eccentricity distribution for exoplanets and stellar binary systems.

An interesting aspect of stellar evolution is the focus of Chap. 3, in which Coel Hellier examines the evolution of close binary stars, concentrating on the test case of cataclysmic variables. Magnetic braking is widely accepted to be the mechanism removing angular momentum from the orbit of close binary stars, so driving their evolution. The best test of this hypothesis is a comparison of the observed orbital-period distribution with the output of population-synthesis codes that incorporate magnetic braking. Cataclysmic variable stars are often used as a test-bed, being one of the best-studied classes of close binary. Yet recent observational results conflict with the assumptions needed to model the orbital-period distribution of these stars. So, even for one of the best-studied classes of close binary, there is no settled understanding of their evolution.

Determining the distances to celestial objects is one of the most important tasks in astrophysics, and in Chap. 4, Stephen Webb surveys recent progress that has been made in the 'distance ladder' and 'geometric' approaches to the clarification of the cosmic distance scale. The impressive successes of these approaches are emphasized, but two intriguing problems are also discussed: firstly, the uncertainty in the distance to the LMC; and secondly, the potential difficulties in reconciling 'local' H_0 estimates with estimates from gravitational lenses modelled within a standard CDM scenario. The chapter also discusses what the near future might hold for our understanding of the cosmic distance scale, including the anticipated contributions of the four major astrometry missions planned for the next decade.

Release of the first data from the Wilkinson Microwave Anisotropy Probe (WMAP) was one of the highlights of early 2003, and these are placed in context by Andrew Jaffe in his discussion of the cosmic microwave background (CMB) in Chap. 5. The CMB provides a direct image of the Universe when it was a few hundred thousand years old, in a hotter, denser and simpler state than today. Fluctuations in the CMB have been mapped by satellites, balloon-borne and ground-based telescopes, allowing measurement of the cosmological parameters describing the contents and evolution of the Universe. The results are consistent with a hot Big Bang and the gravitational growth of large-scale structure in general, and with the inflationary paradigm: the Universe is geometrically flat, with a scale-invariant spectrum of adiabatic initial conditions. The WMAP results and ongoing observations of the polarization of the CMB cement this interpretation.

In recent years there have been important developments in the field of particle astrophysics. In Chap. 6, Lars Bergström and Ariel Goobar review some of the progress relating to the accelerating Universe - in particular, models for the dark energy and ways of distinguishing between them - and to the dark matter problem, where particles beyond the Standard Model are most probably involved. Our knowledge of fundamental cosmological parameters obtained by a combination of observational methods is summarized. The chapter also discusses the present status of neutrino mass and oscillations, and briefly touches upon the question of whether ultra-high energy particles beyond the cut-off provided by scattering on the cosmic microwave background have been detected.

Recent developments in observation and simulation offer new clues to processes during the cosmic 'dark ages', between the epochs of reionization and recombination. These are discussed by William Keel in Chap. 7. The developments inform our picture of galaxy formation and especially our understanding of the intergalactic medium. Studies of QSOs now reveal the final stages of reionization, while detection of enriched material in intergalactic space may trace the overall enrichment from the first generation of very massive stars. The current demographics of supermassive black holes suggests that they grew rapidly during the initial epochs of galaxy building, giving a physical basis for connecting the evolution of galaxies and QSOs.

In Chap. 8, Hyron Spinrad reviews selected current observations of distant galaxies and our interpretation of the fragile and occasionally contradictory data. Galaxies at the 'contemporary limit' of technology and redshift ($z \sim 6$) are difficult to locate, and the large redshift may push some critical confirming and/or interpretative analysis toward unfamiliar IR wavelengths. The chapter concentrates on the observational means and results to explore the early evolution of galaxies. There are biases that seem to intrude on plans for the interpretative aspects of distant galaxy photometry and spectroscopy. The best methods of selection for those very distant systems are discussed; including utilizing strong sub-mm emission from dust, photometry indicating a UV 'spectral break', and finally the signal of a strong Lyman-α emission line, a feature which has now carried us to a galaxy redshift in excess of $z = 6.57$.

Spectroscopy in the optical waveband remains one of the most powerful diagnostic tools available to astronomers. In recent years, as Fred Watson demonstrates in Chap. 9, the technique has undergone a number of fundamental improvements on several different fronts. Perhaps the most spectacular is the coming-of-age of robotic multi-fibre systems, which allow hundreds of objects to be observed simultaneously with a wide-field telescope. Soon, this will be extended to thousands. However, it is not only in instrumental developments that optical spectroscopy has been revolutionized, but also in essential techniques such as sky-subtraction. The chapter outlines recent progress and points towards future developments.

The large scale distribution of galaxies carries information on the nature and evolution of galaxies, as Anthony Fairall describes in Chap. 10. The '2dF' and 'Sloan' surveys are measuring redshifts on a scale never before attempted. The Sloan survey is also producing 5-colour photometric imagery, thereby building a database vastly greater than any previous work. Analyses of their currently available data have already produced a wealth of papers that further quantify the large-scale characteristics of galaxies and their spatial distribution. Correlation functions and power spectra have been estimated, from which cosmological parameters may be extracted. In particular the 2dF survey finds $\Omega_m = 0.29 \pm 0.05$, $\Lambda = 0.7 \pm 0.1$ and $H_0 = 72 \pm 7$, in line with the now accepted standard model.

Active Galactic Nuclei (AGN) and supermassive black holes are the subject of Chap. 11. As Ian Robson shows, while much progress has been made over the past few years in our understanding of the physical processes underway in AGN, there are, nevertheless, many mysteries that still elude us. One of these is the fundamental question of the detailed physical processes in the immediate vicinity of the supermassive black hole, even the supposed accretion disk. Although gravity is acknowledged as the primary energy source, how this energy is transferred into the powerhouse of the AGN and the related observational phenomena remains unclear. This chapter concentrates on the inner workings of the central engine of AGN, utilizing the latest data from space and ground-based facilities.

Since their discovery in 1967, the excitement surrounding Gamma-Ray Bursts (GRBs) has not decreased. In Chap. 12, Gilbert Vedrenne and Jean-Luc Atteia review the observational and theoretical developments between 1973 and 2003 which have finally led to the identification of the origin of GRBs. The early years (1973-1991) brought the excitement of discovery, the study of GRB properties in gamma-rays and the lack of quiescent counterparts. The GRO decade (1991-2000) established for the first time the extragalactic origin of GRBs. The afterglow era began in 1997, when BeppoSAX located GRB970228 within a few hours of the burst. This new era of afterglow observations led to measurement of the distance to sources and identification of their host galaxies. The story of GRBs continues and the advent of a new generation of space missions will surely bring yet more surprises.

The current status of gravitational wave research is reviewed by Leonid Grishchuk in the final contribution, Chap. 13. Recently constructed laser-beam detectors of gravitational waves are approaching the planned level of sensitivity, and in a few years more advanced detectors will become operational. Their sensitivity will be sufficient to meet the most cautious evaluations of the strength and event rates of astrophysical sources of gravitational waves. The current approach to gravitational wave research is very broad. Experimental innovations, source modelling, methods of data analysis, and theoretical issues are all being studied and developed. The race for direct detection of relatively high-frequency waves is accompanied by vigorous efforts to discover the very low-frequency relic gravitational waves. This may be achieved through measurements of the cosmic microwave background radiation. Experimental and theoretical work in connection with space-based laser-beam detectors is also being actively pursued.

This book has benefited from the support and assistance of a large number of people. I would like to offer my sincere thanks to all of the contributing authors for their considerable efforts, perseverance and enthusiasm for this project, and to Professor Sir Martin Rees for so generously agreeing to write the Foreword. I am also most grateful to Brigitte Reichel-Mayer of Springer-Verlag, Heidelberg for her invaluable assistance and advice in the preparation of the LaTeX files, and Professor Wolf Beiglböck, also of Springer-Verlag, Heidelberg, for his constructive comments on the format of the book. Finally, I am indebted to Sue Peterkin of Praxis Publishing for her very considerable assistance at all stages in the organization and coordination of this project, and to Clive Horwood, Publisher, for his encouragement, advice and patience throughout.

Barnham, UK
August 2003

John W. Mason

Contents

**1 Recent Advances in Understanding Star
and Planet Formation**
I.N. Reid, S.L. Hawley 1
1.1   Introduction 1
1.2   Star Formation 1
1.3   Planet Formation 8
1.4   Summary 16
References 17

2 Exoplanets and Their Properties
H.R.A. Jones 21
2.1   Introduction 21
2.2   Finding Exoplanets 22
2.3   Properties of Exoplanets 28
2.4   The Future of Planet Searches 42
References 43

**3 What Drives the Evolution of Close Binary Stars?
The Test Case of Cataclysmic Variables**
C. Hellier 47
3.1   Introduction: The Need for Magnetic Braking 47
3.2   Discussion 57
References 59

4 The Cosmic Distance Scale
S. Webb 61
4.1   Introduction 61
4.2   The Cosmological Distance Ladder 63
4.3   Geometrical Methods 73
4.4   Two Distance-Scale Puzzles 76
4.5   Future Developments 80
References 81

5 The Cosmic Microwave Background
A.H. Jaffe 89
5.1 Introduction 89
5.2 The CMB and the Physics of the Early Universe 94
5.3 CMB Experiments and CMB Data 102
5.4 Current Observations of the CMB 103
5.5 Summary 105
References 107

6 Particle Astrophysics and the Dark Sector of the Universe
L. Bergström and A. Goobar 111
6.1 Introduction 111
6.2 Cosmological Parameters from 'Standard Candles' 113
6.3 Current Results 114
6.4 Cross-Cutting Measurements 114
6.5 How Much Better Can We Do? 116
6.6 The Highest Redshift Supernova: SN1997ff 117
6.7 The Next Generation of SN Experiments 119
6.8 The Quintessence Alternative 119
6.9 The Nature of Dark Matter 123
6.10 More on the Dark Matter 124
6.11 Dark Matter Candidates 125
6.12 Indirect Detection Through Gamma-Rays 127
6.13 Non-Supersymmetric Candidates 130
6.14 Fluxes of Neutrinos in Standard Models 132
6.15 Particles Above the GZK Cutoff? 133
6.16 Summary and Conclusions 133
References 134

7 The Early Universe: From Recombination to Reionization
W.C. Keel 139
7.1 Introduction 139
7.2 Reionization 140
7.3 The First Stars 144
7.4 Where Do Black Holes Come From? 147
7.5 Prospects 150
References 151

8 The Most Distant Galaxies
H. Spinrad 155
8.1 Introduction, Motivations and Questions 155
8.2 Some Issues in the Contemporary Theory
    of Early Galaxy Evolution 158
8.3 A Race for the Maximum Redshift 160

8.4	The Identification of Very Distant Galaxies	160
8.5	The Future	176
References		177

9 Optical Spectroscopy Today and Tomorrow
F. Watson 181
9.1	Introduction	181
9.2	Multi-Object Spectroscopy – Overview	184
9.3	Multi-Object Spectroscopy with Robots	190
9.4	Integral-Field Spectroscopy	195
9.5	Efficiency by Design	198
9.6	Future Challenges in Optical Spectroscopy	202
References		204

10 Large-Scale Structures in the Distribution of Galaxies: The 2dF and Sloan Surveys
A.P. Fairall 211
10.1	Introduction	211
10.2	Target Galaxies	213
10.3	The Sloan Photometric Camera	213
10.4	Sky Coverage	214
10.5	Spectroscopy	215
10.6	Redshift Maps and Large-Scale Structures	217
10.7	Galaxy Properties and Spatial Distribution	219
10.8	Clusters	221
10.9	Redshift Space Distortion, Biasing and the Mass Density of the Universe	222
10.10	Power Spectra	223
10.11	Λ Cosmology	226
10.12	Concluding Remarks	227
References		227

11 Active Galactic Nuclei and Supermassive Black Holes
I. Robson 231
11.1	Introduction and the Big Picture	231
11.2	The Evidence for Supermassive Black Holes	235
11.3	Black Holes and the Relation to the AGN	240
11.4	AGN Evolution and Black Hole Growth	244
11.5	The Central Engine and Accretion Disks	248
11.6	Conclusions	251
References		252

12 The Story of Gamma-Ray Bursts
G. Vedrenne and J.-L. Atteia 255
12.1 Introduction 255
12.2 The Early Times 256
12.3 CGRO: A New Step in Understanding the Origin of GRBs 264
12.4 BeppoSAX: Its Decisive Role in Understanding GRBs 267
12.5 A New Window on Stellar Explosions
 and on the Early Universe 270
12.6 The Story Continues... 276
References 277

13 Update on Gravitational Wave Research
L.P. Grishchuk 281
13.1 Introduction 281
13.2 Elementary Theory of Gravitational Waves 282
13.3 Current Status of Gravitational-Wave Detectors 287
13.4 Gravitational Waves and Astrophysics 291
13.5 Gravitational Waves and Cosmology 296
13.6 Summary 306
13.7 Acknowledgements 307
References 307

Index 311

List of Contributors

Jean-Luc Atteia
Laboratoire d'Astrophysique
CNRS/UPS/OMP
14 Avenue E. Belin
31400 Toulouse Cedex, France

Lars Bergström
Department of Physics
Stockholm University
Alba Nova University Center
S-10691 Stockholm
Sweden

Anthony P. Fairall
Department of Astronomy
University of Cape Town
Rondebosch 7700 South Africa

Ariel Goobar
Department of Physics
Stockholm University
Alba Nova University Center
S-10691 Stockholm
Sweden

Leonid Grishchuk
Department of Physics
and Astronomy
Cardiff University
Cardiff CF24 3YB, UK
and
Sternberg Astronomical Institute
Moscow University
Moscow 119899 Russia

Suzanne L. Hawley
Department of Astronomy
University of Washington
Box 351580 Seattle
Washington 98195, USA

Coel Hellier
Astrophysics Group
Department of Physics
Keele University
Keele
Staffordshire ST5 5BG, UK

Andrew H. Jaffe
Department of Physics
Imperial College London
Blackett Laboratory
Prince Consort Road
London SW7 2AZ, UK

Hugh R.A. Jones
Astrophysics Research Institute
Liverpool John Moores University
Twelve Quays House,
Egerton Wharf
Birkenhead CH41 1LD, UK

William C. Keel
Department of Physics
and Astronomy
University of Alabama
Box 870324 Tuscaloosa
Alabama 35487-0324, USA

I. Neill Reid
Space Telescope Science Institute
3700 San Martin Drive
Baltimore
Maryland 21218, USA

Ian Robson
United Kingdom Astronomy
Technology Centre
Blackford Hill
Edinburgh EH9 3HJ, UK

Hyron Spinrad
Department of Astronomy
University of California
Berkeley
California 94720-3411, USA

Gilbert Vedrenne
C.E.S.R.,
CNRS/UPS/OMP
9 Avenue du Colonel Roche
31028 Toulouse Cedex 4, France

Fred Watson
Anglo-Australian Observatory
Coonabarabran
NSW 2357, Australia

Stephen Webb
LTS, The Open University
Walton Hall
Milton Keynes MK7 1AA, UK

1 Recent Advances in Understanding Star and Planet Formation

I.N. Reid and S.L. Hawley

1.1 Introduction

Star formation is a subject which spans a wide variety of topics, only a limited subset of which can be covered in a review of this length. We have therefore chosen to concentrate on areas closest to our own research interests - low mass stars, brown dwarfs and planets. We also take an empirical approach, discussing primarily the observed consequences of star and planet formation, rather than the possible underlying theoretical mechanisms.

1.2 Star Formation

1.2.1 The present-day and initial mass functions in the field

The strongest empirical constraint on global star formation in the Galaxy is the observed consequence - the present-day mass function. Allowing for stellar evolution, the local mix of stellar populations and the density law perpendicular to the Galactic plane, the observed present-day mass function may be used to infer the initial mass function, the primary result of the star formation process.

The determination of the present-day mass function involves first measuring the luminosity function for a complete sample of stars, and then applying a mass-luminosity relation to convert the number counts per unit magnitude into number counts per unit mass. The luminosity function has been studied for field stars using both nearby star samples, where the distances are known by trigonometric parallax, and pencil-beam photometric surveys which employ a photometric parallax relation to obtain the distances and hence luminosities. The latter investigations are fraught with uncertainty (Reid & Hawley 2000) and we concentrate here on nearby star studies. The most recent determination is that of Reid, Gizis & Hawley (2002), who used a volume complete sample of stars within 25 parsecs of the Sun, including several hundred stars with *Hipparcos* parallax measurements. The results (Fig. 1.1) indicate a slow rise in number density for $0 < M_V < 10$, with a rapid increase to a peak at $M_V \sim 12$ and a shallow decline at fainter absolute magnitudes.

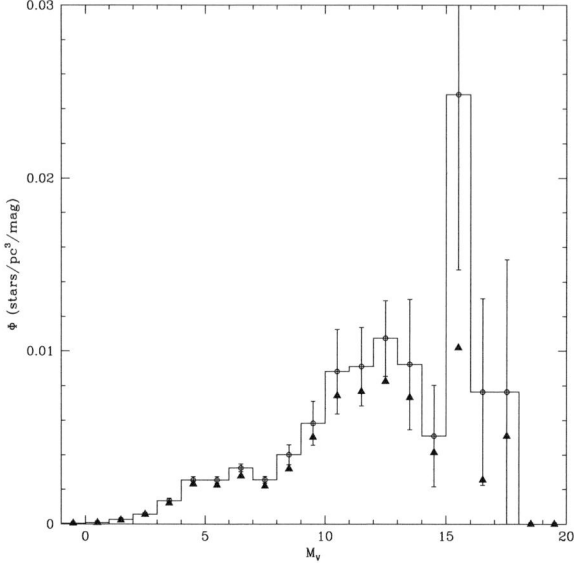

Fig. 1.1. The field star luminosity function derived from the statistics of nearby stars (Reid et al., 2002). The error bars give the Poisson uncertainties; the solid triangles mark the contribution of single stars and primaries in multiple systems.

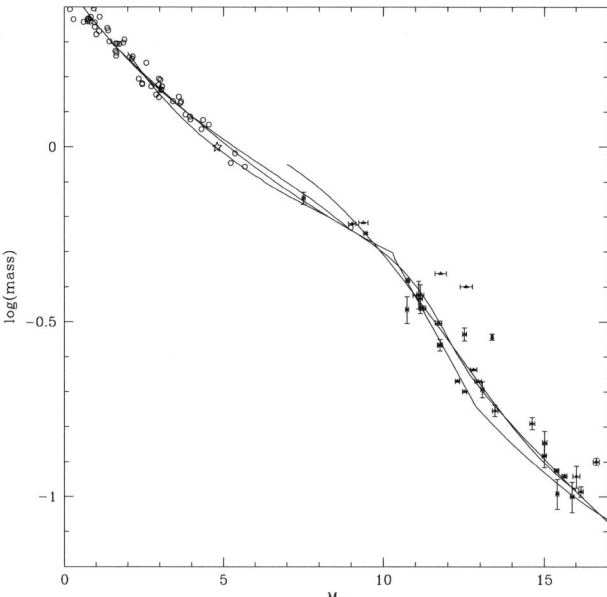

Fig. 1.2. The mass-M_V relation: open circles plot data for eclipsing binaries from Andersen (1991); triangles are low-mass systems from Ségrensan et al (2000); the Sun is plotted as a 5-point star. The mean relations are from Henry & McCarthy (1993), Kroupa, Tout & Gilmore (1993), Delfosse et al. (2000) and from an empirical fit to the Andersen eclipsing systems.

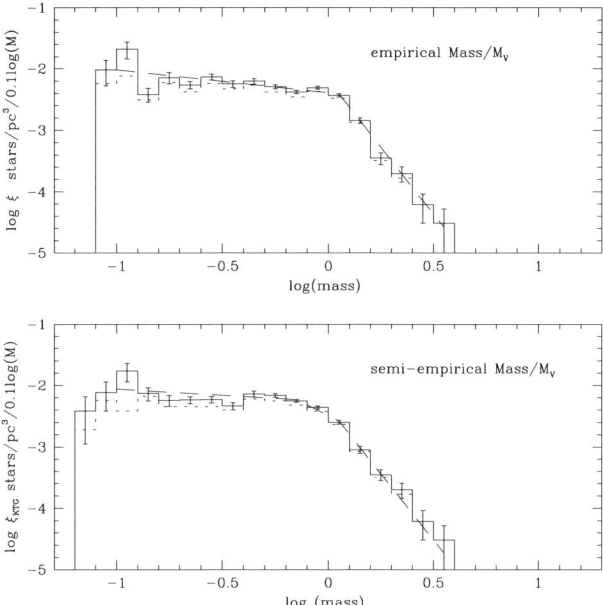

Fig. 1.3. The present-day mass function for field stars, using the empirical and semi-empirical mass/luminosity calibrations described in the text. The dotted histogram shows the contribution from single stars and primaries in multiple systems; the dashed lines are the best-fit power-law relations.

The mass-luminosity relation for high and intermediate mass stars is obtained from observations of eclipsing binary systems; since only four eclipsing M-dwarf systems are known, the calibration at lower masses rests on astrometry of nearby resolved systems. Empirical studies by Henry & McCarthy (1993) and Delfosse et al. (2000) for low mass stars; Andersen (1991) for higher-mass eclipsing systems; and the semi-empirical result of Kroupa, Tout & Gilmore (1993) are compared in Fig. 1.2. It is clear that there is a paucity of data in the range $5 < M_V < 10$ which leads to considerable uncertainty in the mass-luminosity relation over the mass range $\sim 0.5 - 1$ $M_\odot$.

Combining the luminosity function from Fig. 1.1 with these mass-luminosity relations leads to the present-day mass functions illustrated in Fig. 1.3. Results are shown for two composite calibrations: an empirical relation, combining the fit to the Andersen data with the Delfosse et al. relation at $M_V > 9$; and the semi-empirical relation from Kroupa et al. The former favors a two-component power-law mass function, with $\alpha \sim 1.3$ for 0.1 $M_\odot < M < 1$ $M_\odot$ and $\alpha \sim 5.2$ for $M > 1$ $M_\odot$ [1]. The mass function derived from the semi-empirical calibration has breaks at 0.7 $M_\odot$ and 1.1 $M_\odot$, reflecting their dif-

---

[1] Following Salpeter (1955), the mass function is represented as $\psi(M) = M^{-\alpha}$ corresponding to $\xi(M) = M^{-\alpha+1}$ when expressed as a function of log mass, as in Fig. 1.3. $\alpha = 2.35$ is the canonical Salpeter slope.

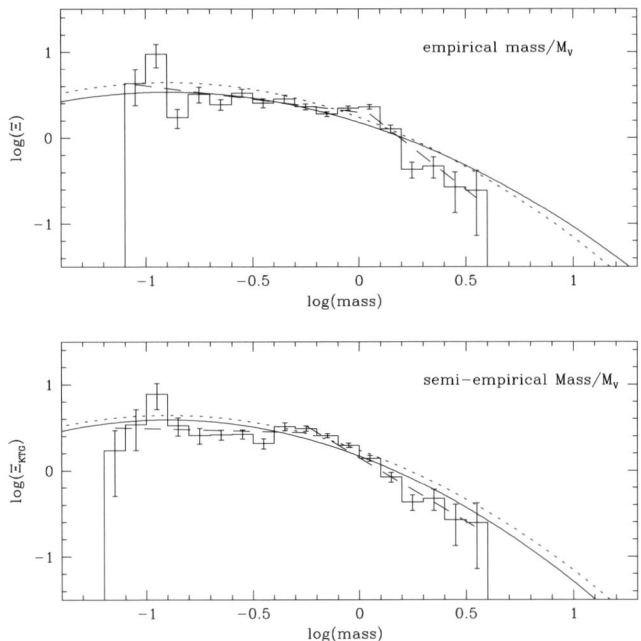

Fig. 1.4. The initial mass function: as in Fig. 1.3, we show the results of alternative mass/M_V calibrations, although the differences are minor. The dashed lines show the best-fit power-law representations; the solid line is the best-fit log-normal relation; the dotted line is the original Miller-Scalo function.

fering fit to the mass-luminosity relation in that mass range (see also Kroupa, 2001). The steep slope evident at $M > 1\ M_\odot$ emphasises that this present-day mass function does not account for higher mass stars which have already evolved off the main sequence.

When the effects due to stellar evolution and galactic structure (the variation in scale height with age) are accounted for, the initial mass function retains its double power law character (Fig. 1.4), with $\alpha \sim 1.1$ for $M < 1.1\ M_\odot$ and $\alpha \sim 2.8$ at higher masses (Reid et al., 2002). A log-normal distribution, suggested originally by Miller & Scalo (1979), does not provide as good a fit to the data, as shown in the figure. The main challenge for star formation theory appears to be an explanation for the change in slope near $M \sim 1\ M_\odot$.

1.2.2 Substellar dwarfs in clusters and the field

The principal difference in the mathematical representations of the IMF described above is manifest at substellar masses, as the power law distribution continues to rise, while the log-normal distribution peaks near 0.1-0.2 $M_\odot$ and declines at lower masses. Observations of field stars and brown dwarfs in the mass range $M < 0.1\ M_\odot$ are still relatively scarce. A preliminary analysis by Reid et al. (1999) found that a simple extension of the low mass power law ($\alpha \sim 1.3$) provided an acceptable fit to the scant data; Chabrier (2002) reached a similar conclusion. The interpretation for field dwarfs is complicated by the rapid luminosity evolution in the brown dwarf régime - the mass-luminosity relation changes radically with the age of the population.

The bulk of the data at substellar masses has therefore been obtained in young open clusters, where the brown dwarfs have primarily M spectral types and are still relatively bright and accessible to observation. In a few cases, observations extend to sufficiently faint magnitudes that they are capable of detecting L dwarfs or even early T-dwarfs (e.g. σ Ori, Zapatero Osorio et al., 2002). If those objects are confirmed as cluster members, the inferred masses are less than $0.015\,M_\odot$ and may even lie below the deuterium burning limit. This clearly suggests that cloud fragmentation can extend to extremely low masses - albeit with the caveat that the mass estimates rest solely on theoretical tracks for an assumed cluster age.

Statistically, the cluster results are mixed, with some studies finding a declining mass function, consistent with a log-normal distribution (e.g. Hillenbrand & Carpenter 2000 for the Orion Nebula Cluster) while others find that a power law mass function with $\alpha \sim 0.5 - 1$ reproduces their results (Béjar et al., 2001; Luhman et al., 2000). Much of the uncertainty lies in the evolutionary models: Luhman et al. found that an $\alpha \sim 1$ power law fit the ONC data when they used a different set of evolutionary tracks. There may also be significant variations in the form of the IMF in different environments: Briceño et al.(2002) suggest that brown dwarfs are less common in the low density Taurus star-forming clouds. Distinguishing between a rising and declining mass function in the substellar régime is an important avenue of continuing research, with implications both for the (dark) baryonic mass content of the Galaxy, and for the theoretical understanding of initial cloud core conditions and the accretion/mass loss processes prevalent during star formation.

1.2.3 Brown dwarf formation

Classically, one of the most important parameters governing star formation is the Jeans mass - the minimum mass for gravitational collapse. At the typical temperatures and densities of molecular clouds, the Jeans mass has a value close to $1\,M_\odot$. This raises two problems for standard collapse models, however: first, how do sub-Jeans mass objects (i.e. over 95% of stars and all brown dwarfs) form in a molecular cloud without requiring extremely high densities ($\sim 10^7$ cm^{-3}); and, second, how do these objects maintain their low masses by avoiding substantial subsequent accretion from remnant cloud material?

Reipurth and Clarke (2001) have suggested one possible mechanism. They propose that stars form in small groups. Dynamical interactions can lead to the ejection of individual stars, truncating accretion by removing the protostar from the local gas reservoir. Ejection should produce both a binary frequency which declines sharply with decreasing mass, and a scarcity of accretion disks around young brown dwarfs. The former prediction is at least qualitatively in agreement with observations (but see further below); the latter appears to be contradicted by line profile measurements and infrared data for very low-mass dwarfs in young star-forming regions (Muzerolle et al., 2000; Natta & Testi, 2001). Moreover, while this scenario can supply brown dwarfs, there is no obvious explanation for the observed form of the mass function for low-mass stars and brown dwarfs.

On the other hand, Nordlund & Padoan (2002) have pointed out that the Jeans mass is valid as a critical limit only in highly idealised situations - quiescent gas clouds, where gravitational energy dominates. Real molecular clouds are highly turbulent, with kinetic energy dominating gravitational energy by two orders of magnitude. Nordlund & Padoan argue that turbulence can drive lower-mass cores to sufficiently high densities for collapse. The resultant IMF is a Salpeter power-law at high masses, and a log-normal at low masses, with the peak dependent on the gas density. This model predicts significantly fewer brown dwarfs in low-density star forming regions (e.g. Taurus) than high-density regions (e.g. Orion). While this is qualitatively consistent with observations, the data in both field and clusters, as discussed above, are a better match to a power-law ($1 > \alpha > 0.5$) at low masses.

1.2.4 Binary statistics and star formation

It has long been an astronomical cliché that most stars are found in binary or multiple systems. Tracking how multiplicity varies as a function of mass, including both the distribution of systemic parameters and the overall multiplicity frequency, can test the viability of different star formation mechanisms. Duquennoy & Mayor's (1991) classic survey of late-F and G stars established a multiplicity fraction (N_{mult}/N_{sys}) of 50 to 70% for solar-type stars. Fischer & Marcy (1993) extended the analysis to early and mid-type M dwarfs, and found that the overall fraction decreased to 30-35%. Reid & Gizis (1997) confirmed this result in the analysis of the volume-limited 8-parsec survey, and also suggested that there was a preference for equal-mass systems at lower masses. These statistics refer to *stellar* companions, so part of the decrease reflects the reduced mass range possible for secondary stars[2]. Statistical sampling alone, however, cannot account for the full effect.

Surveys of binarity in very low-mass stars and brown dwarfs have only become possible within the last few years. Ground-based adaptive optics imaging can achieve resolutions of 0.1 arcseconds at near-infrared wavelengths

---

[2] Brown dwarfs are rarely found as close companions to solar-type stars - the 'brown dwarf desert' - but are not uncommon as wide (> 100AU) systems (Gizis et al., 2001). Indeed, the nearest currently-known brown dwarf, the recently discovered ϵ Indi B (Scholz et al., 2003), lies ~ 1460 AU from the K5 primary.

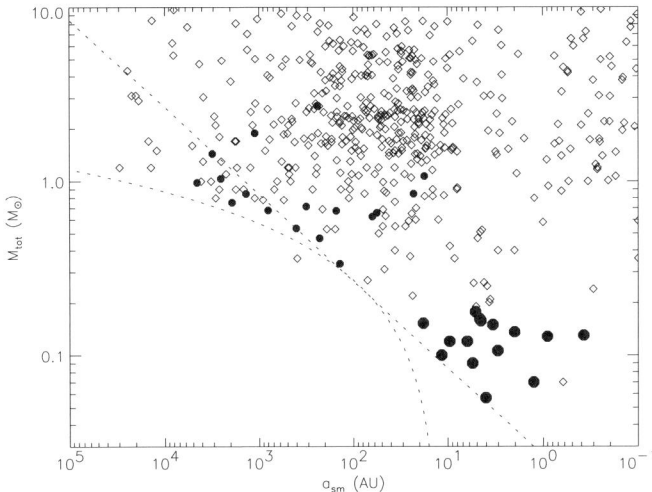

Fig. 1.5. Separation of binary systems as a function of total system mass (from Burgasser et al., 2003). The curved line is the log-normal relation $\log a_{max} \propto M_{tot}$ from Reid et al. (2001); the linear relation is $a_{max} \propto M_{tot}^2$ from Burgasser et al. (2003).

(Close et al., 2001), but remains limited to late-type M dwarfs by the requirement for a relatively bright (R<15) natural guide-star (usually the target itself). Thus, the Hubble Space Telescope remains the most effective means of studying the lowest mass, lowest luminosity dwarfs. To date, both far-red and near-infrared images have been obtained of more than 60 L dwarfs (Martín et al., 1999; Reid et al., 2001; Gizis et al., 2003) and 10 T dwarfs (Burgasser et al., 2003). Allen et al. (2002) have shown that these observations are capable of detecting over 50% of companions with mass ratios ($\frac{m_2}{m_1}$) exceeding 0.4; thus, they extend well into the brown dwarf régime.

The results from all of these surveys are in good agreement:

- The distribution of component separations shows a strong correlation with mass (Fig. 1.5), with an obvious decrease in the maximum separation of binary components, a_{max}, with decreasing mass. Echoing Abt (1988), Reid et al. (2001) pointed out that a_{max} varied linearly with the total system mass in M dwarf systems; Burgasser et al. (2003) have shown that the dependence appears to steepen below $\sim 0.1\,M_\odot$, varying with M_{tot}^2.
- Twenty percent of late-M, L and T dwarfs have resolved companions at separations exceeding $\sim$ 3AU. All of the targets are drawn from magnitude-limited samples, however, so the unresolved binary systems (effectively higher luminosities) are drawn from a larger sampling volume. Thus, the true binary fraction for this range of separation is probably $\sim 10\%$. The relatively sparse radial velocity data available for ultracool dwarfs suggests that a similar proportion are smaller-separation spectro-

scopic binaries, giving a total multiplicity of $\sim 20\%$. This extends the decline in frequency observed between F/G and M dwarfs, despite the fact that these observations include brown dwarf companions.
- Over 75% of the ultracool binaries have mass ratios $\frac{m_2}{m_1} > 0.9$. The lowest mass ratios observed are $\sim 0.7 - 0.75$ (e.g. LHS 2397a, Freed et al., 2003), despite sensitivity limits extending lower by a factor of two.

At present, there is no theoretical model which predicts the observed characteristics. At first sight, the low multiplicity appears consistent with the formation-by-ejection scenario favoured for brown dwarfs by Reipurth & Clark (2001). However, numerical simulations of this mechanism by Bate, Bonnell & Bromm (2001) predict a binary frequency of no more than 5%, significantly below the observed numbers once allowance is made for spectroscopic systems. The $a \propto M^2$ distribution seen at the lowest masses is consistent with the variation in the total binding energy of binary systems, and the upper limit might therefore reflect a critical value. The log-normal relation at higher masses remains unexplained at present.

1.3 Planet Formation

The identification of planetary-mass objects in orbit around a substantial number of nearby stars clearly ranks as one of the most important scientific discoveries of the last century. Moreover, even the initial detection of 51 Peg b (Mayor & Queloz, 1995) confounded the standard paradigm of planet formation, with a jovian-mass planet orbiting a solar-type star at a distance one-third that of Mercury from the Sun. The subsequent years have seen a flood of discoveries, most made by either the Geneva-based team led by Mayor and Queloz, using telescopes at Haute Provence and ESO, or by Marcy, Butler and collaborators, using Lick, Keck and the Anglo-Australian observatories[3] (see also Chap. 2 by H.R.A. Jones in this volume). Detailed observations of these extrasolar planetary systems and of protoplanetary disks, together with theoretical advances, are starting to shed more light on the complex process of planet formation.

1.3.1 The statistics of extrasolar planetary systems

Ideally, we would like to classify objects based on their mode of formation - brown dwarfs, by isolated collapse from the interstellar medium; planets, from fragmentation and accretion within a disk. At present, however, we lack an observational means of distinguishing these modes. For present purposes, therefore, we adopt an empirical definition: we define a planet as a very low-mass object in orbit around an object of significantly higher mass. Without

---
[3] See Jean Schneider's Extrasolar Planets Encyclopedia website (http://cfa-www.harvard.edu/planets/) for a current statistical summary.

formal mass limits, this is a rather loose definition which could be applied to the Earth- Moon system. But the borderlines in mass between brown dwarf and planetary companions are probably intrinsically fuzzy, so we simply suggest applying the above definition with a dose of common sense. Here we have included systems where the companion has a projected mass $M \sin i < 17 M_J$, where M_J is the mass of Jupiter. This specifically excludes all isolated, so-called 'free-floating planets', (i.e. very low-mass brown dwarfs) identified tentatively in some young star clusters and associations (Zapatero Osorio et al., 2000, 2002).

A number of alternate hypotheses to planetary-mass companions have been suggested. Gray & Hatzes (1997) originally proposed that the systematic velocity variations detected in 51 Peg might reflect atmospheric motions due to stellar pulsations. The absence of corresponding photometric variations, together with the poor match between the observed period and any reasonable resonant mode for solar-mass stars, ruled out that possibility. However, activity-driven convective motions can produce the same effect by introducing systematic asymmetries in line profiles as spots rotate in and out of view. The most active stars are excluded *a priori* from the observing programs (eliminating the youngest stars). Even so, at least one star (HD 192263) has recently been discovered to have photometric variations phased with the radial velocity curves, and long-term spot activity is probably responsible for the apparent velocity variations in this case (Queloz et al., 2001).

Finally, Han et al. (2001) have argued that many of the extrasolar planet hosts show residual astrometric motions, indicative of high-mass, brown dwarf companions. Statistically, the existence of so many face-on orbits amongst nearby stars is highly improbable, and in at least one case (ρ^1 Cnc, McGrath et al., 2002) Hubble Space Telescope observations fail to confirm the Han et al. mass estimate. The high masses derived by Han et al. are almost certainly due to incorrect analysis of Hipparcos astrometry (Pourbaix, 2001; Halbwachs et al., 2000; Zucker & Mazeh, 2001).

The discovery that HD 209458 is a transiting system (Charbonneau et al., 2000) provides irrefuteable evidence for the existence of extrasolar planets.[4] The parent star is a near-solar mass main-sequence star, so the depth of the light-curve in mid-transit (1.5%) indicates that HD 209458b has a radius ~ 1.3 times that of Jupiter. With a mass of 0.6 M_J ($i > 89.9°$), this indicates that the planetary atmosphere is inflated by the strong stellar irradiation experienced in the 3-day orbit - consistent with model predictions (Guillot & Showman, 2002). Several attempts have been made to detect spectroscopic signatures of the planetary atmosphere, by observing HD 209458 before and after planetary transits. Terrestrial sky subtraction has proven an

---

[4] In addition, recent HST astrometry of the M4 dwarf, Gl 876 (Benedict et al., 2002) reveals orbital motion, allowing derivation of an inclination of $\sim 84°$ and a mass of $1.8 < M/M_J < 2.5$ for Gl 876b, the longer period low-mass companion in this system. This is obviously well into the planetary-mass régime.

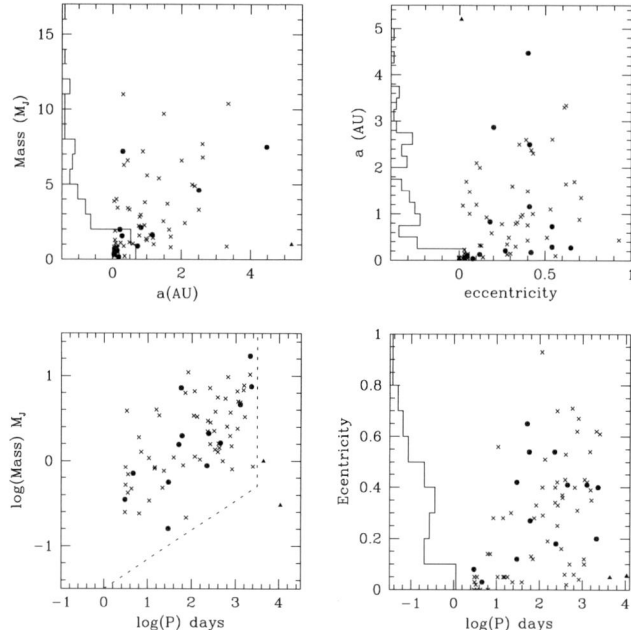

Fig. 1.6. Statistical properties of extrasolar planetary systems: solid points mark planets in multiple systems, and the solid triangles mark the locations of Jupiter and Saturn. The histograms show the distribution of semi-major axis, eccentricity and mass derived from the projected orbits. The dotted lines in the (log(M), log(P)) diagram show the approximate discovery limits of the current radial velocity surveys.

insurmountable barrier to ground-based observations, but Charbonneau et al (2002) succeeded in detecting a significant enhancement of the sodium D lines by using the STIS spectrograph on the Hubble Space Telescope.

As of 15 December 2002, 102 companions in 88 stellar systems (not including the Solar System) meet our working definition of planets [5]. All of these extrasolar planetary (ESP) systems were identified through high-precision radial velocity observations, which leads to significant selection biases. Reflex orbital motion varies with planetary mass and inversely with square of the separation, so current techniques are most effective at identifying high-mass gas giants at relatively small separations. Moreover, the limited heritage of the current observational programs provides an effective barrier to identifying long-period systems - it is difficult to identify a Saturnian analogue with only 5 years of data. Within the last five years, radial velocity observations

---

[5] For up to date summaries, see the Geneva Observatory web site, http://obswww.unige.ch/ naef/$who_discovered_that_planet$.html; the California & Carnegie planet search web-site, http://exoplanets.org/; and Jean Schneider's Extrasolar Planets Encyclopaedia, cited above.

have achieved accuracies close to $1\,\mathrm{m\,s^{-1}}$ (see Chap. 2 in this volume, for more details), and true Jovian analogues (P ~ 5 years), together with more multiple-planet systems, are likely to emerge in the near future.

Even with the current statistical limitations, some trends are evident in the ESP dataset (Fig. 1.6). The lower boundary to the mass distribution in the (mass, a) and (log(mass), log(P)) diagrams reflects the detection limits of the current surveys (velocity precision and time baseline). There is a suggestion that the upper boundary also increases with increasing a and P - that is, hot Jupiters tend to have masses below 3 M_J. The latter systems all have low eccentricity orbits, due to tidal circularisation. At larger radii, however, planetary orbits span a wide range of eccentricity, with a nearly flat distribution between e=0.1 and 0.5. Most significantly, the mass distribution is clearly rising with decreasing mass, and is broadly consistent with a power-law, $N \propto M^{-0.9}$. This is interestingly similar to the form of the mass function for low-mass stars (see above), and may point to a common fragmentation mechanism.

Figure 1.7 shows the companion mass function for G-dwarf spectroscopic binaries, extending coverage from planetary to stellar masses (from Udry et al., 2001). There is a pronounced minimum between ~ 0.02 and $0.08\,\mathrm{M}_\odot$ - the 'brown dwarf desert' identified originally by Marcy. This diagram provides strong qualitative evidence that lower-mass planetary companions are a different type of object. Extrapolating current observations to terrestrial-

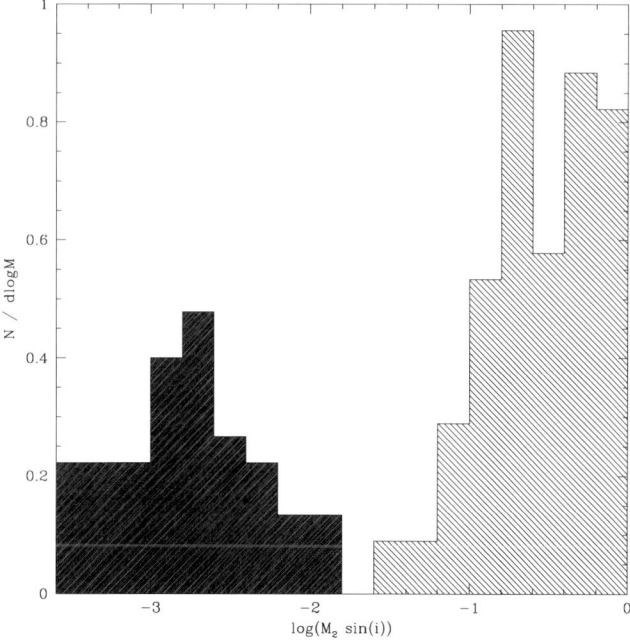

Fig. 1.7. The mass function for G-dwarf companions, from Udry et al. (2001).

planet masses is clearly not valid, but the data suggest that many lower-mass gas giants still await discovery.

The parent stars of the ESP systems are predominantly late F to early K-type main sequence stars (Fig. 1.8), with Gl 876 the only M dwarf (Delfosse et al., 1998; Marcy et al., 2001). This at least partly reflects initial target selection, which concentrates on solar-type stars (e.g. Nidever et al., 2002). Earlier type stars have fewer absorption lines and are generally fast rotators, while later type dwarfs have lower luminosities and correspondingly fainter apparent magnitudes - all factors which tend to decrease radial velocity accuracies. However, the scarcity of M dwarf systems may also reflect decreasing planetary frequency (at least for high mass planets) in lower-mass stars. Several hundred M dwarfs are currently under scrutiny, and this issue should become clearer within the next few years.

Considering only the solar-type stars, the current observations allow us to set a lower limit on the frequency of gas-giant planetary systems. There are 488 stars within 25 parsecs with $2.0 > M_V > 7.0$ and $0.5 < (B-V) < 1.0$ (the box inscribed in Fig. 1.8); $\sim 30\%$ of those systems lack high-precision radial velocity observations, but 25 ($5.1\pm1.0\%$) are known to have planetary-

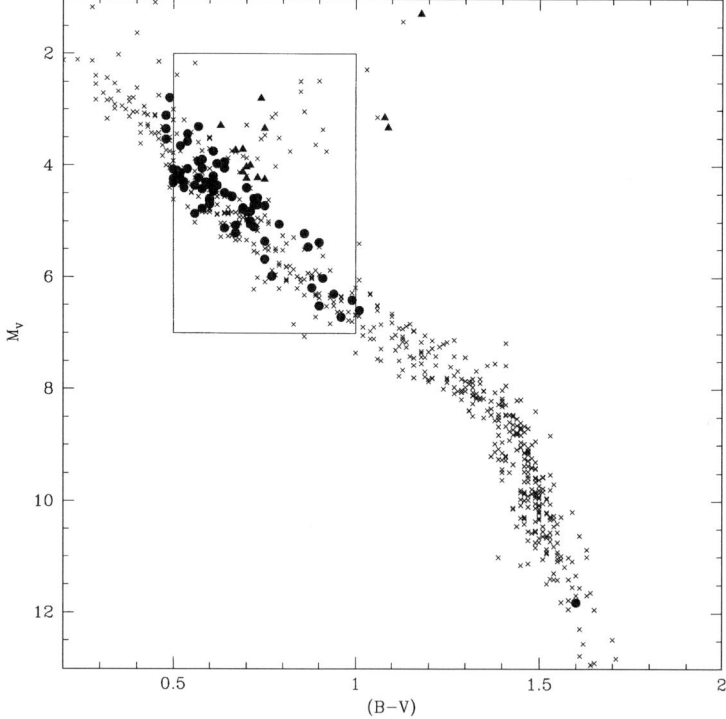

Fig. 1.8. The distribution of ESP host stars (solid points and triangles) in the HR diagram. The reference dataset (crosses) is provided by stars within 25 parsecs of the Sun with Hipparcos parallax measurements accurate to better than 10%.

mass companions. This obviously represents a lower limit, since the current observations are barely capable of detecting solar-system analogues (Fig. 1.6). However, 30% of the planets are 'hot Jupiters' in 3-5 day orbits. These are the easiest planets to detect, with velocity amplitudes rising to 100 m s^{-1}. It is likely that at least 1.5% of solar-type stars possess this type of gas giant companion.

1.3.2 Planet formation and the host star characteristics

With a sample of almost 100 ESP systems, we can start to analyse the characteristics of not only the planets, but also the host stars. The strongest trend which has emerged so far is a clear correlation in planetary frequency with metallicity, as suggested initially by Gonzalez (1997). Later analyses by Laughlin (1999), Gonzalez et al. (2001) and Santos et al. (2001) all confirm the qualitative result that higher metallicity stars are more likely to have planets sufficiently massive to be detected by current techniques. Indeed, these initial studies suggested that planets were almost exclusively found around stars more metal-rich than the Sun. There are complications, however, in quantifying the correlation which were not taken fully into account in these analyses - specifically, measuring abundances on a self-consistent basis, and defining an adequate reference sample.

A wide variety of methods are available for measuring stellar metallicity. Systematic differences, in scale or zeropoint, can exist between different techniques. All of the ESP host stars have been observed at very high spectral resolution, and those data are available for abundance analysis. Such is not the case for the average field star, where metallicities rest primarily on lower resolution data. At least one widely used calibration, Schüster & Nissen's (1986) relations for the *uvby* Strömgren photometric system, systematically underestimates abundances relative to high-resolution analyses (Reid, 2002; Haywood, 2002).

The reference sample should provide unbiased statistics for the local field stars. The simplest method of achieving that goal is to select a complete, distance-limited stellar sample, but this has only become possible with the availability of reliable *Hipparcos* astrometry (see Reid, 2002, for further discussion). That distribution is shown in Fig. 1.9, where we limit the field star sample to disk dwarfs with $d < 25$ parsecs and the (M_V, (B-V)) limits given in §1.3.1. All of the abundances are derived using Haywood's calibration of Strömgren photometry (see also Haywood, 2001). The most significant difference relative to previous analyses is that the Sun lies near the mode, with $\sim 45\%$ of local stars having [m/H]>0 rather than only 5-10%.

Fig. 1.9 matches the field star distribution against the ESP hosts, excluding systems which were selected for observation solely because they are metal rich (e.g. BD-10 3666). We show results for both the full ESP sample scaled to match the 25-parsec volume (i.e. to 5% of the field star sample) and for the ESP hosts within 25 parsecs. The results are statistically identical. There is

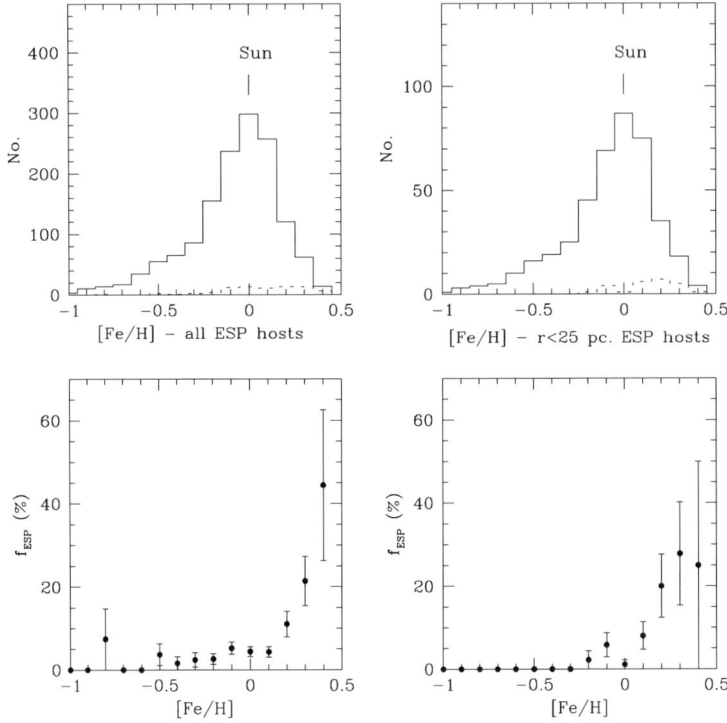

Fig. 1.9. Planetary frequency as a function of metallicity. The left hand panels compare data for F, G and early-K stars within 25 parsecs (solid histogram) against the appropriately-scaled distribution for all ESP host stars (dotted histogram); the right-hand panel shows the same comparison, but limiting the ESP sample to stars within 25 parsecs. The lower panels plot the ratios.

a strong trend for increasing frequency with increasing abundance. However, the frequency is still significant ($\sim 5\%$) for solar analogues, and extends to sub-solar metallicities.

Two mechanisms have been proposed to explain the observed correlation: massive planets form more readily in metal-rich disks; and pollution of the outer envelope by accretion of giant planets. The latter mechanism relies on orbital migration within the protostellar disk. Jupiter is metal-rich relative to the Sun (Z=0.06 vs. Z=0.02), but adding a single Jupiter to the Sun increases the metallicity by only 0.3%. Pollution therefore must occur after the T Tauri phase (when the star is fully convective) and must be confined to the outer convection zone, which ranges in mass from $M_{conv} \sim 0.005$ to 0.1 $M_\odot$ in main-sequence stars. In that case, one would expect a significant decrease in the maximum metallicity with decreasing temperature and increasing M_{conv}. That trend is not observed (Santos et al., 2001; Reid, 2002). Thus, the observations favour nature rather than nurture: super-jovian mass planets form more readily in metal-rich systems.

1.3.3 Dust disks, planet formation and orbital evolution

Planet formation occurs in circumstellar protoplanetary disks, originally detected through the presence of excess flux at infrared wavelengths (Mendoza, 1966). Recent observations have imaged a number of disks in re-radiated thermal emission at mid-infrared wavelengths (Koerner et al., 1998), in scattered light at optical and near-infrared wavelengths (Weinberger et al., 1999) and in molecular (mainly CO) emission at millimetre wavelengths (Beckwith & Sargent, 1993). Table 1.1 summarises the current picture of the main stages of evolution.

The most significant feature of Table 1.1 is the short timescale for gas depletion - scarcely longer than 10^7 years. This sets an upper limit on the formation time of the gas-rich giant planets, and has long posed a problem for conventional, hierarchical models, which envisage accretion of ices onto a $\sim 10^{-3} M_J$ rocky super-embryo (Mizuno, 1980). Direct fragmentation of the protostellar disk due to gravitational instabilities has been suggested as an alternative mechanism (e.g. Boss, 2000); Mayer et al. (2002) have recently derived much shorter formative timescales from full three-dimensional hydrodynamical simulations of gaseous disks. They find that persistent, massive (1-5 M_J) overdense clumps develop in a matter of only a few hundred years. Ice giants, like Uranus and Neptune, can form at large disk radii (>20 AU) for stars forming in clusters, where ultraviolet radiation from luminous stars photoionises the protoplanetary envelopes.

Given the Solar System as a template, the two most unexpected properties of extrasolar planets are the wide range of orbital eccentricities and the existence of hot Jupiters. Both can probably be accounted for through orbital evolution during the early history of the planetary system. The broad distribution in eccentricity was one of the factors which prompted suggestions that these low-mass companions were related to brown dwarfs rather than planets. Stepinski & Black (2001), in particular, have pointed out the strong similarity with the eccentricity distribution of spectroscopic binaries, but this may well reflect similar dynamical histories rather than a common origin. Marzari & Weidenschilling (2002), for example, have shown that gravitational interactions in three-body systems usually lead relatively rapidly ($\sim 10^8$ years) to chaotic orbits. The result is usually ejection of one body, with the remaining

Table 1.1. Evolution of protoplanetary disks

Class	Example	Age yrs	Description
Embedded protostar	L1551-IRS5	$< 10^6$	Gas-rich, $M \sim 0.1 \, M_\odot$, R$\sim$ 100 AU
Classical T Tauri	TW Hya	$few \times 10^6$	Optically-thick, gas-rich, strong CO
Transitional	HR 4796	$\sim 10^7$	dust-rich, gas-depleted, some H_2
Debris disk	β Pic	$\sim 10^8$	optically-thin, gas-free dust disk

planets occupying moderate- to high-eccentricity orbits, often with different inclinations. Alternatively, resonant interactions within the viscous protostellar disk could lead to significant eccentricities while maintaining a co-planar system (Chiang et al., 2002).

It is clear that the temperature profile of protostellar disks does not permit giant planet formation in sub-Mercurian orbits (Guillot et al., 1996). The solution to this dilemma is generally agreed to be orbital evolution (Lin & Papaloizou, 1986; Lin et al., 1996). Giant planets form in the ice zone, beyond ~ 4 AU, but migrate inward rapidly through angular momentum exchange with the disk. This evolution stops when the disk dissipates ($\tau \sim 10^7$ years). An interesting consequence is that the Hill radius decreases with decreasing a, so the outermost satellites of giant planets will be lost during the orbital evolution (Weiss & Stewart, 2002). Moreover, Armitage & Bonnell (2002) suggest that orbital migration might account for the brown dwarf desert in solar-type stars: close brown dwarf companions form coevally with the central star, and are therefore subject to stronger orbital migration in the younger, high density accretion disk than the later-forming gas giant planets.

Trilling, Lunine & Benz (2002) have undertaken a theoretical analysis of migration in disks spanning a range of parameters. Starting from Jupiter's present orbit, they find that a majority of gas giants migrate inward rapidly and merge with the parent star. Most of the survivors lie beyond 1 AU. Planets forming beyond 5 AU and higher mass planets migrate less rapidly, and a larger proportion survive. The latter factor may account for the trend toward lower masses at small semi-major axis evident in Fig. 1.6. Trilling et al. conclude that present planetary catalogues include only one-quarter to one-third of surviving gas giants. This hypothesis will be tested over the next few years as radial velocity surveys extend to longer periods and lower masses.

1.4 Summary

With the observational advances of the last few years, the IMF is now well established for the full range of stellar masses. The main questions facing current theoretical models of star formation are:

1. What is responsible for the change in slope in the IMF at $\sim 1\,\mathrm{M}_\odot$?
2. What is the functional form of the substellar IMF? Are there significant variations dependent on the star forming environment?
3. What mechanism governs the formation of binary and multiple systems at different masses?

The details of planet formation remain less well understood than star formation, although the catalogue of known systems is now advancing to the point where systematic trends and correlations are becoming apparent.

Clearly, the main caveat is the continuing bias towards detecting higher-mass, shorter-period planetary companions. However, it is clear that planetary systems around solar-type stars are both relatively common (a frequency exceeding 5%) and fairly diverse. Extending observations to longer periods and, particularly, lower masses will be a prime research focus for the next 5 to 10 years.

References

Abt, H.A. (1988) Ap. J. **331** 922
Allen, P.R., Koerner, D.W., Reid, I.N., Gizis, J.E., Kirkpatrick, J.D., McElwain, M.W., Murphy, G.R. (2003) IAU Symp. 211 in press
Andersen, J. (1991) Astr. Ast. Rev **3** 91
Armitage, P.J., Bonnell, I.A. (2002) MNRAS **330** L11
Bate, M.R., Bonnell, I.A. & Bromm, V. (2002) MNRAS **332** L65
Beckwith, S.V.W. & Sargent, A.I. (1993) Ap. J. **402** 280
Béjar, V.J.S. et al. (2001) Ap. J. **556** 830
Benedict, G.F. et al. (2002) Ap. J. **581** L115
Briceño, C., Luhman, K.L., Hartmann, L., Stauffer, J.R., & Kirkpatrick, J.D. (2002) Ap. J. **580** 317
Boss, A.P. (2000) Ap. J. **536** L101
Burgasser, A.J., Reid, I.N., Kirkpatrick, J.D., Brown, M.E., Miskey, C.L., Gizis, J.E. (2003) Ap. J. **586** 512
Chabrier, G. (2002) Ap. J. **567** 304
Charbonneau, D., Brown, T.M., Latham, D.W., & Mayor, M. (2000) Ap. J. **529** L45
Charbonneau, D., Brown, T.M., Noyes, R.W., & Gilliland, R.L. (2002) Ap. J. **568** 377
Chiang, E.I., Fischer, D., Thomme, E. (2002) Ap. J. **564** L105–20.
Close, L.M. et al. (2002), Ap. J. **566** 1095
Delfosse, X., Forveille, T., Mayor, M., Perrier, C., Naef, D., & Queloz, D. (1998) A & A **338** L67
Delfosse, X., Forveille, T., Ségransan, D., Beuzit, J.-L., Udry, S., Perrier, C., Mayor, M. (2000) A & A **364** 217
Duquennoy, A. & Mayor, M. (1991) A & A **248** 485
Fischer, D.A. & Marcy, G.W. (1991) Ap. J. **396** 178
Freed, M., Close, L.M., Siegler, N. (2003) Ap. J. **584** 453
Gizis, J.E., Kirkpatrick, J.D., Burgasser, A.J., Reid, I.N., Monet, D.G., Liebert, J., Wilson, J.C. (2001) Ap. J. **551**, L163
Gizis, J.E., Reid, I.N., Knapp, G.R., Liebert, J. Kirkpatrick, J.D., Koerner, D.W., Burgasser, A.J. (2003) A.J. **125** 3302
Gonzalez, G. (1997) MNRAS **285** 403
Gonzalez, G., Laws, C., Tyagi, S., Reddy, B.E. (2001) A. J. **121** 432
Gray, D.F. & Hatzes, A.P. (1997) Ap. J. **490** 412
Guillot, T., Burrows, A., Hubbard, W.B., Lunine, J.I., Sauman, D. (1996) Ap. J. **459** L35-39
Guillot, T. & Showman, A.P. (2002) A & A **385** 156

Halbwachs, J.L., Arenou, F., Mayor, M., Udry, S., & Queloz, D. (2000) A & A **355** 581

Han, I., Black, D.C. & Gatewood, G. (2001) Ap. J. **548** L57

Haywood, M. (2001) MNRAS **325** 1365

Haywood, M. (2002) MNRAS **337** 151

Henry, T.J., McCarthy, D.W. (1993) Ap. J. **350** 334

Hillenbrand, L.A., Carpenter, J.M. (2000) Ap. J. **540** 236

Koerner, D.W., Ressler, M.E., Werner, M.W. & Backman, D. E. (1998) Ap. J. **503** L83

Kroupa, P., Tout, C.A., Gilmore, G.F. (1993) MNRAS **262** 545

Kroupa, P. (2001) MNRAS **322** 231

Laughlin, G. (2000) Ap. J. **545** 1064

Lin, D.N.C., Papaloizou, J. (1986) Ap. J. **309** 846

Lin, D.N.C., Bodenheimer, P., Richardson, D.C. (1996) Nature **380** 686

Luhman, K.L. et al. (2000) Ap. J. **540** 1016

McGrath, M.A. et al. (2002) Ap. J. **564** L27

Marcy, G.W., Butler, R.P., Fischer, D., Vogt, S.S., Lissauer, J.J., & Rivera, E.J. (2001) Ap. J. **556** 296

Marzari, F., Weidenschilling, S.J. (2002) Icarus **156** 70

Martín, E.L., Brandner, W., Basri, G. (1999) Science **238** 1718

Mayor, M. & Queloz, D. (1995) Nature **378** 355

Mayer, L., Quinn, T., Wadsley, J., Stafel, J. (2002) Science **298** 1756–9.

Mendoza, E.E. (1966) Ap. J. **143** 1080

Miller, G.E. & Scalo, J.M. (1979) Ap. J. Sup. **41** 513

Mizuno, H. (1980) Prog. Th. Phys. **64** 544–2.

Muzerolle, J., Briceño, C., Calvet, N., Hartmann, L., Hillenbrand, L., & Gullbring, E. (2000) Ap. J. **545** L141

Natta, A. & Testi, L. (2001) A & A **376** L22

Nidever, D.L., Marcy, G.W., Butler, R.P., Fischer, D.A. & Vogt, S.S. (2002), Ap. J. Sup. **141** 503

Nordlund, A. & Padoan, P. (2003) Lecture Notes in Physics ed. E. Falgarone & T. Passot, in press

Pourbaix, D. (2001) A & A **369** L22

Queloz, D. et al. (2001) A & A **379** 279

Reid, I.N. (2002) PASP **114** 306

Reid, I.N. et al. (1999) Ap. J. **521** 613

Reid, I. N.& Gizis, J.E. (1997) A. J. **113** 2246

Reid, I.N., Gizis, J.E., Kirkpatrick, J.D., & Koerner, D.W. (2001) A. J. **121** 489

Reid, I.N. & Hawley, S.L. (2000) *New light on Dark Stars* (Springer Praxis)

Reid, I.N., Gizis, J.E. & Hawley, S.L. (2002) A. J. **124** 2721

Reipurth, B. & Clarke, C. (2001) A. J. **122** 432

Salpeter, E.E. (1955), Ap. J. **121** 161

Santos, N.C., Israelian, G., Mayor, M. (2000) A & A **363** 228

Scholz, R.-D., McCaughrean, M.J., Lodieu, N. & Kuhlbrodt, B. (2003) A & A **398** L29

Schuster, W.J., Nissen, P.F. (1989) A & A **221** 65

Ségransan, D., Delfosse, X., Forveille, T., Beuzit, J.-L., Udry, S., Perrier, C., & Mayor, M. (2000) A & A **364** 665

Stepinski, T.F. & Black, D.C. (2001) A & A **371** 250
Trilling, D.E., Lunine, J.I., Benz, W. (2002) Ast. & Ap. **394** 241–251.
Udry, S., Mayor, M., Halbwachs, J.-L., & Arenou, F. (2001) ASP Conf. Ser. **239**: Microlensing 2000: A New Era of Microlensing Astrophysics, 91
Weinberger, A.J., Rich, R.M., Becklin, E.E., Zuckerman, B. & Matthews, K. (2000) Ap. J. **544** 937
Weiss, J.W., Stewart, G.R. (2002) DPS **34** 3003
Zapatero Osorio, M.R., Béjar, V.J.S., Martín, E.L., Rebolo, R., Barrado y Navascués, D., Bailer-Jones, C.A.L. & Mundt, R. (2000), Science **290** 103
Zapatero Osorio, M.R., Béjar, V.J.S., Martín, E.L., Rebolo, R., Barrado y Navascués, D., Mundt, R., Eislöffel, J. & Caballero, J.A. (2002), Ap. J. **578** 536
Zucker, S. & Mazeh, T. (2001) Ap. J. **562** 549

2 Exoplanets and Their Properties

H.R.A. Jones

2.1 Introduction

It was the renaissance philosopher Giordano Bruno who first suggested there might be other worlds orbiting the stars of the night sky. Bruno's heretical philosophising came to a fiery end, when in 1600 he was burned at the stake. However, his musings set the stage for one of astronomy's 'Holy Grails' - the search for planets around other stars. Bruno's sad death at the hands of the Holy Inquisition set the stage for fruitless searches over the following 395 years. In 1991 the first definitive extrasolar planet (exoplanet) was discovered around a pulsar using timing measurements (Wolszczan & Frail 1992). A further three planets have now been discovered around PSR 1257+12 as well as a planet around PSR B1620-26 (Backer, Foster & Sallmen 1993). While these are landmark discoveries, the planets' location, next to a stellar remnant, helps little in understanding our own Solar System.

In 1995 the long search ended when Mayor & Queloz (1995) announced the detection of the first exoplanet around a Sun-like star. The radial velocity of the G2V star 51 Peg was used to infer the presence of a Jupiter mass planet in a 4.2 day orbit. The discovery was quickly confirmed and corroborated by Doppler evidence for Jupiter mass planets in orbit around a number of other nearby stars. Although very exciting, the indirect nature of the detection meant that the planet around 51 Peg was controversial for sometime. Evidence was put forward for asymmetric variations in line profiles indicating that the radial velocity variations arose within the photosphere of the star. Whilst such measurements proved incorrect, today the proof of each individual exoplanet still requires substantial scrutiny to ensure that radial velocity variations are not caused by variations in the photosphere of the target star. In the last year, three radial velocity planets have been withdrawn. Nonetheless the number of bona fide objects increases rapidly and today we have an inventory of more than 100 exoplanets. To those outside the field this may sound like a large number, however this is not sufficient a number to understand even the main features of exoplanets. In fact we have only just begun to probe the parameter space. The most pertinent measure of the infancy of this field is that we do not yet have the sensitivity to detect the presence of our own Solar System if it were present around any other star. The detection of Solar Systems like our own seems the vital scientific stepping-stone to the search for other life forms in the Universe.

2.2 Finding Exoplanets

The word planet means 'wanderer' in Greek, thus planets move on the sky relative to the fixed background stars. This definition needs updating since with modern equipment even quite distant galaxies can be seen to move. Discussions about the status of Pluto and 'free-floating planets/brown dwarfs' (McCaughrean et al. 2001) describe the issues. According to the *Oxford English Dictionary* (dictionary.oed.com) the old astronomical meaning of planet 'A heavenly body distinguished from the fixed stars by having an apparent motion of its own among them; each planet, according to the Ptolemaic system, being carried round the earth by the rotation of the particular sphere or orb in which it was placed.' has evolved to:

'The name given to each of the heavenly bodies that revolve in approximately circular orbits round the Sun (primary planets), and to those that revolve round these (secondary planets or satellites)'. More specific definitions are provided elsewhere, e.g., Dictionary.com: 'A non-luminous celestial body larger than an asteroid or comet, illuminated by light from a star, such as the Sun, around which it revolves.' The IAU working group on extrasolar planets provides a half page working definition of extrasolar planet and a working list of candidate extrasolar planets (www.ciw.edu/boss/IAU/div3/wgesp). In this paper the abbreviation exoplanet will be used instead of extrasolar planet.

Fig. 2.1. An image of the 15-70 M_{JUP} (Saumon et al. 2000) brown dwarf Gl229B (centre) 7 arcsec from its M2V companion Gl229A (left), courtesy of Oppenheimer (1999). By contrast Jupiter is approximately ten times closer in and a million times fainter.

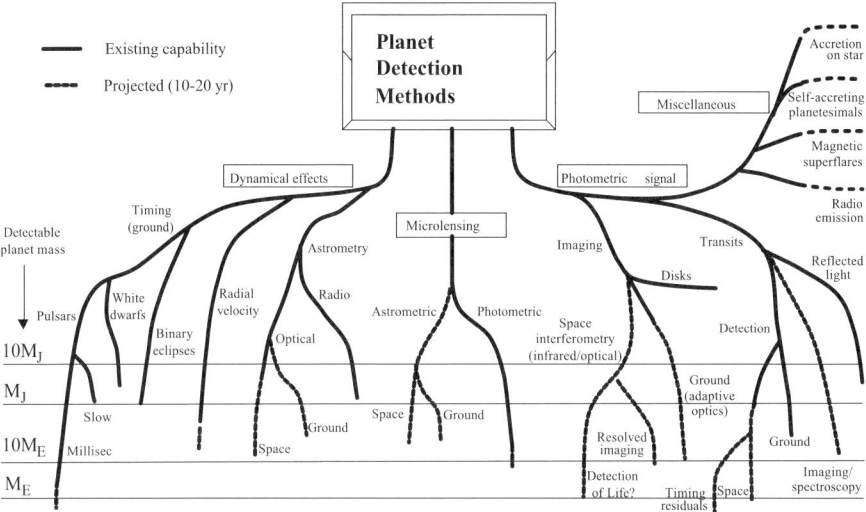

Fig. 2.2. There are many plausible ways to detect exoplanets. This schematic has become known as the 'Perryman-tree' and been adapted from Perryman (2000).

Planetary mass objects may have already have been imaged in young star forming regions (e.g. Lucas & Roche 2000; Zapatero Osorio et al. 1999). Apart from the masses of these objects being very dependent on poorly constrained theoretical models, the 'free-floating' nature of these objects means they fall outside the currently accepted notion of exoplanet. An important strand in most definitions is the concept that to be a 'planet' an object must be in orbit around a 'star'. This proximity to a much brighter object as well as their relative faintness makes planets difficult to find, e.g. Fig. 2.1. Discovery would be simplified were it possible to directly image exoplanets. The best opportunity so far is perhaps the controversial planet around Epsilon Eridani. At only 3.2 pc, it should soon become feasible to image this object although with a separation of 1 arcsecond and a magnitude difference of 15 (a factor of 1,000,000 in brightness) this observation is at the limit of current technology. This difficulty in direct imaging means that a wide range of innovative planet searching ideas have been devised, see Fig. 2.2.

2.2.1 Radial Velocity – Doppler Wobble

So far radial velocity or Doppler wobble has been the most important technique for finding exoplanets, e.g. Fig. 2.3. While the velocities of stars have been measured for many years, the breakthrough in making exoplanet detections by this method has been to observe a stable reference source at the same time as the target star (Butler et al. 1996; Baranne et al. 1996). The current measurement precision of some surveys attains better than 3 metres per second over many years. For comparison Jupiter causes the Sun to wobble

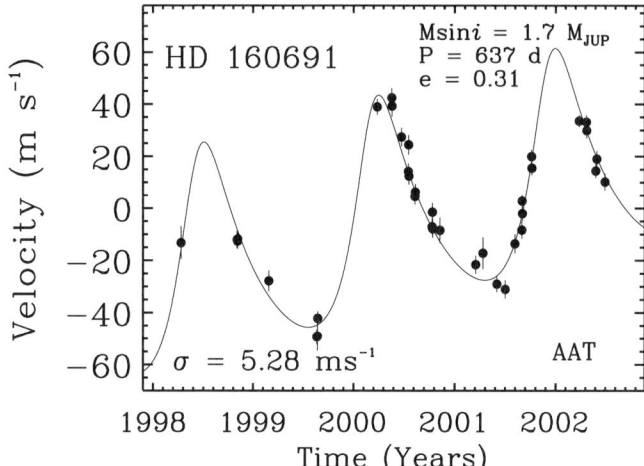

Fig. 2.3. The figure shows the detection of a radial velocity exoplanet around the star HD 160691 by the Anglo-Australian planet search (Jones et al. 2002). Assuming 1.08 ± 0.05 $M_\odot$ for the primary, the minimum (M sin i) mass of the companion is 1.7 ± 0.2 M_{JUP} and the semi-major axis is 1.5 ± 0.1 au. In addition the figure also shows a trend indicating a second companion.

with a velocity of 12.5 metres per second over a 12 year period. Planets one tenth the mass of Jupiter, albeit with much shorter periods than Jupiter, are now detected by the radial velocity technique. The next generation of radial velocity searches can be expected to improve efficiency and bring precisions down to around 1 m s^{-1} by using a new generation of specifically designed spectrometers along with robotic telescopes and automated scheduling.

2.2.2 Astrometric

Since a star and a planet orbit around their common center-of-mass, a large planet will cause movement in the position of a star. Jupiter induces such a motion on Sun. At a distance of 10 parsecs, Jupiter could be seen to give an angular motion of 0.5 milliarcseconds to the Sun. This motion is tantalisingly close to limits of current technology. The last major astrometry mission HIPPARCHOS managed a precision of 2 milliarcsecond. Its successors GAIA and SIM will measure angular motion of microarcsecs and will thus be sensitive to Jupiter mass objects in orbit around all local stars. Figure 2.4 shows astrometric confirmation of the exoplanet around Gl876.

2.2.3 Transits

Around 1 in 3000 stars (e.g. Horne 2003) is expected to have a planet in orbit around it which moves into the line of sight between the star and the Earth.

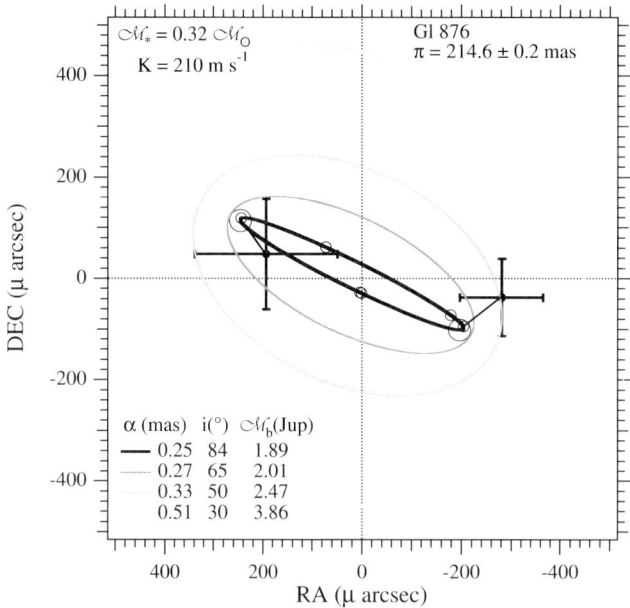

Fig. 2.4. The figure shows the four orbits permitted for the astrometric perturbation of Gl876A due to Gl876b, courtesy of Benedict et al. (2002) using the FGS instrument on the Hubble Space Telescope. The innermost orbit is determined from the simultaneous astrometry-RV solution. Plotted on this orbit are the phases of the astrometric observations, with the two primary phases indicated by large circles. Also plotted (+) are the astrometry-only residual normal points at $\phi = 0.26$ (periastron, lower right) and $\phi = 0.72$. These normal points are connected to the derived orbit by residual vectors.

The drop in brightness of a star can be detected as a cool planet transits across it and blocks some of the light. The detection of planetary tranists requires the highest quality photometry as well as frequent sampling to avoid many of the numerous false positives: grazing/partial eclipses from two stars, reddened A-type stars eclipsed by M dwarfs, giants eclipsed by dwarfs, brown dwarfs, triple systems with pair of fainter eclipse where one is mainly seeing the light of the brighter star. One planet has been seen in transit as predicted from its Doppler wobble measurements, see Fig. 2.5. Konacki et al. (2003) may have made the first direct transit detection and there are many new candidates (Udalski et al. 2002). In addition to more than 20 ground-based projects a number of space missions are also planned (see Table 2.1, Horne et al. 2003). It is nonetheless feasible to search for planetary transits without access to Space Agency budgets (e.g., transitsearch.org).

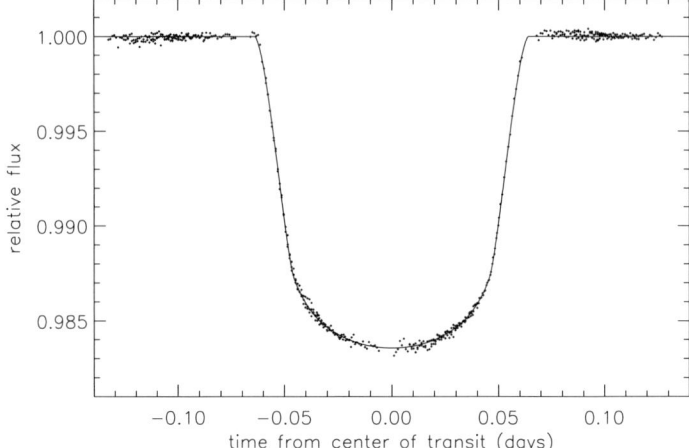

Fig. 2.5. Light curve from Brown et al. (2001) obtained by observing four transits of the planet of HD209458 using the STIS spectrograph on the Hubble Space Telescope. The folded light curve can be fitted within observational errors using a model consisting of an opaque circular planet transiting a limb-darkened stellar disk. In this way the planetary radius is estimated as 1.347 ± 0.060 R_{JUP}, the orbital inclination 86.6 ± 0.14, the stellar radius 1.146 ± 0.050 R_{JUP}. Satellites or rings orbiting the planet would, if large enough, be apparent from distortions of the light curve or from irregularities in the transit timings. No evidence is found for either satellites or rings, with upper limits on satellite radius and mass of 1.2 R_{JUP} and 3 M_{JUP}, respectively. Opaque rings, if present, must be smaller than 1.8 planetary radii in radial extent. The high level of photometric precision attained in this experiment confirms the feasibility of photometric detection of Earth-sized planets circling Sun-like stars.

2.2.4 Gravitational Microlensing

The presence of planets can also be inferred by their focusing and hence amplification of light rays from a distant source by an intervening object. Microlensing (Paczynski 1986) describes the gravitational lensing that can be detected by measuring the intensity variation of a macro-image made of any number of micro-images that are generally unresolved to the observer. Microlensing is capable of probing the entire mass range of planets from Super-Jupiter mass to Earth mass. Figure 2.6 shows the properties of exoplanets currently excluded by the non-detection of planetary mass microlenses. The universality of gravity guarantees that gravitational microlensing is also sensitive to unbound planetary mass objects as well as brown dwarfs. One conventional thought is that planets are formed through accretion and brown dwarfs are formed through fragmentation during star formation. Abundant statistics of microlensing planets can provide an important clue to test this scenario. Microlensing planets are expected to be distant and so follow-up will be limited.

2 Exoplanets and Their Properties

Table 2.1. Properties of radial velocity planets and their hosts.

Planet	$M_{JUP}/\sin i$	T	a	e	K	[Fe/H]	M_{prim}	SpType
HD83443b	0.35	2.986	0.038	0	57	0.33	0.79	K0V
HD46375b	0.25	3.024	0.041	0.02	35.2	0.25	1	K1V
HD179949b	0.93	3.092	0.045	0	112	0.02	1.24	F8V
HD187123b	0.54	3.097	0.042	0.01	72	0.16	1.06	G5
Tau Boo b	4.14	3.313	0.047	0.04	474	0.28	1.3	F7V
BD-103166b	0.48	3.487	0.046	0.05	60.6	0.5	1.1	G4V
HD75289b	0.46	3.508	0.047	0.01	56	0.29	1.05	G0V
HD209458b	0.63	3.524	0.046	0.02	82	0.04	1.05	G0V
HD76700b	0.19	3.971	0.049	0	25	0.1	1	G6V
51 Peg b	0.46	4.231	0.052	0.01	55.4	0.12	1.06	G2IV
Ups And b	0.68	4.617	0.059	0.01	70.2	0.1	1.3	F8V
HD49674b	0.12	4.948	0.057	0	14.3	0.25	1	G5V
HD68988b	1.9	6.276	0.071	0.14	187	0.24	1.2	G0
HD168746b	0.24	6.4	0.066	0	28	-0.07	0.92	G5
HD217107b	1.29	7.13	0.072	0.14	139.7	0.3	0.98	G8IV
HD162020b	13.73	8.42	0.072	0.28	1813	0.01	0.7	K2V
HD130322b	1.15	10.72	0.092	0.05	115	-0.02	0.79	K0V
HD108147b	0.41	10.9	0.079	0.2	40.8	0.2	1.27	F9V
HD38529b	0.78	14.31	0.129	0.28	54.7	0.313	1.39	G4IV
55 Cnc b	0.84	14.65	0.115	0.02	72.2	0.27	1.03	G8V
Gl86	4.23	15.8	0.117	0.04	379	-0.24	0.79	K1V
HD195019	3.55	18.2	0.136	0.02	271	0	1.02	G3IV
HD6434	0.48	22.09	0.154	0.3	37	-0.52	1	G3IV
Gl876c	0.56	30.12	0.13	0.27	81	0	0.32	M4V
Rho Cr b	0.99	39.81	0.224	0.07	61.3	-0.19	0.95	G0-2V
55 Cnc c	0.21	44.28	0.241	0.34	13	0.27	1.03	G8V
HD74156b	1.55	51.6	0.276	0.65	108	0.13	1.05	G0
HD168443b	7.64	58.1	0.295	0.53	470	0.1	1	G2V
Gl876b	1.89	61.02	0.207	0.1	210	0	0.32	M4V
HD121504b	0.89	64.62	0.317	0.13	45	0.16	1	G2V
HD178911b	6.46	71.5	0.326	0.14	343	0.28	0.87	G5
HD16141b	0.22	75.8	0.351	0	10.8	0.22	1	G5V
HD114762b	10.96	84.03	0.351	0.33	615	-0.5	0.82	F9V
HD80606b	3.43	111.8	0.438	0.93	414	0.43	0.9	G5
70 Vir b	7.41	116.7	0.482	0.4	316.1	0	1.1	G4V
HD52265b	1.14	119	0.493	0.29	45.4	0.11	1.13	G0V
HD1237b	3.45	133.8	0.505	0.51	164	0.2	0.9	G6V
HD37124b	0.86	153	0.543	0.2	37	-0.32	0.91	G4V
HD73526b	3.63	188	0.647	0.52	149	0.28	1.02	G6V
HD82943b	0.88	221.6	0.728	0.54	34	0.32	1.05	G0
HD8574b	2.08	228.5	0.77	0.3	65	-0.09	1.17	F8
HD169830b	2.95	230.4	0.823	0.34	83	0.14	1.4	F8V
Ups And c	1.9	241.3	0.829	0.28	53.9	0.1	1.3	F8V
HD89744b	7.17	256	0.883	0.7	257	0.18	1.4	F7V
HD202206b	14.68	258.9	0.768	0.42	554	0.37	1.15	G6V
HD40979b	3.16	260	0.818	0.26	99.9	0.19	1.08	F8V
HD12661b	2.3	263.3	0.823	0.35	74.4	0.29	1.07	G6V
HD150706b	1	264.9	0.82	0.38	33	-0.13	0.98	G0
HD134987b	1.63	265	0.821	0.37	53.7	0.23	1.05	G5V
HD17051b	2.12	312	0.909	0.15	63	0.25	1.03	G0V
HD92788b	3.88	337	0.969	0.28	113	0.04	1.06	G5
HD142b	1.36	338	0.98	0.37	40	0.24	1.1	G1IV
HD28185b	5.7	383	1.03	0.07	161	0.24	0.99	G5
HD177830b	1.24	391	1.1	0.4	34	0	1.17	K0
HD108874b	1.65	401	1.07	0.2	46	0.14	1	G5
HD4203b	1.64	406	1.09	0.53	51	0.22	1.06	G5
HD128311b	2.63	414	1.01	0.21	85	0.08	0.8	K0
HD27442b	1.32	415	1.16	0.06	32	0.2	1.2	K2IV
HD210277b	1.29	436.6	1.12	0.45	39.1	0.16	0.99	G0
HD82943b	1.63	444.6	1.16	0.41	46	0	1.05	G0
HD19994b	1.66	454	1.19	0.2	42	0.23	1.35	F8V
HD20367b	1.12	500	1.28	0.23	27	0.1	1.17	G0
HD114783b	0.99	501	1.2	0.1	27	0.33	0.92	K0
HD147513b	1	540.4	1.26	0.52	31	-0.03	0.92	G3-5V
Hip75458b	8.68	550	1.34	0.71	296	0.03	1.05	K2III
HD222582b	5.2	577.1	1.36	0.76	195.69	-0.01	1	G5
HD23079b	2.76	628	1.48	0.05	62	0	1.1	F8/G0V
HD160691b	1.74	637.3	1.48	0.31	41.1	0.28	1.08	G3IV-V
HD141937b	9.67	658.8	1.48	0.4	247	0.16	1	G2/G3V
16 Cyg b	1.68	798.4	1.69	0.68	50	0.09	1.01	G2.5V
HD4208b	0.81	829	1.69	0.04	18.3	-0.24	0.93	G5V
HD114386b	0.99	872	1.62	0.28	27	-0.03	0.75	K3V
HD213240b	4.49	951	2.02	0.45	91	0.23	1.22	G4IV
HD10697b	6.08	1074	2.12	0.11	114	0.15	1.1	G5IV
47 UMa b	2.56	1090.5	2.09	0.06	49.7	-0.08	1.03	G0V
HD190228b	3.44	1112	1.98	0.52	89	-0.24	1.3	G5IV
HD114729b	0.88	1136	2.08	0.33	19	-0.22	0.93	G3V
HD2039b	5.1	1190	2.2	0.69	136	0.1	0.98	G3IV
HD136118b	11.91	1209	2.39	0.37	212	-0.06	1.24	F9V
HD50554b	3.72	1254	2.32	0.51	78.5	0.02	1.1	F8
Ups And d	3.75	1284	2.52	0.27	61.1	0.1	1.3	F8V
HD216437b	2.09	1293.5	2.38	0.34	37.8	0.13	1.07	G4IV-V
HD196050b	2.81	1300.3	2.41	0.2	49.2	0.15	1.1	G3V
HD216435b	1.23	1326	2.6	0.14	20	0.15	1.25	G0V
HD12661c	1.56	1444.5	2.56	0.2	27.4	0.29	1.07	G6V
HD106252b	6.79	1503	2.53	0.57	150.7	-0.16	1.05	G0
HD37124c	1	1550	2.5	0.4	20	-0.32	0.91	G4V
HD23596b	8	1558	2.87	0.31	125	0.32	1.29	F8
HD30177b	7.64	1620	2.65	0.21	140	0.25	0.95	G8V
HD168443c	16.96	1770	2.87	0.2	289	0.1	1.01	G5
14 Her b	3.9	1775	2.87	0.37	70.4	0.5	0.79	K0V
HD72659b	2.54	2185	3.24	0.18	41.8	-0.14	0.95	G0V
HD38529c	12.78	2207.4	3.71	0.33	169.1	0.313	1.39	G4IV
HD39091b	10.39	2280	3.5	0.63	194	0.09	1.1	G1IV
HD74156c	7.46	2300	3.47	0.4	121	0.13	1.05	G0
HD33636b	9.3	2440	3.5	0.52	164	-0.09	0.99	G0V
Eps Eri b	0.92	2550	3.39	0.43	17.5	-0.1	0.8	K2V
Gl777A	1.15	2613	3.65	0	17.6	0.14	0.9	G6IV
47 UMa c	0.76	2640	3.78	0	11	-0.08	1.03	G0V
55 Cnc d	4.05	5360	5.9	0.16	49.3	0.27	1.03	G8V

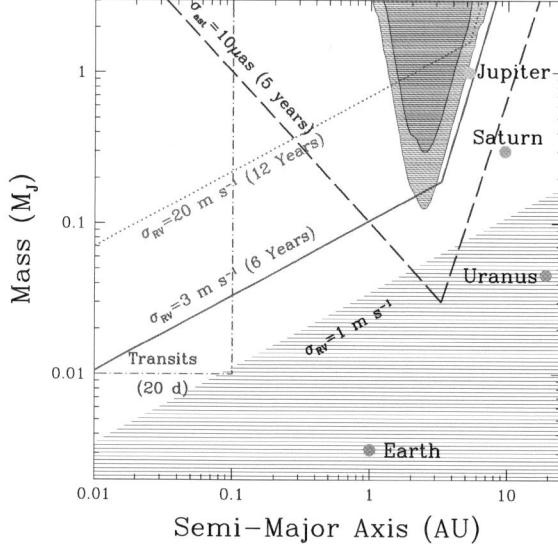

Fig. 2.6. The plot, adapted from Gaudi (2003) shows the current microlensing constraints on exoplanets along with the sensitivities of radial velocity (rv), astrometric (ast) and transit detection methods. The hatched triangular wedge at the bottom of the diagram is used to indicate the expected 1 m s^{-1} precision of the radial velocity technique. The exclusion region from the PLANET microlensing survey based on the 5 year results (mplanet.anu.edu.au) is shown as the triangular hatched regions towards the top right ; $< 45\%, < 33\%$ (outer, inner) of M dwarfs in the bulge have companions in these regions.

2.3 Properties of Exoplanets

Table 2.1 lists the properties of exoplanets and their host stars announced up until the end of 2003 February. This table of exoplanets is based primarily on exoplanets.org/almanacframe.html (Marcy et al. 2003) and makes use of www.obspm.fr/encycl/catalog.html (Schneider 2003) for metallicities [Fe/H], primary masses (M_{prim}) and spectral type (SpType). It is important to note that this is primarily a table of exoplanet *candidates*. They should be considered as *candidates* because apart from Gl876 (Fig. 2.4) and HD 209458 (Fig. 2.5), their orbital inclinations are not well constrained. The exoplanets in orbit around the M dwarf Gl876 are included in the table but generally excluded from further plots and discussions. This is because the rest of the exoplanet hosts are F through K stars. It is expected that the environment for planet formation around M dwarfs may be very different from that of the more massive dwarfs (Trilling, Lunine & Benz 2003). Furthermore Laws et al. (2003) find evidence for a slight decrease in the incidence of radial velocity companions orbiting relatively less massive stars.

The table includes ten multiple systems though there are a number of additional multiples which have not been included since the constraints on their parameters are poor, e.g. the second planet around HD 160691 apparent from the long-period trend in Fig. 2.3. It is expected that as the frequency and quality of radial velocity data continues to improve the number of multiples will rapidly increase. While it is premature to assess the fraction of multiple systems it is important to note that the long running Lick sample of 51 stars has so far found two triples and one double in addition to a further five so-far single planets (Fischer et al. 2003). This means that the multiple fraction is then 3/51 and the current fraction bearing any exoplanet is then 8/51 or 15%. It is vital to note that this is a *lower* limit to the fraction of exoplanets around *suitable* solar-type stars. The original Lick sample numbered 100, though was paired down to 51 by the removal of spectroscopic binaries, chromospherically active, rapidly rotating and evolved stars.

2.3.1 51 Peg's Exoplanet Is Not Typical

While the discovery of 51 Peg was a landmark, as with the earlier discovery of planets around pulsars it was met with scepticism in part because of the difficulties of the measurement but as much because 51 Peg b seemed to have nothing in common with our Jupiter (apart from mass). Its orbit seemed to require radical new ideas about the formation of planets; though in fact all that was required was the rediscovery of the robust theoretical concept of inward planetary migration driven by tidal interactions with the protoplanetary disk (Goldreich & Tremaine 1980; Lin & Papaloizou 1986; Ward & Hourigan 1989). Thus while such a large mass planet could not form in the glare of radiation from its Sun, it was entirely plausible that it had migrated into position through the disk of material around 51 Peg (Lin, Bodenheimer & Richardson 1996). Although 51 Peg-like objects dominated the early discoveries, other types of planets are much more common. Of the 100 or so exoplanets that have been discovered the 51 Peg-like planets (3-5 day orbital periods) comprise 12 and represent a class of planets circling about 0.5% of stars. The 51 Peg class were found first because they are easiest to find. Relatively heavy and close to their stars they exert the largest force on their stars and are thus by far the easiest to detect by the radial velocity method. As more planets are discovered other types of biases in our understanding of exoplanets resulting from our experimental sensitivity are starting to reveal themselves.

Although the exoplanets listed in Table 2.1 were all found using the radial velocity technique they are not discovered from a single well documented and quantified methodology. The compilation relies on a number of different ongoing surveys operating with different samples, sensitivities, instruments, scheduling, strategies and referencing techniques. Cumming et al. (1999, 2003) has thoroughly investigated the observational biases inherent in the Lick and Keck surveys but has yet to report findings for the bulk of detected exoplanets. So far none of the surveys have the 3 m s^{-1} precision over 15 years necessary to detect Jupiter and thus do not yet constrain the

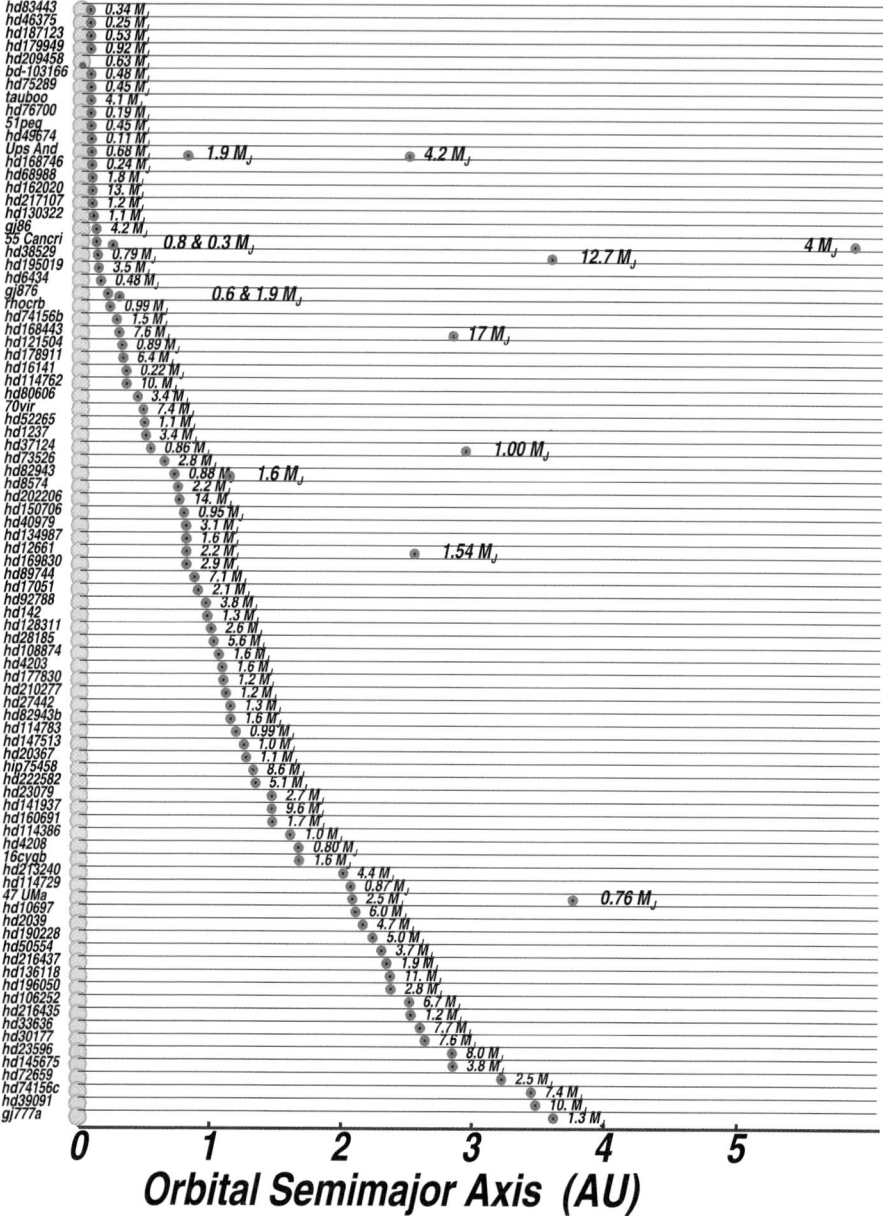

Fig. 2.7. The figure courtesy of exoplanets.org (Marcy et al. 2003) shows the masses and orbital radii of exoplanets found by the radial velocity technique. The left-hand side of the figure indicates the names of primary stars. Masses are labelled in units of Jupiter mass (M_{JUP}) corresponding to 318 Earths.

frequency of Solar Systems analogs. Nonetheless a very wide range of properties have already been found. Figure 2.7 illustrates exoplanets are known with

Fig. 2.8. The figure shows an equatorial density structure for $M_{\mathrm{disc}} = 0.1 M_{\mathrm{star}}$ and is courtesy of Rice et al. (2003). The disc is highly unstable and is fragmenting. The fragments are all gravitationally bound.

- a range of orbital radii from a few days out to beyond the orbit of Jupiter,
- a range of masses from just below Saturn-mass to many times that of Jupiter,
- multiple companions.

The relatively large number and fraction of planets that we have discovered before achieving sensitivity to our own Solar System together with the wide range of parameters discovered suggests that planetary formation and survival are robust. This seems to be borne out by theoretical work, e.g., Fig. 2.8 which indicates that planet formation is an 'easy come, easy go' business, with many planets created and many destroyed, and with an important minority – including our own Jupiter – surviving (Trilling, Lunine & Benz 2003).

2.3.2 Brown-Dwarf Desert/Planet Jungle Around *Solar-Type Stars*

1995 was a watershed year, not just for exoplanets but also for brown dwarf research. A conference on cool stars in Florence saw the announcement of the exoplanet 51 Peg b as well as the brown dwarf Gl229B. Brown dwarfs bridge the gap between stars and planets. Too small and cool to be a star and sustain thermonuclear hydrogen burning but yet too massive to be a planet. (See also Chap. 1 by I.N. Reid and S.L. Hawley in this volume.) Whilst a steady refinement of radial velocity searches means they are spectacularly sensitive to exoplanets, searches are actually far more sensitive to the presence of brown dwarfs which have hardly been found. For many years it was expected that planets would be found by extending radial velocity searches of brown dwarfs

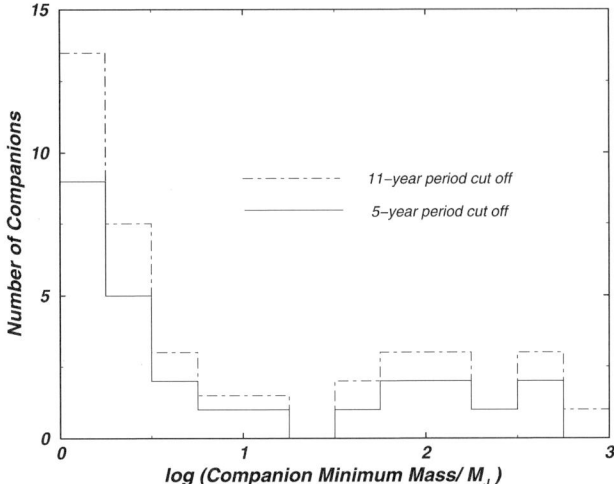

Fig. 2.9. The distribution of companion minimum mass uncovered by the Anglo-Australian planet search by Blundell et al. (2003, MNRAS, submitted) is shown by a dotted line; the solid line shows the distribution corrected for completeness after 11 years.

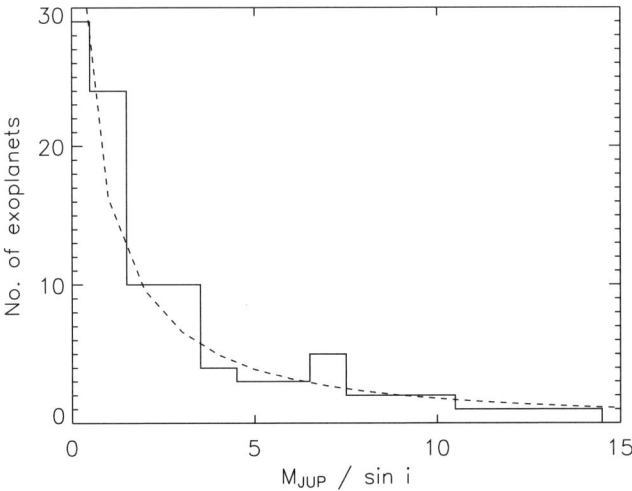

Fig. 2.10. The number of exoplanets per unit mass are shown as a solid line. The dotted line depicts number of exoplanets proportional to $\text{Mass}^{-1.3}$.

to lower masses. Figure 2.9 shows the problem. While stellar radial velocity companions are relatively abundant and easy to detect, there is a relative deficit of objects from around 100 to 10 M_{JUP}, approximately the brown dwarf regime. Once sensitivity to below 10 M_{JUP} is achieved the number of exoplanets rapidly increases. The increase in the mass function towards lower masses for all exoplanets is shown in Fig. 2.10. This sharply rising detection rate at low masses is found against a sensitivity function for finding planets that falls in proportion to mass.

Before the lack of brown dwarfs around solar type stars can be concluded it will be necessary to acquire orbital inclinations for a reasonable sample of the objects in Table 2.1, although Jorissen, Mayor & Udry (2001) have made a statistical deconvolution of known exoplanets and find very few of them to lie above 10 M_{JUP}. While the distribution of binary mass ratios for very different masses is poorly constrained, the dramatic turnaround in the mass function between brown dwarfs and exoplanets seems good evidence for planets and brown dwarfs having different formation and survival mechanisms. Brown dwarfs are modelled to form as part of the star forming process (Bate, Bonnell & Volker Bromm 2003), whereas planets arise from the circumstellar disc left around stars after they have formed. Regardless of exact formation mechanism, Armitage & Bonnell (2002) find that the 'brown dwarf desert' is a consequence of rapid orbital migration of brown dwarfs within an evolving protostellar disk.

2.3.3 Exoplanet Mass Function Rises Towards Low Masses

Figure 2.11 indicates how the discovered exoplanets compare with a two different mass functions. With the assumption that the mass function is constant with period it appears that a steep mass function is favoured. The simulations

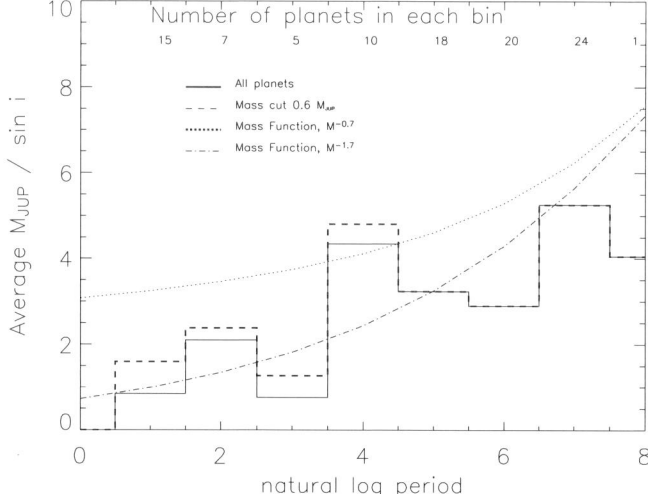

Fig. 2.11. The figure investigates the average mass of exoplanets per period bin. The histogram shows the uncorrected distribution and the dashed histogram shows the correction when the planets below 0.6 M_{JUP} are removed. The dashed and dotted lines indicate the mass functions for mass functions of -0.7 and -1.7 for the assumption of a mass function from 0.6-10 M_{JUP}. At the top of the plot is shown the number of exoplanets in each period bin.

of Tabachnik & Tremaine (2002) and Zucker & Mazeh (2002) using around 60 planets favour a much flatter mass distribution. The Bayesian approach of Cumming et al. (2003) promises to incorporate a knowledge of detection sensitivities for a single sample. So far, the relative small numbers of objects as well as the selection biases preclude much confidence in a particular value of the mass function. Zucker & Mazeh (2002) note the probable lack of high mass exoplanets at short periods. Based on Fig. 2.11 it seems that the expected lower average mass of close-in planets persists to periods beyond the 51 Peg-type objects.

2.3.4 The Parent Stars of Exoplanets Are *Very* Metal-Rich

An important characteristic of a star is the fraction of 'metals' (elements heavier than Helium) it contains - the metallicity. Gonzalez (1998) found exoplanet host stars to be metal-rich. This conclusion has been confirmed by many authors with different samples, methodologies and spectral synthesis codes (e.g., Reid 2002; Santos et al. 2003). Figure 2.12 shows just how metal-rich the exoplanet primaries are. Only a single bin is anywhere close to solar metallicity. All other bins are at least 0.1 dex above the solar; whereas the Sun and other solar type dwarf stars in the solar neighbourhood have an average metallicity of 0 or even slightly less (Reid 2002). Above solar metallicities, where most of the known exoplanets are hosted, the probability of detecting a exoplanet is proportional to its metallicity. By a metallicity of +0.3 dex the frequency of stars with exoplanets is effectively 100%. This re-

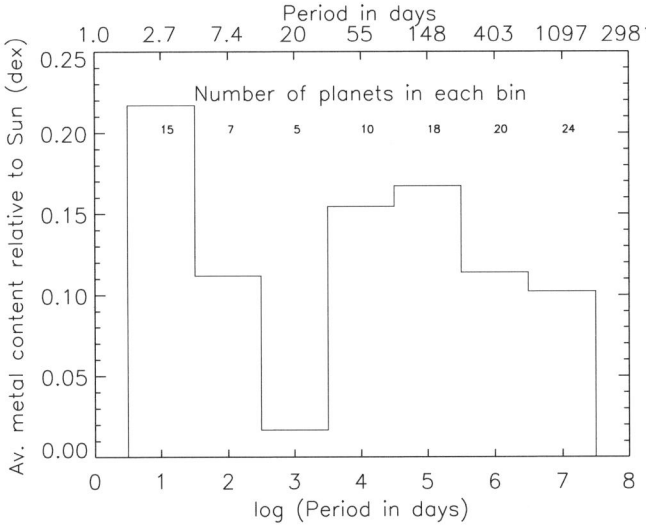

Fig. 2.12. Average metallicity of the primaries of exoplanets plotted as a function of period. The last bin at a natural log Period of 8 has been omitted because it contains a single object.

sult, representing the only link between the presence of planets and a stellar photospheric feature, has been given two main explanations. The first is based on the classical view that giant planets are formed by runaway accretion of gas on to a 'planetesimal' having up to 10 Earth masses. In such a case, we can expect that the higher the proportion is of dust to gas in the primordial cloud (i.e. metals), and consequently in the resulting protoplanetary disc, the more rapidly and easily may planetesimals, and subsequently the observed giant planets be built.

Opposing this view, it had been proposed that the observed metallicity 'excess' may be related to the 'pollution' of the convective envelope of the star by the infall of planets and/or planetesimals. So far evidence for this has only been found in one of the many systems investigated. While such pollution may play a small role, giants stars with planets all have high metallicities

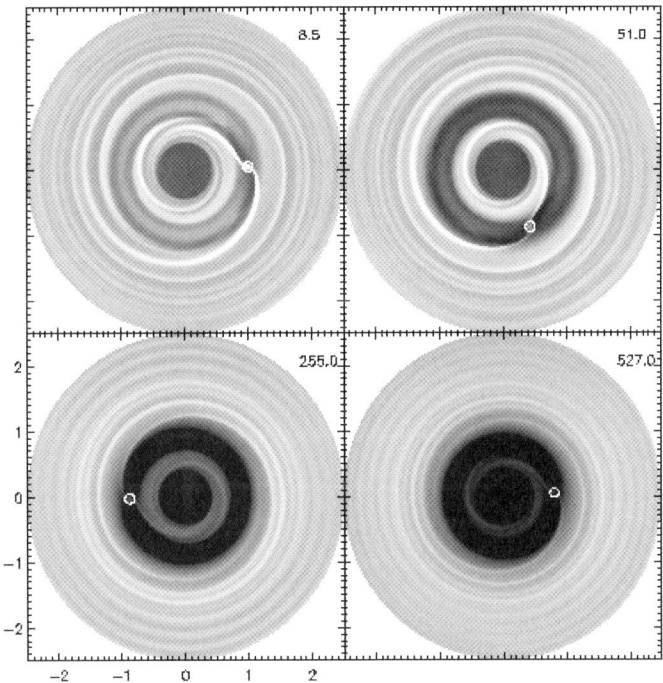

Fig. 2.13. This figure (courtesy of Nelson et al. 2000) shows the evolution of a protoplanet embedded in a protostellar disc for calculation. The relative surface density of disc material is represented by the grey-scale. The white circle represents the position of the protoplanet. The disc is initially unperturbed at the outset of the simulation. The number of orbits are depicted in the top right-hand corner of each snapshot. The transfer of angular momentum between the disk and the protoplanet leads to the formation of an annular gap, or surface density depression, in the vicinity of the planet's orbit, which is cleared after about 200 orbits for a Jupiter-mass planet.

and show no evidence for pollution (Santos et al. 2003). These are crucial systems because they are well mixed and not subject to the mixing and surface uncertainties that limit our confidence in measuring 'pollution' in Sun-like stars. Besides the [Fe/H] differences, there is currently some debate about possible anomalies concerning other elements. But the relatively low number of exoplanets known, and possible systematics with respect to the samples do not permit firm conclusions to be reached (Santos et al. 2003). So far, it seems that the higher metallicity of most planet-harbouring stars arises because high metallicity environments have a higher probability of planet formation and migration. The relatively low metal content of the Solar System may be consistent with the relative lack of migration (Lineweaver 2001). Thus it appears that like the early detections of 51 Peg-type exoplanets we are finding a surfeit of exoplanets around metal-rich stars because they are easier to detect.

Figure 2.12 shows that the parent stars of 51 Peg-like exoplanets have the highest metallicity compared with the rest of the sample suggesting that exoplanet migration is more efficient in metal-rich systems. The idea that migration is tied to metallicity also seems to be borne out by the overall decreasing metallicity for increasing period. Following the peak of 51 Peg-like metallicities there is a curious drop in average metallicity. Perhaps the average distribution of planets in different metallicity environments is quite different. The migration of large planets is modelled to be accompanied by gap formation in the protostellar disc (e.g. Fig. 2.13). Perhaps in the highest metallicity systems, the efficiency of migration allows detectable planets to

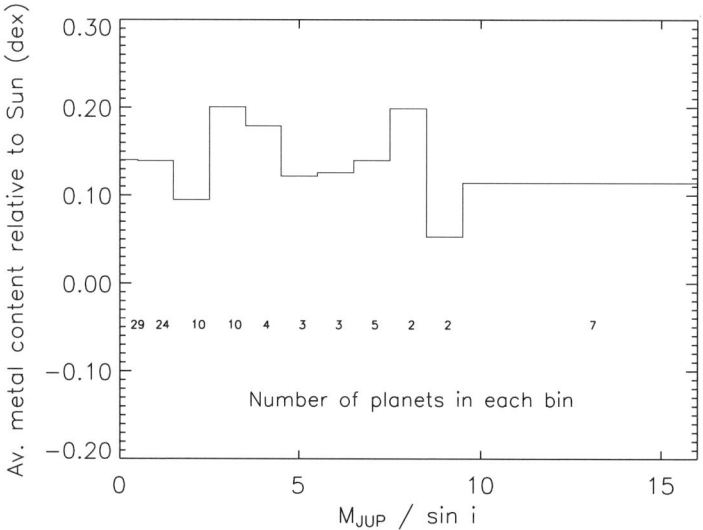

Fig. 2.14. Average metallicity versus mass for exoplanets. The high mass exoplanets (10–16 $M_{JUP}/\sin i$ have been combined into a single bin).

Fig. 2.15. In the centre of the image, courtesy of Grosso et al. (2003), is a young stellar object HH 30 IRS at the edge of the Rho Ophiuchi dark cloud. The light from the central star is blocked out by dust grains in the disk. Other grains below and above the disk midplane scatter the stellar light, producing a typical pattern of a dark lane between two reflection nebulae. This system known as the 'Flying Saucer' should provide an excellent target for studies of the protoplanetary disk and the nascent planets in the system.

move-in to 51 Peg type orbits or otherwise accrete onto their host star but setup a gap region at slightly larger radii inhibiting planets in this regime. Then planets are seen in this 'gap' region only in lower metallicity systems where migration properties are different. One might expect to see a relationship between metallicity and mass of exoplanets though yet none is readily apparent, e.g. Fig. 2.14. Millimetre observations with array instruments such as ALMA may enable the spatial resolution to discern the various components of protostellar disks such as that in Fig. 2.15.

2.3.5 Exoplanet Atmospheres

Future space missions such as DARWIN and TPF should enable direct exoplanet spectra to be acquired rather than making inferences about exoplanets from studying their host stars. Synthetic spectra for exoplanet atmospheres have made rapid enhancements through modelling of brown dwarfs and the incorporation of stellar irradiation leading to detailed predictions for observable exoplanet polarisations. Their first direct test has come from HD209458. In addition to the exquisite light curve (Fig. 2.5), Charbonneau et al. (2002) have made a detection of the atmosphere of HD209458b. They

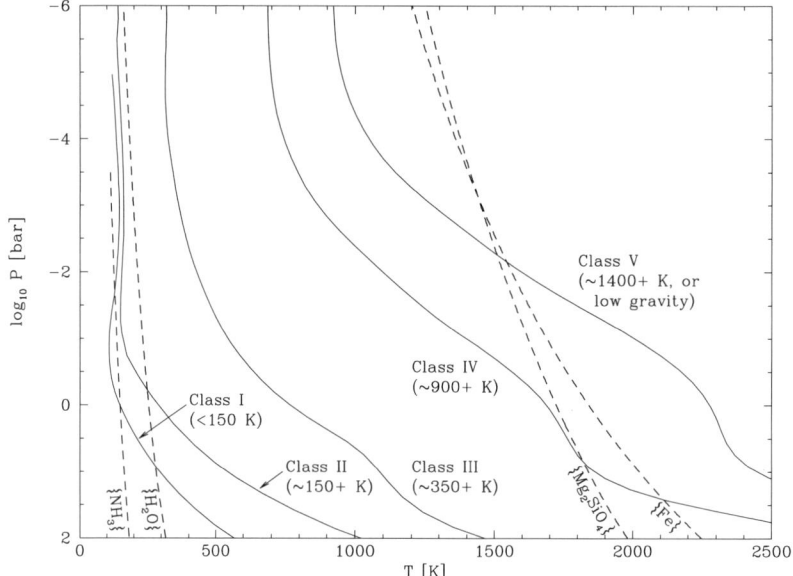

Fig. 2.16. The figure, courtesy of Sudarsky, Burrows & Hubeny (2003), depicts sample temperature-pressure structures for Classes I through V. Also shown are the condensation curves for the principal condensates, ammonia, water, forsterite (a representative silicate), and iron.

detected additional dimming of sodium absorption during transit due to absorption from sodium in the planetary atmosphere. The observed dimming is reasonably well modelled by planetary atmosphere models that incorporate irradiation and allow for sodium to be out of thermal equilibrium (Barman et al. 2002). Sudarsky et al. (2000, 2003) have classified exoplanets into 'composition classes'. Figure 2.16 shows the exoplanet types I to V. The Class I profile intersects the ammonia condensation curve at roughly one bar. At greater pressures, the temperature is too high for ammonia condensation to occur. This profile also intersects the water condensation curve at nearly 10 bars, indicating the presence of water clouds at this depth. Of course, a whole variety of Class I objects with differing temperature-pressure profiles will exist, and the ammonia and water condensation curves will be intersected at different pressures for each object, but by definition Class I EGPs have an ammonia cloud that is well above a water cloud. The warmer Class II profile crosses only the water condensation curve, indicating the existence of a water cloud in the upper atmosphere. Classes IV and V intersect the silicate and iron condensation curves, but at quite different pressures.

2.3.6 Why Don't Exoplanets Have Circular Orbits Like Our Solar System?

Apart from the short-period exoplanets whose orbits are circularised by the tidal pull from their parent star, Fig. 2.17 shows the eccentricity of exoplanet orbits is much higher than in the Solar System. Fischer et al. (2003) find that it is rather close to that observed for stars. According to our paradigm of planetary formation a planet (formed in a disk) should keep a relatively circular (low eccentricity) orbit. In order to boost exoplanet eccentricities it is necessary to imagine interactions between multiple planets in a disk and between a planet and a disk of planetesimals and perhaps the influence of a distant stellar companion. In fact dynamical interactions between planets seem inevitable since even with the fairly poor sampling of known exoplanets, 10 multiple systems have already been found. Of these a number are in resonant orbits. Thus 'dynamical fullness' is probably important and suggests that interactions play a vital role in determining the properties of many exoplanets. These orbital complexities mean that to understand exoplanets more generally, it will be necessary to find all the main components in planetary systems. This will require using results from different techniques, particularly radial velocity and astrometric, to disentangle the various planetary components in orbit around nearby stars.

Fischer et al. (2003) suggest that selection effect may play a role in the high eccentricities observed among the single exoplanets discovered to date. Most known exoplanets reside within 3 AU due to the limited duration (10 yr)

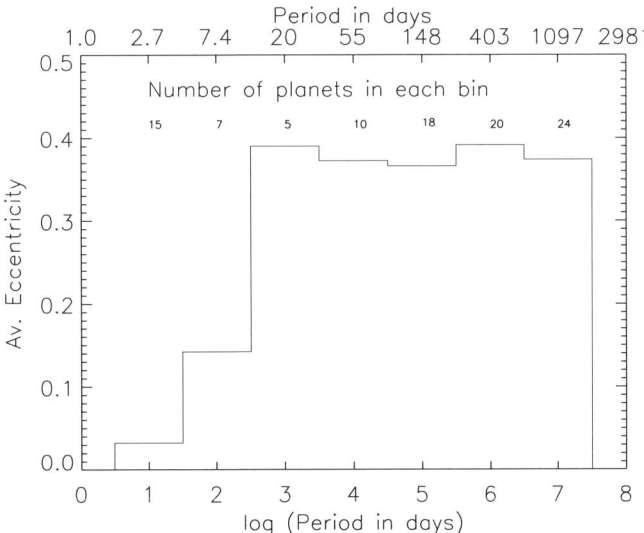

Fig. 2.17. Eccentricity versus natural log period for exoplanets. In the Solar System, Pluto is the only planet with an eccentricity of greater than 0.1. HD80606 b ($e = 0.93$) is in a stellar binary system and has been omitted from this plot.

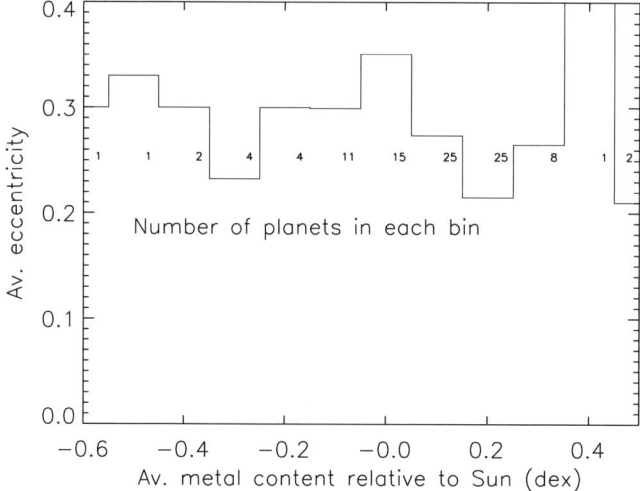

Fig. 2.18. Eccentricity versus metallicity for exoplanets.

of the Doppler surveys. Thus the planets detected to date represent a subset that ended up within 3 AU. Giant planets within 3 AU may systematically represent the survivors of scattering events in which the other planet was ejected while extracting energy from the surviving planet and throwing it inward. This would give rise to us systematically detecting the more massive, surviving planet residing in an orbit with a period less than 10 years. At least Fig. 2.18 suggests that metallicity does not play an important role.

Various mechanisms have been proposed to explain the orbital eccentricities, namely planet-planet interactions, planet-disk interactions, and planet migration leading to resonance capture. A comprehensive model for orbital eccentricities may include more than one process. Two or more planets will migrate in a viscous disk at different rates, allowing them to capture each other in mean-motion resonance. Such resonances can then pump the eccentricities of both planets up to values as high as 0.7 (Chiang 2003) as observed. The growth in eccentricities, especially as the disk dissipates, can render the two orbits unstable as close passages occur. Ford et al. (2003) show that such close passages often lead to ejection of the less massive planet, leaving one behind. This sequential set of processes provides a natural explanation for the resonances and eccentricities observed among single and multiple planets. Models combining dynamical scattering and tidal interactions can now naturally account for the observed range of periods and eccentricities (Adams & Laughlin 2003).

2.3.7 The Two Component Period Distribution

The distribution of exoplanets with period in Figs. 2.11 and 2.12 suggest that exoplanets show key differences with period. Figure 2.19 indicates two

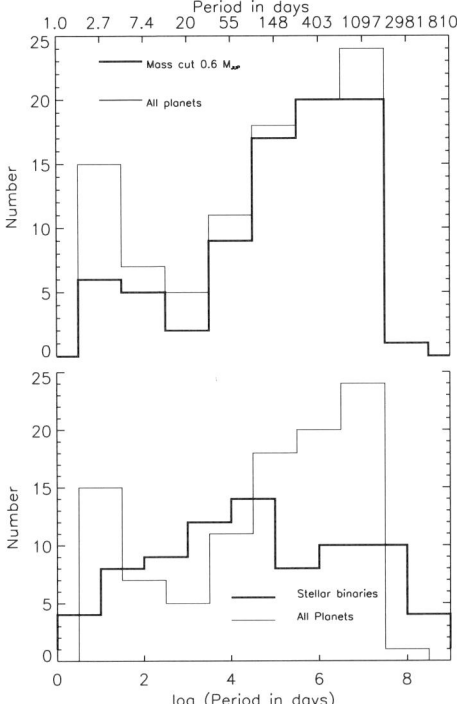

Fig. 2.19. The number of exoplanets discovered within natural logarithm period bins is shown (Jones et al. 2003). The top part of the figure compares exoplanets with masses less than 10 $M_{\rm JUP}/\sin i$ and greater than 0.6 $M_{\rm JUP}/\sin i$. The 0.6 $M_{\rm JUP}/\sin i$ cut is chosen to allow for the incompleteness introduced by including lots of more readily detected low-mass short-period exoplanets (Armitage et al. 2002). The bottom part compares exoplanets with the stellar binaries as found by Duquennoy & Mayor (1991).

separate features in the exoplanet period distribution. A peak of short-period exoplanets is seen in the 51 Peg-type objects, then a dearth, followed by an smooth rise in the number of exoplanets toward longer periods. The short-period peak can be explained by a stopping mechanism such as Lindbad resonances between the planet and disk (Kuchner & Lecar 2002) and photoevapouration (Matsuyama, Johnstone & Murray 2003) limiting exoplanet migration to periods of greater than 3 days. Once there exists a sample of short-period exoplanets derived from a wider range of primary masses it should be possible to distinguish between different stopping mechanisms by comparing the orbital characteristics of the exoplanet with those of the primary.

For example, Kuchner & Lecar predict that planets around early A-type main sequence stars will collect at a radius much further from the star (0.3 AU) than the radius where solar-type stars collect. Inspection of Table 2.1 suggests that there may indeed be a small trend toward higher mass stars with circularised planets occurring at slightly larger radii. The rise in the number of exoplanets towards longer periods is becoming more apparent are discovered and is very well reproduced by exoplanet migration scenarios which envisage planets migrating inwards (e.g., Armitage et al. 2002; Trilling et al. 2003) as well as outwards (Masset & Papaloizou 2003).

2.4 The Future of Planet Searches

The impetus of discovery and characterisation means that exoplanets discovery should continue to increase as objects are found from a wide range of techniques. The power of characterisation using several techniques has already been proven for Gl876 and HD209458. A much deeper understanding of exoplanets should become apparent once such data exists for a large sample. The 'dynamical fullness' seen for our own Solar System and the multiple exoplanets, suggests that the characterisation of exoplanets will increasingly be made using a variety of techniques. In the short-term the combination of radial velocity, astrometry, transit and interferometric imaging measurements seem most promising. Notwithstanding an increasing rate of discovery and characterisation of an even broader spectrum of exoplanets, the next few years should continue to bring dramatic improvements in the realism of exoplanet formation and evolution simulations.

The revolution in exoplanet science is just beginning. At the moment we are sensitive to a small fraction of the planet parameter space and in particular are not yet sensitive to our own Solar System going around a nearby star. Despite having plenty of examples of exoplanets and of higher mass brown dwarfs, we are still challenged by the definition of a planet. We need a better idea of planets as a function of host star properties, planet mass, composition (gas, ice, rock, metal) and orbit parameters and the intercorrelation of parameters. It will be fascinating to see the importance of environment on exoplanets, not only on the major planets that we detect over the next few years but also the minor constituents such as comets and asteroids. Substantial research into chemical differentiation will be necessary, do other terrestrial planets have non-equilibrium atmospheres? The range of ground and space based endeavours suggests that by the time current generation of astrophysically inclined school leavers have finished their PhDs, the research edge may have moved to the characterisation of more Earth-like planets. By then the technical and scientific resources should exist to start to constrain a range of scenarios for exolife.

References

Adams F.C., Laughlin G., 2003, Migration and dynamical relaxation in crowded systems of giant planets, Icarus, in press

Armitage P.J., Bonnell I. A., 2002, The brown dwarf desert as a consequence of orbital migration, MNRAS, **330**, L11

Armitage P.J., Livio M., Lubow S.H., Pringle J.E., 2002, Predictions for the frequency and orbital radii of massive extrasolar planets, MNRAS, **334**, 248

Backer D.C., Foster R.S., Sallmen S., 1993, A second companion of the millisecond pulsar 1620-26, Nature, **365**, 817

Bate M.R., Bonnell I.A., Bromm V., 2003, The formation of a star cluster: predicting the properties of stars and brown dwarfs, MNRAS, **339**, 577

Baranne A. et al., 1996, ELODIE: A spectrograph for accurate radial velocity measurements, A&AS, **119**, 373

Barman T. et al., 2002, Non-LTE Effects of Na I in the Atmosphere of HD 209458b, ApJ, **569**, 51

Benedict et al., 2002, A Mass for the Extrasolar Planet Gl 876b Determined from Hubble Space Telescope Fine Guidance Sensor 3 Astrometry and High-Precision Radial Velocities, ApJ, **581**, L115

Brown T., Charbonneau D., Gilliland R., Noyes R., Burrows A., 2001, HST Time-Series Photometry of the Transiting Planet of HD 209458, ApJ, **552**, 699

Butler R.P., Marcy G.W., Williams E., McCarthy C., Dosanjh P., Vogt S.S., 1996, Attaining Doppler Precision of 3 m s^{-1}, PASP, **108**, 500

Charbonneau D., Brown T.M., Noyes R.W., Gilliland R.L., 2002, Detection of an Extrasolar Planet Atmosphere, ApJ, **568**, 377

Chiang E.I., 2003, Excitation of Orbital Eccentricities by Repeated Resonance Crossings: Requirements, ApJ **584**, 465

Cumming A., Marcy G.W., Butler R.P., 1999, The Lick Planet Search: Detectability and mass thresholds, ApJ, **526**, 896

Cumming A., Marcy G.W., Butler R.P., Vogt S.S., 2003, The Statistics of Extrasolar Planets: Results from the Keck Survey, ASP Conference Series 294: Scientific Frontiers in Research on Extrasolar Planets, eds. D. Deming and S. Seager, in press

Duquennoy, A., Mayor, M., 1991, Multiplicity Among Solar Type Stars in the Solar Neighbourhood, A&A, **248**, 485

Fischer D.A. et al., 2003, A Planetary Companion to HD 40979 and Additional Planets Orbiting HD 12661 and HD 38529, ApJ, **586**, 1394

Ford E.B., Rasio F.A., Yui K, 2003, Dynamical Instabilities in Extrasolar Planetary Systems, ASP Conference Series 294: Scientific Frontiers in Research on Extrasolar Planets, eds. D. Deming and S. Seager, in press

Gaudi B.S., 2003, Microlensing Searches for Extrasolar Planets: Current Status and Future Prospects, ASP Conference Series 294: Scientific Frontiers in Research on Extrasolar Planets, eds. D. Deming and S. Seager, in press

Goldreich P., Tremaine S., 1980, Disk-Satellite Interactions, ApJ, **241**, 425

Gonzalez G., 1997, The Stellar Metallicity-Giant Planet Connection, MNRAS, **285**, 403

Grosso N., Alves J., Wood K, Neuhaeuser R., Montmerle T., Bjorkman J.E., 2003, Spatial study with the VLT of a new resolved edge-on circumstellar dust disk discovered at the periphery of the rho Ophiuchi dark cloud, A&A, **586**, 296

Horne K., 2003, Status and Prospects of Planetary Transit Searches: Hot Jupiters Galore, ASP Conference Series 294: Scientific Frontiers in Research on Extrasolar Planets, eds. D. Deming and S. Seager, in press

Jones H.R.A., Butler R.P., Tinney C.G., Marcy G.W., Penny A.J., McCarthy C., Carter B.D., 2002, Extrasolar planets around HD 196050, HD 216437 and HD 160691, MNRAS, **337**, 1170

Jones H.R.A., Butler R.P., Tinney C.G., Marcy G.W., Penny A.J., McCarthy C., Carter B.D., 2003, An exoplanet around Tau1 Gruis, MNRAS, **341** 948

Jorissen A., Mayor M., Udry S, 2001, The distribution of exoplanet masses, A&A, **379**, 992

Konacki M., Torres G., Jha S., Sasselov D., 2003, A new transiting extrasolar planet, Nature, **421**, 507

Kuchner N., Lecar M., 2002, Halting plant migration in the evacuated centres of protoplanetary disks, ApJL, **574**, 87

Laws C. et al., 2003, Parent Stars of Extrasolar Planets VII: New Abundance Analyses of 30 Systems, AJ, **125** 2664

Lineweaver C.H., 2003, An Estimate of the Age Distribution of Terrestrial Planets in the Universe: Quantifying Metallicity as a Selection Effect, Icarus, **151**, 307

Lin D.N.C., Bodenhimer P., Richardson D.C., 1996, Orbital migration of the planetary companion of 51 Pegasi to its present location, Nature, **380**, 606

Lin D.N.C., Papaloizou J.C.B., 1986, On the Tidal Interaction Between Protoplanets and the Protoplanetary Disk. III Orbital Migration of Protoplanets, ApJ, **309**, 846

Lucas P.W., Roche P.F., 2000, A population of very young brown dwarfs and free-floating planets in Orion, MNRAS, **314**, 858

Marcy G.W., Butler R.P., Vogt S.S., Fischer D.A., McCarthy C. et al., 2003, The California & Carnegie Planet Search, http://exoplanets.org

Masset F.S., Papaloidzou J., 2003, Runaway migration and the formation of hot Jupiters, ApJ, **588** 494

Matsuyama I., Johnstone D., Murray N., 2003, Halting planet migration by photoevaporation from the central source, ApJL, **585**, 143

Mayor M., Queloz D., 1995, A Jupiter-mass companion to a solar-type star Nature, **378**, 355

McCaughrean M.J., Reid I.N., Tinney C.G., Kirkpatrick J.D., Hillenbrand L.A., Burgasser A., Gizis J., Hawley S.L., 2001, What is a planet?, Science **291**, 1487

Nelson R., Papaloizou J.C.B., Masset J., Kley W., 2000, The Migration and Growth of Protoplanets in Protostellar Discs, MNRAS, **318**, 18

Oppenheimer B., 1999, Brown dwarf companions of nearby stars, PhD, California Institute of Technology

Paczynski B., 1986, Gravitational Microlensing at Large Optical Depths, ApJ, **301**, 503

Perryman M., 2000, Extra-Solar Planets, Rep. Prog. Phys., **63**, 1209

Reid I.N., 2003, On the Nature of Stars with Planets, PASP, **114**, 306

Rice W.K.M., Armitage P.J., Bate M.R., Bonnell I.A., 2003, The effect of cooling on the global stability of self-gravitating protoplanetary discs, MNRAS, **339**, 1025

Santos N.C., Israelian G., Mayor M., Rebolo R., Udry S., 2003, Metallicity, orbital parameters, and space velocities, A&A, **398**, 363

Saumon D. et al., 2000, Molecular Abundances in the Atmosphere of the T Dwarf Gl229B, ApJ, **541**, 374

Schneider J., 2003, The Extrasolar Planets Encyclopaedia, http://www.obspm.fr/encycl/encycl.html

Sudarsky D., Burrows A., Pinto P., 2000, Albedo and Reflection Spectra of Extrasolar Giant Planets, ApJ, **538**, 885

Sudarsky D., Burrows, A., Hubeny P., 2003, Theoretical Spectra and Atmospheres of Extrasolar Giant Planets, ApJ, **588** 1121

Tabachnik S., Tremaine S., 2002, Maximum-likelihood method for estimating the mass and period distributions of extrasolar planets, MNRAS, **335**, 151

Trilling D., Lunine J.I., Benz W., 2003, Orbital migration and the frequency of giant planet formation, A&A, **394**, 241

Udalski A. et al., 2002, The Optical Gravitational Lensing Experiment. Planetary and Low-Luminosity Object Transits in the Carina Fields of the Galactic Disk, Acta Astronomica, **52**, 317

Ward W.R., Hourigan K., 1989, Orbital Migration of Protoplanets: The Inertial Limit, ApJ, **347**, 490

Wolszczan A., Frail D., 1992, A Planetary System around the Millisecond Pulsar PSR1257+12, Nature, **255**, 145

Zapatero Osorio M.R., Bejar V.J.S., Rebolo R., Martin E.L., Basri G., 1999, An L-Type Substellar Object in Orion: Reaching the Mass Boundary between Brown Dwarfs and Giant Planets, ApJ, **524**, L115

Zucker S., Mazeh T., 2002, On the Mass-Period Correlation of the Extrasolar Planets, ApJ, 568, 113L

3 What Drives the Evolution of Close Binary Stars? The Test Case of Cataclysmic Variables

C. Hellier

3.1 Introduction: The Need for Magnetic Braking

The interaction of two stars in a close binary gives rise to a diversity of variable-star types. Some are rare curiosities, such as the strange binaries SS433 and Her X-1, while others are of major astrophysical importance. For example the progenitors of type 1a supernovae are thought to lurk amongst the close binaries, though their identity is unclear. But what drives the evolution of such binaries?

Mass transfer, occurring when one component of a binary overfills its Roche lobe, is one possibility, but this can only drive evolution that is so rapid as to be rarely seen. To appreciate this, consider moving material from the more massive to the less massive star. The material has to move further from the center of mass, and thus gains angular momentum. The process must therefore remove angular momentum from the orbit, bringing the stars closer together and so shrinking their Roche lobes, leading to more mass transfer. The process is thus unstable, either on the dynamical timescale, or, depending on how the radius of the donating star responds to mass loss, on the Kelvin–Helmholtz timescale, both of which are short compared to a binary lifetime.

The movement of material from the less massive to the more massive star will, by the same logic, increase the orbital separation and cause the donating star to detach from its Roche lobe, halting mass transfer. Hence, stable mass transfer via Roche-lobe overflow needs to be driven by a mechanism that removes angular momentum from the binary.

In higher-mass stars other effects can play a role, such as the expansion resulting from nuclear burning, or mass transfer from strong stellar winds, but these are less significant in low-mass stars ($\lesssim 1$ $M_\odot$) since the timescale to evolve off the main sequence is long and the stellar winds are feeble.

One angular-momentum-loss mechanism that will always occur is gravitational radiation (GR). But this scales as a^{-4} (where a is the binary separation) and so is rarely important for binaries with orbital periods exceeding a few hours, since the GR timescale will exceed the Hubble time, though it can dominate in ultra-short-period binaries.

The answer to "what drives the evolution of low-mass binaries?" is thus — by general acceptance — magnetic braking. The theory of magnetic braking, dating back to Weber & Davis (1967), invokes a stellar wind that is forced by the star's magnetic field to flow along field lines and thus to corotate with the star out to radii far exceeding the stellar radius. When the wind particles eventually break free, the long lever arm ensures that they carry away significant angular momentum, so braking the star's rotation. Further, if the star is in a close binary, and so distorted from spherical, tidal forces will ensure that it corotates with the orbit. Ultimately, the angular momentum is lost from the orbit (the largest reservoir of angular momentum), thus driving the binary to shorter periods (and so, paradoxically, causing the secondary to spin faster). The evolution to shorter orbital periods shrinks the Roche lobes and so effects mass transfer.

In principle, one of the best tests of the magnetic braking hypothesis would be to measure the rate of orbital-period change by timing the eclipses of high-inclination binaries. In practice, though, such data are dominated by short-term fluctuations on timescales of decades, most likely resulting from solar-type magnetic cycles of the component stars (e.g. Warner 1988), and so tell us nothing about secular evolution.

As a result, the crucial test is a comparison of the observed orbital-period distribution with the output of population-synthesis codes that incorporate magnetic braking. One of the best-studied classes of close binaries is the cataclysmic variables (CVs), in which a white dwarf (the primary) accretes from a low-mass, Roche-lobe-filling red dwarf (the secondary). This group is notable for showing a multitude of observational clues on a wide range of astrophysical topics and so is an excellent test-bed [see Warner (1995) for a thorough review of CVs, or Hellier (2001a) for an introductory account].

Modelling the orbital-period (P_{orb}) distribution of CVs has been a major industry (e.g. Paczynski & Sienkiewicz 1981; Rappaport, Joss & Webbink 1982; Rappaport, Verbunt & Joss 1983; Spruit & Ritter 1983; McDermott & Taam 1989; Kolb 1993; Howell, Nelson & Rappaport 2001; see King 1988 for an accessible review). At times, these endeavours have led to close agreement, suggesting that the problem was generally solved, with only details remaining to be clarified. Recently, though, new observational results have scuppered the observational underpinning of the magnetic-braking models used to reproduce the P_{orb} distribution, and have thrown into doubt the whole picture of CV evolution.

It is thus timely — and the purpose of this article — to rehearse the observational facts presented by CVs, and review our current understanding of them. To keep the article coherent, I don't deal with the double-degenerate AM CVn stars, where the different mass–radius relation of the helium-rich secondaries results in a very different evolution; nor do I address the CVs with highly magnetic white dwarfs, where the interplay of the primary's and secondary's magnetic fields might radically change the evolution (e.g. Webbink & Wickramasinghe 2002).

The CV P_{orb} distribution is shown in Fig. 3.1. One of the biggest difficulties in interpreting it is that it results from the convolution of several factors including: (1) the evolution of CVs, (2) the initial P_{orb} distribution they are born with, and (3) age effects, such as changes in the CV birth rate over galactic history, that would mean we don't see an equilibrium distribution.

We know very little about the last two. We think that CVs are a product of common-envelope evolution — the short orbital periods and thus small separations imply that the red-dwarf secondary would have been engulfed in the red-giant envelope when the white dwarf formed. During the common-envelope phase the friction would have caused the stars to spiral in, reducing the orbital period by a factor of ~ 100, and expelling the envelope (e.g. Livio 1996), but the uncertainties in this process make predictions of the resulting P_{orb} distribution difficult. Even then, the post-common-envelope binary is likely to be detached, and will have to lose further angular momentum (presumably by magnetic braking) before becoming a CV. If the mass ratio is too high, the system might then undergo a phase of rapid thermal-timescale mass transfer before settling down to the sedate mass transfer of the observed CV population (e.g. Schenker & King 2002).

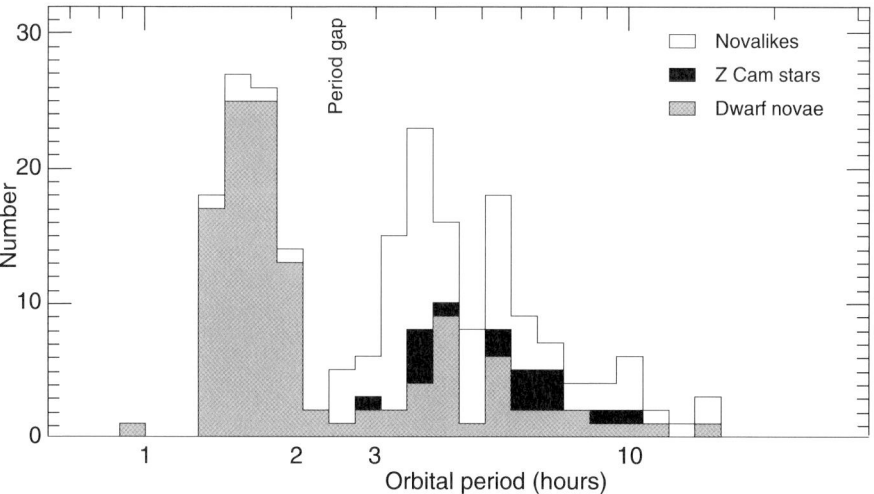

Fig. 3.1. The orbital-period distribution of cataclysmic variables, based on a compilation by Ritter & Kolb (figure from Hellier 2001a).

3.1.1 Fact 1: The number of CVs declines at orbital periods greater than ~ 7 hrs, and only a few are seen beyond 12 hrs

The existence of a maximum period is readily explained. The fact that stable mass transfer requires the secondary to be lighter than the white dwarf, and the existence of the Chandrasekhar limit on white-dwarf masses, places a limit on the mass of the secondary. In turn, this places a limit on the secondary's radius, assuming it to lie on the main-sequence, and thus on the size of a Roche lobe with which the star can make contact, and thus on the orbital period of the CV. These factors lead to a limit of $P_{\rm orb} \lesssim 12$ hrs, as observed, although the shape of the long-period tail of the distribution will depend on the distribution of white-dwarf masses in addition to the factors discussed in the last paragraph.

The occasional long-period CV, such as GK Per with $P_{\rm orb} = 48$ hrs, is then explained as a system that came into contact when the secondary had left the main sequence and begun ascending the giant branch.

3.1.2 Fact 2: We see a large spread of mass-transfer rates at all orbital periods

The existence of a spread in mass-transfer rate ($\dot{M}$) is based on two lines of argument. First, there is a critical $\dot{M}$ (a few 10^{-9} $M_\odot$ yr^{-1}) below which accretion discs undergo a hydrogen-ionization instability, producing dwarf-nova outbursts (e.g. Osaki 1996). We see both dwarf novae and systems with stable discs (called novalikes) at all orbital periods, implying a spread of $\dot{M}$.

Further, although distance estimates are always difficult, one can use the absolute magnitudes to estimate the mass-flow through the disc. One finds systems with $\dot{M}$ as high as a few 10^{-8} $M_\odot$ yr^{-1}, and others as low as a few 10^{-11} $M_\odot$ yr^{-1}, a rate compatible with that from GR (e.g. Patterson 1984; Warner 1995, Fig 9.8).

In principle, the magnetic-braking $\dot{M}$ at a given orbital period should be a function of the stellar masses and little else. This is because the rate of angular-momentum loss ($\dot{J}$) depends primarily on the rotation rate of the secondary, and on the secondary's field strength and stellar wind, which are thought to be dynamo driven and so also set by the rotation rate. Since the secondary's rotation is locked to the orbit, the braking $\dot{J}$ should be primarily a function of $P_{\rm orb}$.

Further, since the red-dwarf mass is set by the Roche-lobe size and thus by the orbital period, and since the white-dwarf mass varies by only a factor ~ 2, the dependence on stellar masses should have little effect. Thus, the observed $\dot{M}$ spread (factors of ~ 100 at any given orbital period) cannot be a feature of magnetic braking.

A possible solution is to invoke the nova explosions — the thermonuclear runaways of accreted material on the white-dwarf surfaces — that are thought to occur at intervals of $\sim 10^5$ yrs in all CVs. The ejection of 10^{-4} $M_\odot$ as a nova shell will cause an instantaneous change in $P_{\rm orb}$, which will alter the contact with the Roche lobe and hence change $\dot{M}$.

The most recent investigation (Kolb et al. 2001; Kolb 2002) suggests that this can produce an $\dot{M}$ spread of ~ 100, but only under restricted circumstances ($3 \lesssim P_{\rm orb} \lesssim 5$ hr, with a moderate white-dwarf mass, and a high ejecta mass). Thus the effect of nova explosions is probably not the whole story.

Another possibility results from the fact that mass-transfer produces accretion luminosity which irradiates the secondary and so might induce further mass transfer, with the feedback leading to an $\dot{M}$ well above the secular mean. This could only be sustained for a limited time, and would be followed by an episode of abnormally low $\dot{M}$ (Wu, Wickramasinghe & Warner 1995). The validity of the idea has been debated (e.g. King et al. 1995; 1996), but something similar is required to explain the observed $\dot{M}$ spread.

3.1.3 Fact 3: Few CVs are found in the 'period gap' between 2 and 3 hrs

The period gap is easily seen in Fig. 3.1 and is statistically highly significant (e.g. Hellier & Naylor 1998). We discuss this fact along with a companion statement:

3.1.4 Fact 4: Below the period gap most systems have a low mass-transfer rate; above the gap the mass-transfer rate is generally much higher

Figure 3.1 shows that nearly all the CVs below the gap are the lower-$\dot{M}$ dwarf novae, whereas most higher-$\dot{M}$ novalikes are above the gap. Further, the critical $\dot{M}$ for stability increases with orbital period ($\dot{M}_{\rm crit} \propto P_{\rm orb}^{1.7}$; e.g. Osaki 1996) and we find that the dwarf novae above the gap generally have a higher $\dot{M}$ than those below the gap, including some that are on the verge of stability (the Z Cam stars). Overall we find that $\dot{M}$ is typically 10^{-10} $M_\odot$ yr^{-1} below the gap and typically 10^{-9}–10^{-8} $M_\odot$ yr^{-1} above the gap.

The traditional explanation of these facts has been as follows. It is thought that the secondary's magnetic field is generated by an α–Ω dynamo, in which differential rotation winds up the internal poloidal field, amplifying it and converting into a toroidal field. Turbulent convection then converts the toroidal field back to a poloidal field, completing the feedback loop of the dynamo (e.g. Eggleton 2001).

Further, this dynamo is thought to operate most efficiently at the boundary between the star's radiative core and its convective outer envelope. The traditional picture proposes that this α–Ω dynamo operates efficiently in CVs above the period gap, driving $\dot{M}$ rates several orders of magnitude above that expected from GR alone. Note that CV secondaries, being locked to orbital cycles of a few hours, have rotation rates much faster than typical for lone stars, greatly boosting any dynamo effect. Also, we have both direct and indirect evidence that their surfaces are highly spotted (e.g. Hessman, Gänsicke & Mattei 2000; Webb, Naylor & Jeffries 2002), a strong indicator of magnetic activity.

The traditional picture then points out that at orbital periods of ~ 3 hrs the secondary will have been whittled down to a mass (~ 0.3 M$_\odot$) at which it will be fully convective. It is suggested that the disappearance of any boundary between convective and radiative regions shuts down the α–Ω dynamo, so shutting down magnetic braking (Rappaport et al. 1983; Spruit & Ritter 1983).

Since the secondary has been losing mass at a high rate as it approaches 3 hrs, it will be out of thermal equilibrium. Thus, when braking shuts off, it will relax to equilibrium and shrink within its Roche lobe, reducing $\dot{M}$ to zero. Such a star will then be in the period gap, but be too faint to be noticed.

The lower $\dot{J}$ resulting from GR will continue to reduce the orbital period, bringing the secondary back into contact at ~ 2 hrs. The star will then be seen below the gap, transferring mass at the lower rate driven by GR. The width of the gap will be determined by the degree to which the star had been out of thermal equilibrium, and thus by the value of $\dot{M}$ above the gap.

The above picture was consistent with the Weber & Davies (1967) theory which predicts $\dot{J} \propto \Omega^3$, where Ω is the stellar rotation rate. Since CV secondaries are rapidly rotating, this dependence implies that they brake strongly. The theory was tested using the rotation rates of single stars in open clusters of different ages, from which one deduces the braking they experience. The resulting slow-down with age was found to be in line with the Ω^3 law (Skumanich 1972).

It was noted, though, that the model had not been tested on the very fastest rotators, which would be directly comparable to CVs, and evidence accumulated that braking saturated at the highest rotation rates. More recent, extensive study of spin-down rates of stars in open clusters, including faster rotators than in the Skumanich study, concurs that $\dot{J} \propto \Omega^3$ for slower rotators but finds $\dot{J} \propto \Omega$ above a critical rotation rate, $\Omega_{\rm crit}$, where the threshold $\Omega_{\rm crit}$ is of order a few days (see Sills, Pinsonneault & Terndrup 2000 and refs therein). This therefore implies much lower $\dot{J}$ values for the rapidly rotating CV secondaries.

Figure 3.2, from work by Andronov, Pinsonneault & Sills (2003), makes clear the discrepancy between the data and the $\dot{J}$ required by traditional models of the period gap. Andronov et al. find (1) the braking $\dot{J}$ above the gap is orders of magnitude lower than needed; (2) there appears to be a smooth decline in $\dot{J}$ through periods corresponding to the period gap, contrasting with the abrupt shut-off required; and (3) any change from a shell dynamo at the radiative–convective boundary to a distributed dynamo appears to occur near 0.6 M$_\odot$, not at the 0.3 M$_\odot$ required by the period-gap theory. With Andronov et al.'s prescription for $\dot{J}$, there would be no period gap, and nor would there be any novalikes.

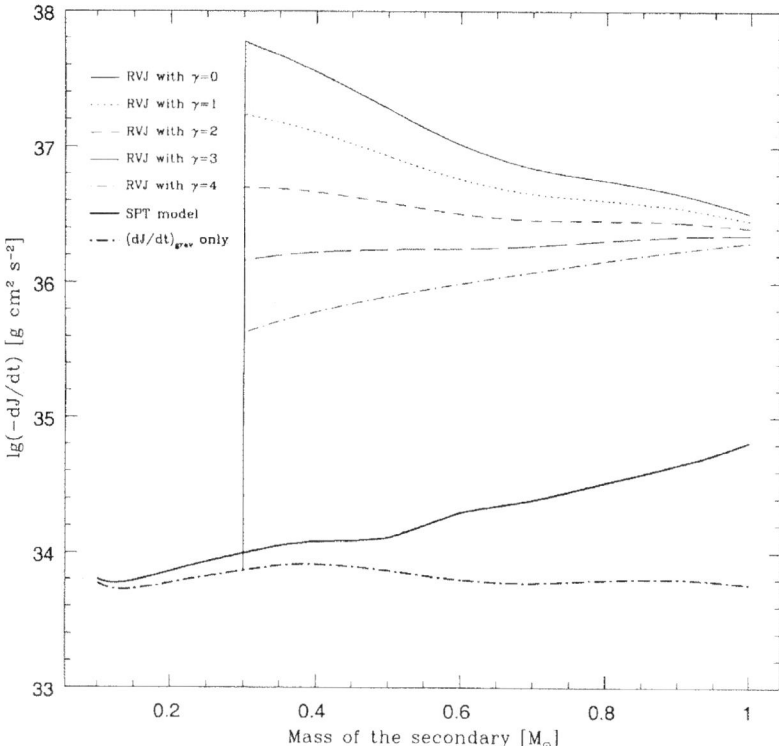

Fig. 3.2. The observed angular-momentum-loss rate (bold line) from the prescription of Sills et al. (2000), and the loss-rate from GR alone (lowest line). The other lines show the loss rate required to reproduce the period gap, from Rappaport, Verbunt & Joss (1983), with different values of a model parameter γ. Note the logarithmic scale, revealing a marked discrepancy. Figure from Pinsonneault, Andronov & Sills (2002); note that a version of this figure in Andronov et al. (2003) contains an error in the GR $\dot{J}$ rate.

It is too early to judge the full impact of these results on CV evolutionary theory — it remains possible that CV secondaries behave differently from single stars in some way — but at the present time the theory of the period gap must be regarded as unfounded.

3.1.5 Fact 5: Between 3 and 4 hrs, there are few systems with a medium mass-transfer rate, but here, and only here, we see systems that alternate between very high and very low mass-transfer rates

In the region 3–4 hrs, just above the 2–3-hr period gap, we see fewer dwarf novae than expected (e.g. Shafter 1992), but we do see novalikes with abnormally high $\dot{M}$ (the SW Sex stars). We also see VY Scl behaviour, a term used to label stars which spend half their time as high-$\dot{M}$ novalikes and half their time in low states with little or no mass transfer, but none of their time as medium-$\dot{M}$ dwarf novae. These VY Scl stars are seen only in the 3–4-hr period range.

These facts suggest that some mechanism discourages CVs in the 3–4-hr range from having medium $\dot{M}$ rates. One suggestion is that irradiation-feedback cycles (Wu et al. 1995) would boost any medium $\dot{M}$ to a high $\dot{M}$. Note that, of the systems above the gap, the 3–4-hr binaries have the smallest separations and so irradiation-feedback would be most effective.

The question then occurs, does the development of new phenomena in the 3–4-hr range have any bearing on the existence of the 2–3-hr period gap? There is no substantive model for this, but with the theory of the period gap up in the air, this question, first asked by Robinson et al. (1981), should be reopened.

3.1.6 Fact 6: The orbital-period distribution shows a minimum period at 77 mins. The cutoff is abrupt with no prior increase or decrease in the number of systems

The basic principles of the period minimum have been understood since papers by Paczynski & Sienkiewicz (1981) and Rappaport, Joss and Webbink (1982) (see King 1988 for a review). Remembering that the natural response to mass transfer is an increase in orbital period, we realise that as the secondary mass decreases it will, at some point, cease to generate significant nuclear energy and begin to obey the inverted mass–radius relation of a white dwarf. This means that the secondary will expand on further mass loss, and the system will evolve to longer periods again.

Also significant is that the Kelvin–Helmholtz timescale lengthens as the secondary mass decreases ($\tau_{KH} \propto M^{-3}$), and will become comparable to the mass-loss timescale (which cannot be greater than the GR timescale). Approaching the minimum, the secondary is increasingly unable to shrink in response to mass loss, and becomes oversized for its mass. Thus, the actual value of the minimum period depends on the $\dot{M}$ rate at which the system approaches the minimum.

Traditional models have assumed that $\dot{J}$ below the period gap is that from GR alone. However, this leads to several problems (see Fig. 3.3). First, the low $\dot{J}$ leads to a period minimum about 10 mins lower than that observed. Second, since $\dot{P}_{orb} = 0$ at the minimum, we expect systems to pileup at that value, and hence to see a spike in the P_{orb} distribution, which we don't. Thirdly, beyond the period minimum, the $\dot{J}$ driven by GR should diminish rapidly (since the period is lengthening again, and the secondary is now very light) and thus there should be large numbers of faint systems just beyond the minimum (e.g. Kolb 1993). We have evidence that some of the observed systems are turning the corner (Fig. 3.4), but not a single one is convincingly post-minimum (Patterson 1998; 2001).

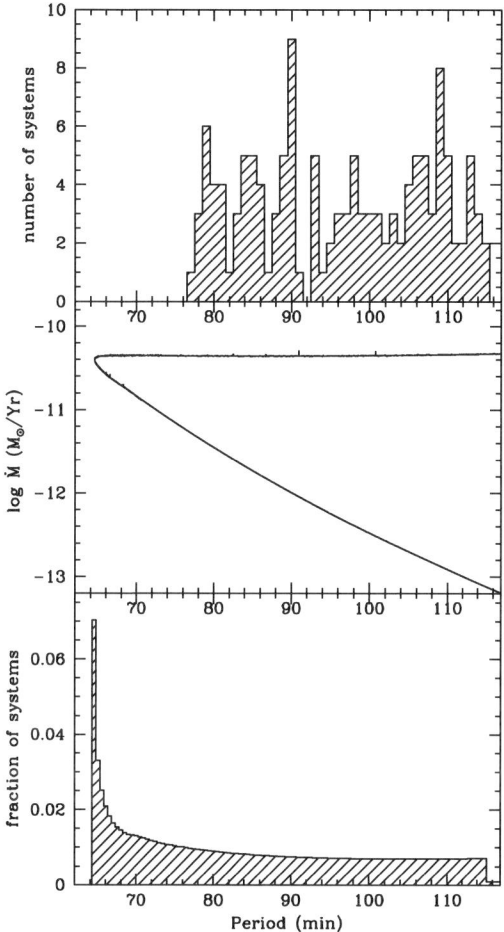

Fig. 3.3. *Top:* The observed number of systems near the period minimum. *Middle:* The predicted $\dot{M}$ from gravitational radiation alone. *Bottom:* The predicted distribution of systems based on $\dot{J}$ from GR alone. Figure from Barker & Kolb (2003).

The revised $\dot{J}$ prescription of Andronov et al. (2003) helps to some extent, since there is still some magnetic braking below the gap (Fig. 3.2); however the addition is probably insufficient to solve the problem, since we require roughly twice the GR rate to increase the minimum period to the observed 77 mins.

Additional $\dot{J}$ will mean that systems arrive at the minimum with their secondaries further out of equilibrium, and thus will drop in $\dot{M}$ more precipitously while turning the minimum; further, extra $\dot{J}$ could drive systems round the period minimum faster, and destroy them more rapidly afterwards. These effects could help to reduce the period spike (though not remove it entirely,

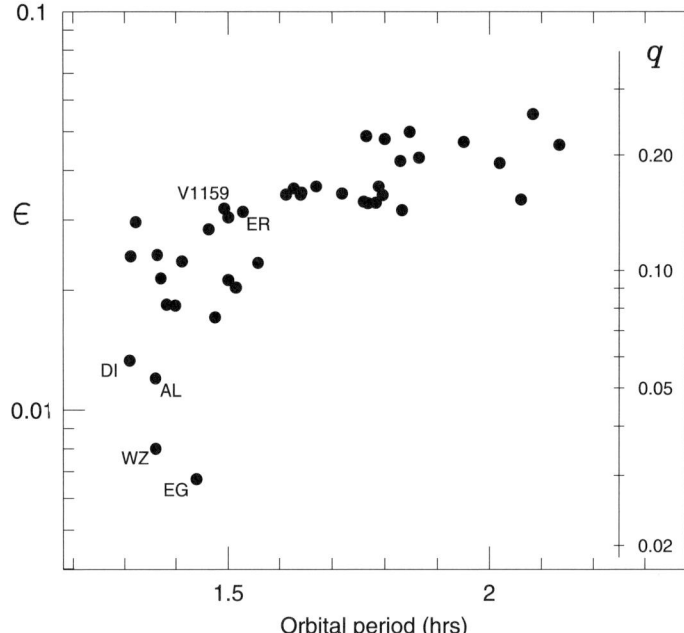

Fig. 3.4. The mass ratio ($q = M_{\rm sec}/M_{\rm wd}$) for dwarf novae below the period gap, deduced from the superhump period excess (ϵ). The existence of four systems at very low q (DI UMa, AL Com, WZ Sge and EG Cnc) gives the impression that CVs are 'turning the corner' at the period minimum. The ER UMa systems (DI UMa, V1159 Ori & ER UMa itself) all have $\dot{M}$ values at least an order of magnitude greater than those of the WZ Sge stars (EG Cnc, AL Com and WZ Sge), illustrating the spread in $\dot{M}$ within systems of similar period. Figure from Hellier (2001b), based on work by Patterson (1998, 2001).

since $\dot{P}$ still goes to zero) and explain the absence of post-minimum systems. The end products would not be CVs so much as white dwarfs orbited by planet-like objects (e.g. Patterson 1998), and such systems would not draw attention to themselves, though they could appear in white-dwarf surveys.

Thus an additional $\dot{J}$, beyond that prescribed by Andronov et al. (2003), appears necessary to explain the period minimum, as well as being required to explain the period gap.

3.1.7 Fact 7: Near the minimum period there is an exceptionally large range of mass-transfer rates and secondary star masses

Below the gap we still see the occasional high-$\dot{M}$ novalike (Fig. 3.1), while near the period minimum we see several ER UMa stars, whose frantic outbursting behaviour is thought to be driven by $\dot{M}$ rates ($\sim 6 \times 10^{-10}$ M$_\odot$ yr^{-1}) well above those of normal dwarf novae at the same orbital period ($\sim 1 \times 10^{-10}$ M$_\odot$ yr^{-1}; see Osaki 1995). These facts are important because they confirm the existence of much higher $\dot{M}$ than driven by GR, even near the period minimum.

Clustering around the period minimum are the WZ Sge stars. They are characterized by very low intrinsic luminosities and long intervals between outbursts, implying low $\dot{M}$ rates of 3×10^{-11} $M_\odot$ yr^{-1}, compatible with $\dot{J}$ from GR alone (e.g. Patterson 1998). Further, some have exceptionally small secondary-star masses. We can deduce this from studying 'superhumps' — tidal instabilities in accretion discs caused by the perturbation of the secondary star. The period of the superhump depends on the mass-ratio of the system, and hence leads to the secondary mass. Extensive work on these systems by Patterson and the CBA network (see Patterson 1998) shows that at the period minimum, and only at this period, we see systems dropping down to exceptionally low mass-ratios (see Fig. 3.4), implying secondary masses as low as 0.02 $M_\odot$.

This is important because it supports the idea that the observed minimum is a true minimum, where something happens to CVs, and that there aren't numbers of shorter-period systems that we are overlooking. Further, it argues against models in which the minimum is merely the product of the finite age of the galaxy, being the limit to which CVs have so-far evolved (see King & Schenker 2002 for models of this type).

While many of the ultra-low-mass systems have ultra-low $\dot{M}$ values, compatible with GR alone, at least one high-$\dot{M}$ ER UMa star (DI UMa), and possibly a second (RZ LMi), is among their number. This is another manifestation of the spread in $\dot{M}$, and suggests that even CVs turning the period minimum are capable of high-$\dot{M}$ episodes.

3.2 Discussion

The cataclysmic-variable $P_{\rm orb}$ distribution is determined by the angular-momentum loss thought to result from magnetic braking of the secondary star. In principle, and assuming that magnetic braking in CV secondaries is the same as that in single stars, the $\dot{J}$ can be determined by studies of open clusters. However, the open-cluster $\dot{J}$ deduced by Andronov et al. (2003) is orders of magnitude below that required to explain the 2–3-hr period gap, the main feature of the CV $P_{\rm orb}$ distribution, and also too low to explain the period minimum.

An alternative is that the systems on either side of the gap represent different CV populations, perhaps those with nuclear-evolved and unevolved secondaries (Andronov et al. 2003). However, this still wouldn't explain the fact that the majority of CVs above the gap are observed to have $\dot{M}$ rates much higher than the Andronov et al. prescription.

Perhaps the observed sample is just the 'tip of the iceberg', boosted to short-lived, high-$\dot{M}$ states, with the mean $\dot{M}$ for the class being much lower. However, systems with the Andronov et al. $\dot{M}$ would be regularly outbursting dwarf novae and therefore more likely to be noticed than the higher-$\dot{M}$ nova-likes. Thus, the high $\dot{M}$ systems would have to be complemented by systems of such a low $\dot{M}$ that they never undergo dwarf-nova outbursts.

Irradiation-induced feedback cycles are one explanation for a bimodal distribution of $\dot{M}$. However, King et al. (1996) argue that they are unlikely to be significant for the majority of CVs, unless their effect is boosted by a mechanism by which higher $\dot{M}$ can itself cause higher $\dot{J}$. There have been suggestions of such an effect (Livio & Pringle 1994) but no settled model (King & Kolb 1995; King et al. 1996).

Note that $\dot{M}$ cycles which retain the mean $\dot{M}$ rate of Andronov et al. would still leave us without an explanation for the period gap. There are also several lines of evidence that the high-$\dot{M}$ states of the observed systems are long-lived (see Kolb 2002). These include the fact that spectral types of CV secondaries above the gap are later than expected for their size, implying a departure from thermal equilibrium caused by high $\dot{M}$ (e.g. Baraffe & Kolb 2000), and the fact that white dwarfs in CVs above the gap are generally much hotter, implying high long-term $\dot{M}$ (e.g. Townsley & Bildsten 2002).

It is thus tempting to take at face value the observed $\dot{M}$ values above the gap and stick stubbornly to the idea that $\dot{J}$ above the gap is at least an order of magnitude higher than the Andronov et al. (2003) prescription. This has the additional benefit that, provided the extra $\dot{J}$ shuts off at 3 hrs, it can explain the period gap in the traditional way. We can also ponder whether the additional $\dot{M}$ is related to the VY Scl behaviour and dearth of dwarf novae at 3–4 hrs.

However, the above stubbornness doesn't answer the question of why CV secondaries would behave differently from the open-cluster stars on which Andronov et al. base their work. There are some differences: CV secondaries are Roche-lobe shaped whereas lone stars are axisymmetric, and there might be systematic differences in field topology that would influence the braking, but these currently seem insufficient to explain discrepancies of orders of magnitude.

We therefore need a much better understanding of (1) the spread in $\dot{M}$ among systems at the same orbital period and the possible role of $\dot{M}$ cycles, and (2) the enhanced $\dot{J}$ that appears to exist for most CVs, and particularly those with periods >3 hrs. Until then, when answering the question with which we started: "What drives the evolution of close binaries?", we have to admit that, at least in the test case of cataclysmic variables, the answer is very unclear.

Acknowledgement

I thank Ulrich Kolb for valuable comments on this work.

References

Andronov N., Pinsonneault M., Sills A., 2003, Cataclysmic variables: an empirical angular momentum loss prescription from open cluster data, ApJ, **582**, 358

Baraffe I., Kolb U., 2000, On the late spectral types of cataclysmic variable secondaries, MNRAS, **318**, 354

Barker J., Kolb U., 2003, The minimum period problem in cataclysmic variables, MNRAS, **340**, 623

Eggleton P.P., 2001, The Braking of Wind, in Evolution of Binary and Multiple Star Systems, ASP Conf. Ser., Vol. 229. Podsiadlowski, Ph, Rappaport S., King A.R., D'Antona F., Burderi, L., eds, p.157

Hellier C., 2001a, Cataclysmic variable stars: how and why they vary, Springer–Praxis, Heidleberg

Hellier C., 2001b, On echo outbursts and ER UMa supercycles in SU UMa-type cataclysmic variables, PASP, **113**, 469

Hellier C., Naylor, T., 1998, On the orbital period distribution of cataclysmic variables, MNRAS, **295**, L50

Hessman F.V., Gänsicke B.T., Mattei J.A., 2000, The history and source of mass-transfer variations in AM Herculis, A&A, **361**, 952

Howell S.B., Nelson L.A., Rappaport S., 2001, An exploration of the paradigm for the 2–3 hour period gap in cataclysmic variables, ApJ, **550**, 897

King A.R., The evolution of compact binaries, QJRAS, **29**, 1

King A.R., Frank J., Kolb U., Ritter H., 1995, Mass transfer cycles in cataclysmic variables, ApJ, **444**, L37

King A.R., Frank J., Kolb U., Ritter H., 1996, Global analysis of mass transfer cycles in cataclysmic variables, ApJ, **467**, 761

King A.R., Kolb U., 1995, Consequential angular momentum loss and the period gap of cataclysmic variables, ApJ, **439**, 330

King A.R., Schenker K., 2002, A new evolutionary picture for CVs and LMXBs, in The Physics of Cataclysmic Variables and Related Objects, ASP Conf. Ser., Vol. 261, Gänsicke B.T., Beuermann K., Reinsch, K., eds, p.233

Kolb U., 1993, A model for the intrinsic population of cataclysmic variables, A&A, **271**, 149

Kolb U., 2002, Braking and bouncing, in The Physics of Cataclysmic Variables and Related Objects, ASP Conf. Ser., Vol. 261, Gänsicke B.T., Beuermann K., Reinsch, K., eds, p.180

Kolb U., Rappaport S., Schenker K., Howell S., 2001, Nova-induced mass transfer variations, ApJ, **563**, 958

Livio M., 1996, Common envelope evolution in binary systems, in Evolutionary processes in binary stars, Wijers R.A. M. J., ed, NATO Advanced Science Institutes, Vol. 477, Kluwer Academic Publishers, Dordrecht, p.141

Livio M., Pringle J.E., 1994, Star spots and the period gap in cataclysmic variables, ApJ, **427**, 956

McDermott P.N., Taam R.E., 1989, On the disrupted magnetic braking model for the period gap of cataclysmic variables, ApJ, **342**, 1019

Osaki Y., 1995, A model for a peculiar SU Ursae Majoris-type dwarf nova ER Ursae Majoris, PASJ, **47**, L11

Osaki Y., 1996, Dwarf-Nova Outbursts, PASP, **108**, 39

Paczynski B., Sienkiewicz R., 1981, Gravitational radiation and the evolution of cataclysmic binaries, ApJ, **248**, L27

Patterson J., 1984, The evolution of cataclysmic and low-mass X-ray binaries, ApJS, **54**, 443

Patterson J., 1998, Late evolution of cataclysmic variables, PASP, **110**, 1132

Patterson J., 2001, Accretion-disk precession and substellar secondaries in cataclysmic variables, PASP, **113**, 736

Pinsonneault M.H., Andronov N., Sills A., 2002, Cataclysmic variables: An empirical angular momentum loss prescription from open cluster data, in The Physics of Cataclysmic Variables and Related Objects, ASP Conf. Ser., Vol. 261, Gänsicke B.T., Beuermann K., Reinsch, K., eds, p.208

Rappaport S., Verbunt F., Joss P.C., 1983, A new technique for calculations of binary stellar evolution, with application to magnetic braking, ApJ, **275**, 713

Rappaport S., Joss P.C., Webbink R.F., 1982, The evolution of highly compact binary stellar systems, ApJ, **254**, 616

Robinson E.L., Barker E.S., Cochran A.L., Cochran W.D., Nather R.E., 1981, MV Lyrae – Spectrophotometric properties of minimum light, or on MV Lyrae off, ApJ, **251**, 611

Schenker K., King A.R., 2002, A new evolutionary picture for CVs and LMXBs. II. The impact of thermal-timescale mass transfer, in The Physics of Cataclysmic Variables and Related Objects, ASP Conf. Ser., Vol. 261, Gänsicke B.T., Beuermann K., Reinsch, K., eds, p.242

Shafter A.W., 1992, The role of the dwarf nova period distribution in understanding the evolution of cataclysmic variables, ApJ, **394**, 268

Sills A., Pinsonneault MH., Terndrup D.M., 2000, The angular momentum evolution of very low mass stars, ApJ, **534**, 335

Skumanich A., 1972, Time scales for Ca II emission decay, rotational braking, and lithium depletion, ApJ, **171**, 565

Spruit H.C., Ritter H., 1983, Stellar activity and the period gap in cataclysmic variables, A&A, **124**, 267

Townsley D.M., Bildsten L., 2002, Compressional heating of accreting white dwarfs in CVs, in The Physics of Cataclysmic Variables and Related Objects, ASP Conf. Ser., Vol. 261, Gänsicke B.T., Beuermann K., Reinsch, K., eds, p.31

Warner B., 1988, Quasiperiodicity in cataclysmic variable stars caused by solar-type magnetic cycles, Nature, **336**, 129

Warner B., 1995, Cataclysmic variable stars, CUP, Cambridge

Webb N.A., Naylor T., Jeffries R.D., 2002, Spectroscopic evidence for starspots on the secondary star of SS Cygni, ApJ, **568**, L45

Webbink R.F., Wickramasinghe D.T., Cataclysmic variable evolution: AM Her binaries and the period gap, MNRAS, **335**, 1

Weber E.J., Davis L., 1967, The angular momentum of the solar wind, ApJ, **148**, 217

Wu K., Wickramasinghe D.T., Warner B., 1995, Feedback mass transfer in cataclysmic variables: an explanation of VY Sculptoris, PASA, **12**, 60

4 The Cosmic Distance Scale

S. Webb

4.1 Introduction

Determining the distances to celestial objects is perhaps the most fundamental task in astronomy. It is also the most difficult. Consider, for example, the interesting compact object RX 185635–3754, which is one of the closest isolated neutron stars [121, 122]. Its relative brightness, its isolation and its apparently thermal spectrum from X-ray to optical wavelengths makes this object ideal for the study of the surface emission properties of neutron stars. If we knew its distance then, in combination with estimates of its angular diameter coming from models of the spectral energy distribution [90], we might be able to constrain the dense matter equation of state on the interior composition of neutron stars. Two recent parallax determinations by a team using the *HST*, however, have produced inferred distances varying by almost a factor of two! (61 pc [119] and 117 pc [120]; see also [50].) The first estimate leads to a radius and mass for the neutron star that is smaller than can be accommodated by any reasonable equation of state; the refined measurement allows for reasonable equations of state, and no new physics need be contemplated. Many other examples could be given of the controversy surrounding various aspects of the cosmic distance scale, ranging from the distance to γ-ray bursters (the confirmation that the progenitor events of at least some GRBs lie at cosmological distances [75] took some 30 years after the discovery of GRBs [55]) to the value of H_0 itself (it is not so long ago that astronomers were engaged in heated debate over whether $H_0 = 50$ or $H_0 = 100$ [*note*: all references to H_0 in this review are in units of $\mathrm{km\,s^{-1}\,Mpc^{-1}}$]).

Given that distance determination is so difficult, the interested layperson might reasonably ask how much confidence we can have in any of our astrophysical and cosmological models. The best answer we can presently give is that, over many decades, astronomers have painstakingly constructed a cosmological 'distance ladder'. By identifying local objects whose absolute magnitude and size is known, we can search for them at larger distances and use them as 'standard candles' and 'standard rods'. This lets us identify and calibrate further candles and rods, and the process can be repeated. Figure 4.1 shows an abbreviated version of the distance ladder approach.

In the interests of clarity, Fig. 4.1 omits several rungs in the distance ladder approach, as well as the interconnections between rungs that must be made in order to obtain self-consistency. Nevertheless, it is clear that the distance ladder is a delicate construction; revisions in the lower rungs of the distance scale propagate through. There are certain key distance indicators, and it is important that distances derived from them are cross-checked by other methods; furthermore, we must allow for the possibility that the calibration sample of key indicators might be affected by local conditions. The book by Webb [123] contains a detailed discussion of the cosmic distance scale as it appeared in 1999. One purpose of the present review is to offer an elementary survey of the recent progress made in refining the accuracy of various rungs of the distance ladder; this is done in Sect. 4.2.

In some situations it is possible to bypass the distance ladder by identifying direct, geometrical distance indicators (Sect. 4.3). These indicators are important in their own right. They are also useful in offering cross-checks on distances derived from the ladder approach, and for calibrating key distance ladder indicators. The advantage of these geometric indicators is clear: they are free from the 'accumulation of errors' that plagues the ladder approach. It must be remembered, however, that one must struggle to minimize the model dependence of any derived distances; furthermore, one must strive to understand and document realistic systematic errors. The reliability of geometric distances is sometimes unclear.

In Sect. 4.4 we discuss in a little more detail two of the many intriguing aspects of the cosmic distance scale: the distance to the Large Magellanic

Estimate H_0 and then use redshift as a distance indicator (for a particular cosmological model).
↗
Calibrate extragalactic distance indicators (PNLF, SBF, Tully–Fisher, FP, Type I SNe, etc) and penetrate into the Hubble flow.
↗
Determine distances to nearby galaxies using standard candles; cross check with other candles (e.g. TRGB and RC).
↗
Use all previous techniques (and others) to calibrate stellar standard candles: e.g. Cepheid, RR Lyrae and Mira variables.
↗
Calibrate open cluster main sequences so that we can use main sequence fitting.
↗
Use secular and statistical parallax on groups of stars that lie beyond the range of trigonometric parallax.
↗
Use trigonometric parallax to find distances to as many different types of nearby stars as possible.

Fig. 4.1. A simplified view of the ladder approach to the problem of the cosmic distance scale.

Cloud (LMC) and the possible discrepancy in H_0 estimates derived from gravitational lenses compared with those based on 'local' approaches to the estimate of H_0. No attempt is made to resolve the issues under discussion. We conclude in Sect. 4.5 with a look at what the short-term future might hold for our understanding of the cosmic distance scale.

4.2 The Cosmological Distance Ladder

This section presents a general introduction to the cosmological distance ladder couched in terms of recently published results. If the purpose of the distance ladder is to provide a robust estimate of the Hubble parameter H_0 (so that we may use this value in our standard cosmological models in order to map the Universe), then it is clear that tremendous progress has been made in recent years. Even if astronomers can still disagree about the precise value of H_0 – and it is clear that some work still remains in pinning down its value – nearly all astronomers would agree about the *meaning* of H_0. Nevertheless, a small number of astronomers continue to publish papers questioning the received wisdom regarding cosmic distances, and in particular the notion that redshift can be used as a distance indicator [2, 3, 6, 7, 125]. Except for the five references cited immediately above, the present review concentrates solely upon the standard interpretation of distance indicators.

4.2.1 Parallax

Historically, the most important astronomical distance indicator has been trigonometric parallax. Parallax does not take us far along the distance ladder, but it remains of fundamental importance. The various methods used to determine the distance to remote galaxies, and ultimately therefore the fundamental parameters of the universe itself, such as its geometry, deceleration and energy content, all depend upon our knowledge of the distance to local objects.

Although ground-based determinations are still important (see, e.g. [18]), a revolution in parallax measurements occurred with the advent of space-based telescopes. The *HST* and the *Hipparcos* astrometry mission have collected precision parallaxes on over 10^5 objects, and less precise parallaxes on many more objects. These fundamental measurements have improved our understanding of all aspects of the cosmic distance scale. (Note that, even though good parallax measurements are now available, one must be careful to avoid biases when analysing the data. The Lutz–Kelker bias [66], which depends upon the relative parallax error $\delta\pi/\pi$ and the spatial distribution, causes us to underestimate the distance to stars. The Malmquist or incompleteness bias, which is of particular significance when considering extragalactic distance indicators, is due to the various apparent magnitude cut-offs as they are progressively imposed on a catalog. See [101] for a useful tutorial to gain an understanding of bias problems as they pertain to parallax measurements.) Analysis of the *Hipparcos* catalog continues to provide surprises; it has recently revealed, for example, previously unrecognized members of the nearby stellar population [59]. The planned successors to *Hipparcos* (see Sect. 4.5) will no doubt produce even more surprises.

The methods of secular and statistical parallax, which use as a baseline the motion of the Sun relative to the local standard of rest (LSR), enable us to estimate the mean distance to a carefully chosen group of stars. The Sun moves towards the solar apex with a velocity of about 2.83 AU per year, relative to the LSR, so measurements accumulated over tens of years enable us to measure distances of the order of 500 pc – and thus help calibrate candles beyond the reach of trigonometric parallax. Once again, the *Hipparcos* mission proved its value in improving the definition of the LSR.

4.2.2 Main Sequence Fitting

An important step on the distance ladder is the Hyades cluster, since it defines the position of much of the main sequence (MS) in the HR diagram. Once we have the distance to the Hyades, we can calibrate the MS – and thus we can use the MS as a standard candle. The Hyades distance can be determined by the moving cluster method [25] and directly using *Hipparcos* parallax measurements [87]. The best estimate of the distance to the cluster centre is 46.34 ± 0.27 pc. The method of MS fitting enables the determination of distances out to about 7 kpc (with the important caveat that the Hyades MS seems to be rather brighter than the MS defined by the Pleiades and the open cluster NGC 2516; the source of this discrepancy has yet to be satisfactorily resolved – see [69] for a recent discussion of this point).

A recent example of the MS-fitting distance technique appears in [85]. The authors use an expanded sample of local subdwarfs with *Hipparcos* parallax errors of less than 13% to perform MS fitting to the globular cluster 47 Tuc. They obtain a dereddened distance modulus to the cluster of $\mu_0 = 13.25^{+0.06}_{-0.07}$ mag. This result is in agreement with previous distance moduli derived from white dwarf fitting, and thus reduces the 3 Gyr uncertainty in the cluster age that had been apparent in previous comparisons of MS and white dwarf fitting.

4.2.3 Variable Stars

Variable stars retain their historic importance as distance indicators. Herein we discuss RR Lyrae, Mira and Cepheid variables, and note only in passing that a relationship between the rate of decline in brightness and peak absolute magnitude allow novae to act as a useful cross-check on distances derived by other means.

RR Lyrae Variables

RR Lyrae stars can be used to estimate distances to globular clusters, which in turn can help us calibrate other distance indicators. Furthermore, as we shall see later, they can also be used as extragalactic distance indicators.

There has been some dispute about the absolute magnitude of RR Lyrae stars. Recently, astrometric data from FGS 3, a white-light interferometer on the *HST* [8, 16] has been used [9] to find $\pi_{\text{abs}} = 3.82 \pm 0.2$ mas for RR Lyrae itself. A weighted average of *HST*, *Hipparcos* and previous ground-based determinations gives $\pi_{\text{abs}} = 3.87 \pm 0.19$ mas. (Encouragingly, this value is in good agreement with the recent pulsation parallax derived for the same object [14]: $\pi_{\text{puls}} = 3.858 \pm 0.131$ mas.) Adopting an average line-of-sight extinction to RR Lyrae of $\langle A_V \rangle = 0.07 \pm 0.03$ gives $M_V^{\text{RR}} = 0.61^{-0.11}_{+0.10}$.

The functional dependence of M_V^{RR} on metallicity remains controversial (see, e.g. [99]), with the relation usually expressed in the form

$$\langle M_V^{\text{RR}} \rangle = a \left[\frac{\text{Fe}}{\text{H}} \right] + b. \tag{4.1}$$

Mira Variables

It has long been known from LMC observations [28] that there is an infrared period–luminosity (PL) relationship for Mira variables, with average absolute magnitudes at K (2.2μm) of the form

$$M_K = -3.47 \log P + \gamma \tag{4.2}$$

where P is the period of oscillation in days. Data from *Hipparcos* have been useful in providing more precise calibrations of this relation. Recently, a Mira PL zero-point of 0.84 ± 0.14 was obtained from local Miras with *Hipparcos* parallaxes [127]. An independent zero-point of 0.93 ± 0.14, based on Miras in globular clusters using a cluster distance scale set via the *Hipparcos* parallaxes of subdwarfs, has recently been obtained [30].

Cepheid Variables

Since Cepheids are intrinsically bright the *HST* can in principle detect them out to 30 Mpc. Furthermore, although Cepheids are not superabundant, neither are they rare. Finally, they obey a PL relationship, the theoretical reasons for which are well understood. Cepheids therefore retain their position as our most important and our best extragalactic distance indicator. Nevertheless, problems remain with this venerable distance indicator.

It is now almost 100 years since Leavitt first noticed that bright Cepheids in the Small Magellanic Cloud (SMC) had longer periods than dim Cepheids in the Cloud, an observation that led to a PL relationship of the form

$$\langle M_V \rangle = a + b \log P. \tag{4.3}$$

To determine the slope we need a set of Cepheids with a large range in period and whose relative magnitudes are known; the best place to establish the parameter b is the LMC, which is better for this purpose than the SMC because it is less extended along the line of sight. (Note that, when Cepheids are used as a distance indicator, the slope of the PL relation is commonly assumed to be universal; but perhaps it must be adapted to the morphological type of each host galaxy [83]?) To establish the zero point we must study Cepheids at a known distance.

Early attempts at calibrating the PL relationship used the methods of statistical parallax, since the closest Cepheids were, until recently, tantalizingly out of reach of trigonometric parallax. (Except for Polaris, all local Cepheids are more than 250 pc distant. Most determinations of M_V have thus relied upon large-number statistics. It is only recently, using *Hipparcos*, that precision trigonometric parallaxes have been available for Polaris and for the prototype δ Cep. Recently, FGS 3 on *HST* has been used to measure the parallax of δ Cep with significantly greater precision [10].) Table 4.1 presents some relatively recent and influential calibrations.

The galactic calibration of Cepheids has sometimes been viewed with suspicion, in part due to the effects of reddening; applying the calibration to LMC Cepheids is even more tricky. The reddening of Cepheids in the Local Group (LG) spiral M 33 similarly requires better determination; a recent Cepheid distance determination to this galaxy [61] (which combines single-epoch *I*-band *HST* observations with ground-based period observations to provide an efficient distance-measurement technique) is about 0.3 mag smaller than the same team's distance using TRGB and RC methods (see later), which are themselves in good agreement with other recent estimates [102].

Other methods of deducing the luminosity of Cepheids include MS fitting and the Baade–Wesselink method (see, e.g. [123] for details). The Barnes–Evans semi-geometric method [4], a variant of the Baade–Wesselink method, combines physical displacements (obtained by integrating the radial velocity curve) with angular diameters (obtained from a surface-brightness–color relation) to infer the distance and mean radius values that make these sets

Table 4.1. Some recent Cepheid P–L calibrations

Calibration	Date	Ref.
$\langle M_V \rangle = -1.40 \; - 2.76 \log P$	1991	[67]
$\langle M_V \rangle = -1.38 \; - 2.77 \log P$	1997	[107]
$\langle M_V \rangle = -1.43 \; - 2.81 \log P$	1997	[29]
$\langle M_V \rangle = -1.294 - 2.77 \log P$	1998	[36]
$\langle M_V \rangle = -1.458 - 2.76 \log P$	2001	[35]

of numbers most consistent. In [78], a new calibration of the Barnes–Evans relation is presented. In [36, 82], the Barnes–Evans method is employed to determine the distances to nearby galaxies in an attempt to bypass the LMC calibration.

The effect of metallicity on the Cepheid P–L relation also remains controversial. Recent theoretical work [32] suggests that, for Cepheids with periods longer than about 10 days, a correction to LMC-based distance moduli may be required, with the sign and the amount of this correction depending upon the helium and metal content of the Cepheids.

There are reasons to suppose that first overtone (FO) Cepheids can provide more precise distances than fundamental (F) Cepheids [15]. (At $\log P = 0.3$, the width of the FO instability strip is only 400 K; compare this with a width of 900 K for F Cepheids at $\log P = 1$.) Unfortunately, the detection of FO Cepheids in external galaxies is more difficult than for F Cepheids, since they are fainter and their luminosity amplitudes smaller. However, complete and accurate samples are available for LMC FO Cepheids (see Sect. 4.4.1).

Care must also be taken regarding bias effects; in [109] the authors provide arguments that there is a Cepheid population-incompleteness bias that tends to produce too small distances (and thus too high values of H_0).

The principle behind the modern distance ladder approach is to use well understood and well calibrated standard candles – of which Cepheids are the best example – and use them to calibrate secondary distance indicators, which can then be employed at distances beyond the Cepheid horizon. Given the above-mentioned uncertainties in the Cepheid distance scale (and we have yet to mention the possible discrepancy between the Cepheid distance to NGC 4258 and a direct distance measurement to that galaxy – see Sect. 4.3), it is vital that we have other extragalactic distance indicators – both to act as a check on Cepheid distances, and to determine distances to early-type galaxies (classical Cepheids are Population I objects; there are 'Population I' and 'Population II' strands to the distance ladder, and it is important – if only as a sanity check – to be able to tie them together). We should perhaps also bear in mind a remark made in 1935 by Hubble regarding the Cepheid distance scale: "further revision is expected to be of minor importance". This was before Baade made his famous correction for Cepheid population type [31]!

4.2.4 Tip of the Red Giant Branch (Population II)

A plot of red giant stars on a HR diagram shows a discontinuity, which marks the helium core flash and shows the maximum luminosity of red giant stars. This tip of the red giant branch (TRGB) is potentially a reliable distance indicator to galaxies with resolved stellar populations because of the very weak dependence of the TRBG on metallicity [60]; furthermore, the method requires an observation at only a single epoch. The method has therefore been widely used to confirm distances within the Local Group, and more recently to determine distances to galaxies within nearby groups.

The TRGB method has been used on observations made by WFPC2 on *HST* [54] to determine a distance to the important LG spiral M 33 (one of the main calibrators for several secondary distance indicators, including some mentioned below) of $d_{\mathrm{TRGB}} = 916 \pm 17 (random)$ kpc. The same paper also presents a distance to M 33 based on the helium-burning red clump (RC) method [80], which has gained considerable attention recently as a potential standard candle. The zero point of this method is set geometrically – about 10^3 such stars in the solar neighborhood had parallaxes measured by *Hipparcos*. However, controversy still surrounds the appropriate treatment of possible metallicity and age effects on the *I*-band absolute magnitude of the red clump [68, 96] (though IR measurements can circumvent these problems). The RC distance of $d_{\mathrm{RC}} = 912 \pm 7 (random)$ kpc is in good agreement with the TRGB value, and with other recent estimates [102]. (Interestingly, these results are about 0.3 mag larger than Cepheid distances, a finding that the authors attribute to the uncertainty in reddening for M 33 Cepheids.) The TRGB method has been used to find precise distances to seven nearby galaxies [73]: Leo I, Sextans B, NGC 1313, NGC 3109, UGC 3755, UGC 6456 and UGC 7577. The method has also been used to determine the most precise distances yet to NGC 4214 and UGC 685 [68]. In [51, 52], new distances are presented to galaxies in the nearby Centaurus A and M 81 groups of galaxies; and in [53] distances to 18 nearby galaxies are presented (which, incidentally, leads to a radius of the zero-velocity surface of the LG of 0.94 ± 0.1 Mpc).

These results all shed light on the kinematics and structure of the closest groups, which will in turn aid our understanding of the origin of the small-scale structure of the Universe. One interesting finding coming from these improved distances is that the local Hubble flow is cold, with a 1D mean random motion of about 30 km s^{-1} – a somewhat paradoxical result, given the lumpy nature of the LG. Why should the local rate of expansion be so close to the global rate? Are we seeing here, as some authors suggest [5], a signature of dark energy?

4.2.5 Tully–Fisher Method (Population I)

The Tully–Fisher (TF) method of determining the distances to spiral galaxies relies on an empirical relation between a galaxy's luminosity and its rotational velocity (which can be determined by Doppler broadening of the 21-cm emission line). For a spiral at an angle of inclination i to our line of sight, and with a 21-cm linewidth W measured at 20% of the peak, the magnitude is given by

$$M = a \log\left(\frac{W}{2 \sin i}\right) + b \qquad (4.4)$$

where a and b are constants. Modern applications of the method generally employ infrared measurements. The typical uncertainty in the TF method is large for an individual galaxy, but the method works well for groups of galaxies and it applies out to distances as far as 100 Mpc if care is taken to account for Malmquist bias.

Many studies have looked at how to best employ the TF relation as a distance indicator (see e.g. [48, 106]), and this work continues. In [49], for example, the physical sources of scatter in the TF relation are considered. In [104], TF data on 161 Virgo spirals are used to map the 3D structure of this important galaxy cluster; the results appear to confirm that Virgo has a considerable line-of-sight depth – with a depth about four times larger than its width.

4.2.6 The Fundamental Plane and D_n–σ (Population II)

For clusters of elliptical galaxies, distances may be estimated using the fundamental plane (FP), which is a relationship between three parameters – the effective radius R_e of a galaxy, the average surface brightness I_e within that radius, and the velocity dispersion σ of stars in the central region – of the form

$$R_e = k\sigma^{1.36} I_e^{-0.85} \tag{4.5}$$

where k is a constant. The related D_n–σ method (D_n is the physical size of the galaxy defined at specified surface brightness level), which might be considered as essentially an approximation to the fundamental plane, may also be used to estimate distances. In [11], 452 ellipticals and S0 galaxies in 28 clusters were used to construct a Malmquist-corrected template D_n–σ relation of the form

$$\log D_n = 1.203 \log \sigma + 1.406. \tag{4.6}$$

Distances to individual galaxies are uncertain to about 20% using this method. Encouragingly, however, peculiar velocities obtained using this method are in agreement with determinations based on spiral TF distances.

4.2.7 Surface Brightness Fluctuations (Population II)

The surface brightness fluctuation (SBF) method pioneered by Tonry and Schneider [111], relies on the RMS variation in flux from pixel to pixel when a galaxy is imaged by CCD: surface brightness fluctuations scale with distance as d^{-1}. The method works best on early-type galaxies [110].

In principle, using the *HST*, the method can be applied to 125 Mpc; in practice, the method has been used much more locally. Furthermore, the SBF method has employed a smaller number of researchers than either the TF or FP methods. Most SBF data has been collected as part of a ground-based survey initiated by Tonry and co-workers; the full SBF survey data, comprising about 300 galaxies, is described in [112]. In [13], the distances to 170 galaxies are compared using three methods: SBF, FP, and distances predicted from the observed galaxy density field. The analysis, which highlights the value of the SBF method for early-type galaxies, suggests $H_0 = 73 \pm 4$ (internal error), with a further 15% error coming from the Cepheid calibration. Deep I-Band SBF distances to NGC 564 and NGC 7619 are reported in [72]. When added to the SBF sample in [35], the authors derive $H_0 = 71 \pm 4 \pm 6$.

4.2.8 Planetary Nebula Luminosity Function (Population I *and* II)

Planetary nebulae (PNe) afford an excellent opportunity to tie together the Population I and II sides of the distance ladder. Trigonometric parallax has been used to measure the distance to several nearby PNe, and expansion parallax has been used to measure the distance to several more [63]. Early studies of M 31 implied that the upper limit of absolute magnitude, M^*, was -4.48, although recent work has brightened this to -4.53. Rather than use individual PNe as standard candles, however, a better approach is to define the planetary nebula luminosity function (PNLF). By using a narrowband filter to photograph the sky at 5007 Å – the [O III] emission line – and then subtracting a similar image taken just away from the $\lambda 5007$ line, many PNe may be detected in galaxies out to about 20 Mpc. As a result of this work, Jacoby and Ciardullo propose a PNLF of the form

$$N(M) \propto e^{0.307M}\left(1 - e^{3(M^* - M)}\right). \tag{4.7}$$

By detecting enough PNe in the 'flat' part of the distribution, we can use eqn. 4.7 to determine the position of the cut-off. The characteristic cut-off in the PNLF thus potentially provides an excellent standard candle: a comparison of the PNLF in a remote galaxy with that in M 31 gives the distance to the galaxy (in terms of the M 31 distance). Although M 31 remains the main calibrator for the method, there are now several galaxies with a Cepheid distance in which the PNLF method also applies.

In an intriguing paper [22], Ciardullo, Jacoby and their colleagues report the results of surveys for PNe in six galaxies, and derive their distances using the PNLF; they then compare PNLF, SBF and Cepheid distances. There are now 28 large galaxies with both SBF and PNLF distance measurements, and the analysis in [22] seems to suggest that Cepheid-calibrated PNLF distances are on average about 0.3 mag smaller than Cepheid-calibrated SBF distances. The reason for this discrepancy remains a matter for debate, but it seems possible that internal extinction in the six SBF calibrating galaxies can explain the result. If this is indeed the explanation, then the SBF estimate of H_0 must be increased by 7%.

An extremely interesting aspect of [22] is that one of the six galaxies with a new PNLF distance measurement is NGC 4258 (see Sect. 4.3). The PNLF distance is $7.6^{+0.2}_{-0.3}$ Mpc, which is about 1σ larger than the geometric distance to the galaxy coming from megamaser emission. This is not *necessarily* a serious problem; however, by taking a short LMC distance (see Sect. 4.4.1) then both Cepheid and PNLF distances are in perfect agreement with the geometric distance. This finding would raise the *HST* Key Project result to $H_0 = 78 \pm 8$.

4.2.9 Supernovae

Supernovae (SNe) of all types are among the most interesting distance indicators. The expanding photosphere method (EPM) of Type II SNe (which, since these objects occur primarily in spiral arms, is a Population I indicator) has been used to produce distances to about 20 galaxies. The method can be applied as a primary indicator, but it is usually calibrated against galaxies with a known Cepheid distance, in particular LMC (which hosted SN1987A), M 81 (which hosted SN1993J) and M 101 (which hosted SN1970G). Recently, the EPM has been used on the Type II plateau SN1999gi to derive a distance of $11.1^{+2.0}_{-1.8}$ Mpc to its host galaxy NGC 3184 [62]. This value is in agreement with a recent TF distance estimate [84], but is about 2 Mpc closer than Cepheid-based distances to two galaxies commonly assumed to be in the same cluster [35]. The scatter in the Hubble diagram from the EPM is of the order of 20%; Hamuy and Pinto argue [42] that more precise distances can be inferred from a correlation between the envelope expansion velocities of the Type II SNe plateau subclass and the bolometric luminosities during the plateau phase. Their analysis leads to $H_0 = 55 \pm 12$. Unfortunately the authors of this method must use SN1987A (assuming an LMC distance) to calibrate the Hubble diagram; furthermore, SN1987A is not a prototype of the plateau class. Future observations – including a forthcoming Cepheid-based distance to SN1999em and the possibility of observing high-z Type II SNe during the plateau phase with 8 m class telescopes — may improve the effectiveness of this method. At present, however, Type Ia SNe are perhaps more persuasive distance indicators than Type II SNe.

Type Ia SNe would seem to be ideal standard candles. First, they are luminous; various lines of evidence suggests that their peak absolute magnitude is about −19.5. Second, they are reliable; the dispersion in peak magnitude seems to be about 0.2 mag [17]. Some Type Ia SNe may be anomalous – for example SN1991bg seems to have been intrinsically red, SN1991T intrinsically brighter than normal – and we need to take account of reddening in the host galaxy. But these factors can be dealt with. Until relatively recently, the main drawback with using Type Ia SNe as a distance indicator was their extreme scarcity and the fact that they did not stay long at their peak brightness. The recent advent of programs like the Supernova Cosmology Project and the High-z SN Search, which can find large numbers of these objects 'on demand', makes Type Ia SNe excellent probes of the distant universe. The discovery that distant SN are dimmer than they should be has led to one of the most exciting discoveries in recent astronomy: the acceleration of the universal expansion [86, 94].

4.2.10 Redshift–Distance Relation

Now that we are in the era of 'industrial' redshift detection, the most effective extragalactic distance indicator is simply the Hubble redshift–distance relation. Astronomers are continuing to push back the distance boundaries, with the detection of quasars at $z \sim 6$ [71] and even a galaxy at $z = 6.56$ [46]. (These discoveries are not mere exercises in 'record breaking'; they can tell us about conditions in the first few percent of the age of the Universe, and the mere presence of these high-redshift objects suggests that the reionization epoch lies beyond $z = 6.6$.) In order to convert these redshifts into distances, of course, we need to know H_0 (and at these redshifts one must also specify the particular cosmological model one is using and, indeed, what definition of distance one prefers). So we return to one of the most controversial questions in recent astronomical history: what is the value of the Hubble parameter?

By resolving Cepheids in 28 spirals within 25 Mpc, the *HST* Key Project were able to calibrate some of the secondary distance indicators we have so far discussed, and then penetrate so far out into space – in the 60–400 Mpc range – that perturbations in the Hubble flow became a minor component of the error budget [35]. Their final value for H_0 ($= 72 \pm 8$) is perhaps the consensus view among the astronomical community. (The uncertainty in the Key Project value is in large part due to the uncertainty in the LMC distance (see Sect. 4.4.1).) However, using very similar data, Sandage and co-workers [95] have recently obtained $H_0 = 58.7 \pm 6.3$. Part of the discrepancy is explained by the different Cepheid relations that are employed in the analyses; the Key Project relies on the large sample from the OGLE-II microlensing group [115], whereas the Sandage team employs an older calibration from Madore and Freedman [67]. Not all of the difference can be explained away by discrepant Cepheid calibrations, however; the Sandage team points to more subtle differences in an attempt to bridge the gap between the two approaches [95]. Furthermore, in the latest of a series of papers, Sandage continues his examination of bias properties in extragalactic distance indicators [100]. He considers a simplified galaxy cluster model, and shows that a Teerikorpi cluster incompleteness bias [108] will exist unless the cluster catalog is sampled completely over the entire cluster luminosity function. He applies his analysis to the actual complete catalogs of the Virgo and Ursa Major galaxy clusters to derive bias-free distance moduli. The subsequent value of H_0 using this analysis? He gets $H_0 = 58 \pm 6$.

Using a new statistical analysis of Huchra's compilation of nearly all estimates of H_0 [47], Gott and co-workers [38] find a median value of $H_0 = 67$. The analysis is intriguing, and the authors claim that this is arguably the best summary of current knowledge because it uses all available data and makes no assumption about the distribution of measurement errors.

The *Wilkinson Microwave Anisotropy Probe* (*WMAP*) recently published the most precise value yet for H_0: 71 ± 4 [105]. This is in striking agreement with the Key Project value, particularly since the two teams use different observables, different underlying physics, and different model assumptions. Although recent estimates for H_0 range, within published quoted errors, between 43 and 86, there are surely new signs of an emerging consensus for the value of the Hubble parameter – and indeed, with the new *WMAP* results, for the parameters of a standard cosmological model.

4.3 Geometrical Methods

Faced with the sometimes conflicting results emanating from the distance ladder approach, it is tempting to look for the the certainty that can come from geometric methods. Different geometric methods can bypass some or all of the ladder's rungs. Unfortunately, geometric methods are in short supply (of individual galaxies, only LMC and NGC 4258 have had their distance measured precisely with geometric techniques) and the precision of 'cosmological' measurements is still not high. But all of these methods will, in the future, increase in importance.

4.3.1 SN Light Echo

The distance to SN1987A (and thus one object in the LMC; note, though, that the precise location of the object with respect to the LMC plane is uncertain) can be determined simply by comparing the angular and physical size of the circumstellar ring around the SN. The angular size of the ring can be measured directly from *HST* images; its physical size can be calculated from the observed light travel time to the ring, as measured by *IUE* lightcurves of UV emission lines seen between 8–700 days after the explosion. The method sounds straightforward, but there are several questions to consider. Does the visible structure have a smooth, ring geometry? Do the *IUE* lightcurves and *HST* images correspond to gas that is in the same location in the structure? Can we neglect the delay time between the UV pulse first hitting the ring and the subsequent appearace of UV line emission? And so on.

The first estimate of the geometric distance to SN1987A was 51.2 ± 3.1 kpc [81]. The most recent and comprehensive treatment, by Gould and Uza [39], gives a distance of 48.8 ± 1.1 kpc.

4.3.2 Megamasers

One of the most important contributions to the whole cosmic distance scale debate was the precise determination of the distance (7.2 ± 0.3 Mpc) to NGC 4258 from observations of a circumnuclear maser [44, 76]. This is now perhaps the best-known distance in astronomy! These strong water masers are bright, small and have very narrow line widths – features that make them ideal probes of galactic structure. By tracking spectral features it is possible to measure maser position and acceleration relative to the centre of the disk (in the case of NGC 4258 there is a measured acceleration of $9.3 \pm 0.3 \,\mathrm{km\,s^{-1}\,yr^{-1}}$). Assuming circular orbits and the validity of Newtonian mechanics, one can easily derive the physical extent of the maser from the galactic centre and thus determine the distance.

4.3.3 Eclipsing Binaries

Well detached main sequence eclipsing binary (EB) systems have been proposed as being virtually ideal standard candles for determining distances as far as the LMC. The method is mainly a geometric technique; some radiative physics is involved, but no knowledge is required of distances to calibrating objects. (See [128] for further technical details.)

Observationally, the method requires monitoring of the light curves (preferably in more than one filter) and the radial velocity. The light curves provide relative star dimensions (R_1/a and R_2/a, where the R are the main radii and a is the orbit size) while the radial velocities provide a, from which information the radii can be derived. With the absolute radii known, the luminosities follow if we know the emission per unit surface area. The latter can be predicted from stellar atmosphere models that are fitted to the spectral energy distributions of the EB system of interest. It is clear, therefore, that the EB method can skip many of the calibrating rungs of the distance ladder; it is worth emphasizing, however, that large systematic errors might be introduced if our modelling of stellar atmospheres is uncertain.

As we shall see in Sect. 4.4.1, the method has been used recently to determine the distance to EB systems in the LMC, with intriguing results.

4.3.4 Sunyaev–Zel'dovich Effect

The Sunyaev–Zel'dovich (SZ) effect involves the interaction of the hot intergalactic gas found in rich galaxy clusters with the Cosmic Microwave Background. CMB photons can scatter off high-energy electrons in the gas (the inverse Compton effect), causing a deficit of background radiation at $\lambda > 1.38\,\mathrm{mm}$ and an excess of background radiation at $\lambda < 1.38\,\mathrm{mm}$. Most work focuses on the decrement in the background. We can use the SZ effect to determine the distance to a cluster of galaxies in the following way. The X-ray brightness of the cluster at a particular frequency, $I(\nu)$, depends upon the line-of-sight depth R of the cluster, the electron temperature T_e and density N_e through:

$$I(\nu) = a\, N_e^2 T_e^{-1/2} e^{-h\nu/kT_e} R \qquad (4.8)$$

where a is a constant that can be calculated. The drop in the background radiation temperature due to the SZ effect is given by:

$$\frac{\Delta T_{\rm r}}{T_{\rm r}} = -bN_{\rm e}T_{\rm e}R \qquad (4.9)$$

where b is a constant that can be calculated. We can combine these two equations to find R; and if we assume that the cluster is spherical, we have the linear diameter. In other words, we have a standard rod. (Clearly, the possibility of cluster asphericity leads to an uncertainty, and it is unclear how to treat this in individual cases.) The SZ effect has been used in this way to measure the distance to several clusters [12].

Recent SZ observations of the massive cluster Abell 1413 at $z = 0.143$, using the Ryle telescope, have been combined with X-ray data from *ROSAT* and *ASCA* [40]. The resulting estimate for the Hubble parameter is $H_0 = 57^{+23}_{-16}$ for a cosmology with $\Omega_M = 1$, $\Omega_\Lambda = 0$. The largest errors in the error budget arise from uncertainty in the X-ray temperature, noise in the SZ measurement, and uncertainty in the line-of-sight depth of Abell 1413.

4.3.5 Gravitational Lenses

The light from the distant object like Q0957+561 reaches us by two different paths because the light is bent by an intervening galaxy. If the quasar flickers in brightness, then we see Q0957+561B flicker more than a year after Q0957+561A has flickered. By modelling the mass distribution of the lensing galaxy the time delay Δt between the two paths gives us an estimate of H_0, since

$$H_0 \Delta t \propto \frac{d_{\rm L} d_{\rm S}}{d_{\rm S} - d_{\rm L}} \qquad (4.10)$$

where the d are the normalized angular diameter distances of the lens (L) and source (S). This technique, originally proposed by Refsdal [91, 92], works over cosmological distances.

With more than 50 multiply imaged systems now known, the prospect of determining an accurate estimate of H_0 from gravitational lenses is improving. Two recent papers describe observations of the time delay in gravitational lenses using the Nordic Optical Telescope, from which H_0 is estimated using the Refsdal method. In [45], observations are described of the quadruple quasar RX J0911.4+0551 (at $z = 2.80$), which is lensed by a galaxy and a cluster at $z = 0.77$. The time delay between the A and B components is measured to be 146 ± 8 days (2σ). After modelling the mass distribution of the lensing system, and adopting a cosmology with $\Omega_M = 0.3$, $\Omega_\Lambda = 0.7$, the authors derive $H_0 = 71 \pm 4$ (random, 2σ) ± 8 (systematic). In [19], the time delay of the doubly-imaged quasar SBS 1520+530 (at $z = 1.86$), lensed by a galaxy (which the authors have discovered to lie at $z = 0.717$) is found to be 128 ± 3 days (1σ). Adopting the same procedure as in [45], the authors derive $H_0 = 51 \pm 9$ (1σ error).

A major uncertainty in the standard method of determining H_0 from gravitational lenses is the modelling of the lens. A new method for obtaining

H_0 from quadruple images, which assumes only that the lens potential is split into a radial part described by a simple power-law profile and an angular part described by a quite general shape function [21], when applied to the quadruple lenses PG 1115+080 and B1422+231 gives $H_0 = 56^{+22}_{-26}$ (90%CL) for a $\Omega_M = 0.3$, $\Omega_\Lambda = 0.7$ cosmology.

4.4 Two Distance-Scale Puzzles

The progress made over the past decade in the illumination of the cosmic distance scale has been undeniably impressive. New techniques have been developed and old techniques have been improved. Nevertheless, the essence of the ladder approach is to 'mix and match' in order to obtain consistent results; the geometric methods are either rare or else imperfectly understood. It is thus not surprising that astronomers can (and do) argue endlessly about the details. In this section, a brief introduction is given to two of the more intriguing aspects of the distance scale debate (with no attempt made to resolve the issues).

4.4.1 Clouding the Issue: The LMC Distance

The most important (and controversial) rung on the distance ladder is the distance to the LMC. The importance of the LMC lies in the fact that it is distant enough for us to consider its constituents, at least to a first-order approximation, as being at a constant distance from us; at the same time it is close enough so that the several types of distance indicator it contains can be measured with high accuracy, and thus provide us with valuable consistency checks for the whole ladder approach. Furthermore, the *HST* Key Project group adopted a particular LMC distance and used this as their zero point for the Cepheid PL relation, from which they calibrated their secondary distance indicators [35]. (They adopted an LMC distance of 50.1 ± 2.5 kpc, i.e. a distance modulus μ_0 of 18.50 ± 0.10 mag.) The quoted uncertainty in their final value for H_0 ($= 72 \pm 8$) was in large part due to the uncertainty in the LMC distance. If our value for the LMC distance changes, so does our *HST*-based estimate of cosmological distances.

Prior to *Hipparcos*, the LMC distance modulus appeared to be converging on $\mu_0 = 18.5 \pm 0.1$ mag (i.e. a 5% error), and a discrepancy with the distance based on RR Lyrae determinations could be blamed upon calibration errors for the RR Lyrae stars. It was hoped that *Hipparcos* measurements would secure the LMC distance. If anything, however, the LMC distance is more uncertain than before. Existing determinations of the LMC distance now span a wide range, with values of μ_0 between 18.1 to 18.8 mag appearing in the recent literature. These values are often grouped into 'long' distance scale results ($\mu_0 \approx 18.7$ mag; $d > 50$ kpc) and 'short' distance scale results ($\mu_0 \approx 18.3$ mag; $d < 50$ kpc). In some cases, the same distance-estimating technique can be made to support either of these scales, depending upon one's assumptions and preferred method of analysis (see, e.g. [24, 118, 124] for recent discussions).

An early indication of the LMC distance problem arising from *Hipparcos* measurements came when parallaxes were used to determine the Pleiades distance directly. This led to a downward revision in the Pleiades distance modulus of 0.25 mag (12%) [74, 117]. (This was an important result, since it implied either that the *Hipparcos* catalog contains systematic errors [77, 89] or that our understanding of FGK stellar evolution is imperfect.) Now, compared to the Hyades, the Pleiades are more similar in age and metallicity to the few Cepheid-containing open clusters and associations that can be used for calibration by MS fitting. So authors who preferred a calibration anchored to the Pleiades found a large movement in their derived LMC distances. (There was no change in Hyades-based zero-points. See also, however, [114].) The PL relation was also calibrated directly by Feast and Catchpole [29] using *Hipparcos* parallaxes of 26 Cepheids, but this was at the very limit of what can be extracted from the data. Nevertheless, using that calibration they derived an LMC distance modulus of 18.7 ± 0.1 mag — clear support for the long scale.

Recent Cepheid-based techniques have provided estimates of the LMC distance modulus that *are* consistent with the *HST* KP calibration. In [15], for example, FO Cepheids are used to get 18.53 ± 0.08 mag (theory) and 18.48 ± 0.13 mag (empirical). A new and more precise measurement of the parallax of δ Cep has been used [10] to obtain $\mu = 18.50 \pm 0.13$ mag (or, applying a rather uncertain correction for metallicity, $\mu = 18.58 \pm 0.15$ mag). In [26], however, *Hipparcos* parallaxes of 219 Cepheids are used to produce an absolute calibration of the Galactic distance scale sampled by a Baade–Wesselink distance indicator, and $\mu_0 = 18.59 \pm 0.04$ mag is argued for; this would correspond to a 5% decrease in the *HST* Key Project value of H_0.

Using the initial Key Project calibration of the Cepheid PL relation we get a Cepheid-based distance to NGC 4258 of 8.1 ± 0.4 Mpc [70] – 13% longer than the maser distance. The final Key Project calibration of the relation suggests a distance of 7.8 ± 0.3 Mpc – in closer agreeement with the maser distance, but still 1.2σ away. As mentioned earlier, the PNLF distance to NGC 4258 is $7.6^{+0.2}_{-0.3}$ Mpc [22] – about 1σ away. Interestingly, if a *short* LMC distance were employed, then Cepheid, PNLF and geometric distances would be in complete agreement. (On the other hand, it is claimed that pulsation models of Cepheid behaviour can be used to derive a Cepheid distance to NGC 4258 of 7.3 ± 0.6 Mpc while keeping $\mu_0 = 18.5$ mag for the LMC distance modulus [20].) Clearly, though, more than one maser example is required before we can make definite conclusions.

What of RR Lyrae stars? A 1998 paper [65] suggested $\mu_0 \approx 18.3$ for both the RR Lyrae and Cepheid distances, based on *Hipparcos* data. In [27], evidence is presented for classifying the RR Lyrae star CM Leo as a regular FO pulsator with an absolute magnitude of $M_V = 0.47 \pm 0.04$. Using this value as a standard candle, and after correcting for evolutionary and metallicity effects, the authors derive an LMC distance modulus of $\mu_0 = 18.43 \pm 0.06$ mag. In [9], the *HST* parallax of RR Lyrae itself leads to $\mu_0 = 18.38$–$18.53^{-0.11}_{+0.10}$ mag (with an attendant significant uncertainty due to the extinction of the RR Lyrae population local to the LMC). The result marginally supports the short scale. The authors of a recent paper [23] use new photometry and spectroscopy for more than a hundred RR Lyrae stars in two fields near the LMC bar to derive new estimates of the average magnitude, the local reddening, the luminosity–metallicity relation, and the distance to the LMC. They claim that, using these values, it is possible to reconcile the short- and long-distance scales on a common value of $\mu_0 = 18.515 \pm 0.085$ mag.

The situation with other methods is little clearer. It is claimed, for example, that well detached EBs are ideal standard candles for obtaining not only the distance to the LMC, but also for probing its structure. In an early paper [41], the EB method applied to HV 2274 generated an adjusted LMC distance of 45.7 ± 1.6 kpc ($\mu_0 = 18.30 \pm 0.07$ mag), a result that appeared to argue strongly for the short distance scale. Recent work [33] by the same authors on the EB HV 982, however, implies a distance to the optical centre of the LMC bar of 50.7 ± 1.2 kpc. The results thus differ by 5 kpc! Treating these results as marginally consistent, independent measures of *the* LMC distance implies a mean LMC distance of between 46 kpc and 51 kpc. (A further result on the EB EROS 1044, at 47.5 ± 1.8 kpc [93], implies a mean distance of about 48 kpc.) An alternative explanation, however, is that there is a significant depth to the stellar distribution in the LMC. There are other indications of a significant line-of-sight structure (the initial distance found for SN1987A in [81] is consistent with the distance to HV 982; perhaps there is a distinct region located behind the main mass of the LMC?). As the EB programme develops – there are plans to include 20 systems – we may be able to answer this question. At present, however, the matter is far from resolved.

In [96], the results of a theoretical analysis of the behaviour of M_K^{RC}, the red clump K-band absolute magnitude, are applied to determine the distances to the globular cluster 47 Tuc (in agreement with the *Hipparcos* MS-fitting distance of [85]) and to the OGLE II field. The result confirms the RC LMC distance modulus of $\mu_0 = 18.5$ found by [1]. Near-infrared RC distances to two LMC clusters, Hodge 4 and NGC 1651, also support the Key Project adopted value; based on these two clusters, a mean distance modulus of 18.54 ± 0.10 mag is found [103]. This agrees with another recent near-infrared RC distance: 18.471 mag [88].

As well as possessing a PL relationship for Miras (see (4.2)), we also know [126] that in the LMC the bolometric PL relation extends up to at least ~ 1500 days. We saw in Sect. 4.2 that two new calibrations of the zero point have been found; so we can use LMC field Miras to estimate the distance modulus. This was done in [30], which obtained an LMC true modulus of 18.60 ± 0.1.

Geometric methods are no less confusing. The original geometric determination of the distance to SN1987A [81], combined with an estimate of the position of the SN within the LMC, gave $\mu_0 = 18.50 \pm 0.13$ mag. A later determination of the distance [39] gave a much smaller *upper* limit of $\mu_0 < 18.37 \pm 0.04$ mag.

An interesting graph in [9] is that of 80 recent determinations of the LMC distance modulus, based on 21 independent methods (including their own RR Lyrae distance). The weighted average of these is 18.47 ± 0.04 mag.

The bottom line? Some recent measurements that were not included in the Key Project compilation (coming from RC, RR Lyrae and TRGB methods, for example) are consistent with the value adopted by the Key Project. Yet there are systematic differences between several other results from different techniques that exceed the formal errors. We still do not understand the distance to our closest major neighbor. Perhaps to resolve the LMC distance issue we will have to wait for the planned astrometric missions (see Sect. 4.5).

4.4.2 H_0 from Geometry: A Case for Corrective Lenses?

There are now nine gravitational lens systems with accurate time delay measurements; four of these are unsuitable for an accurate determination of H_0 since they have poorly determined properties (either lens position or overly complicated lenses). That leaves five systems with good time delay measurements, simple environments and a dominant lens galaxy with regular isophotes. Kochanek [56–58] has analysed the five systems, and found estimates for H_0 in the two limiting cases for physically possible galactic mass distributions: dark matter models with flat rotation curves, and constant M/L models. The former leads to $H_0 = 48^{+7}_{-4}$ (95%CL), the latter to $H_0 = 71 \pm 6$ (95%CL). (The average value, including uncertainties in the mass distribution, is $H_0 = 62 \pm 7$.)

The Hubble parameter from constant M/L models is in excellent agreement with the Key Project and *WMAP* results; but our expectation is that galaxies possess massive, extended dark-matter haloes. The models with flat rotation curves are in agreement with the estimates coming from the Sandage team; but these estimates appear to be in conflict with many other lines of evidence (and, as we have seen, there are even indications that the Key Project 'local' estimate of H_0 may require a slight upward revision). Of course, it is possible that Kochanek's analyses produce estimates of H_0 that are systematically low; but other recent independent analyses of gravitational lens systems produce low estimates of H_0 (for example, in [113], the lens system PG1115+080 is analysed to produce an estimate of H_0 that is consistent with [95] but only marginally consistent with the final Key Project result [35]).

Some part of our understanding would seem to be faulty – either our understanding of the dark matter distribution in galaxies, the consensus value of H_0 coming from the 'local' ladder approach, or our models of gravitational lensing systems. It is too early to say where the misunderstanding lies – but future work in this area will prove fascinating!

4.5 Future Developments

At the start of 2002 it was hoped that four major astrometry missions planned for the next decade would provide us with detailed knowledge of the cosmic distance scale. (The four missions were *FAME* (Full-Sky Astrometric Mapping Explorer), *DIVA* (Deutsches Interferometer für Vielkanalphotometrie und Astrometrie), *GAIA* (originally Global Astrometric Interferometer for Astrophysics, though now in search of an acronym) and *SIM* (Space Interferometry Mission).) Unfortunately, in January 2002, NASA cited budget cuts and opted to cancel *FAME*; at the time of writing, it is unclear whether the programme can continue in some altered form. Nevertheless, analyses that have been published regarding the *FAME* catalog (such as the possibility of calibrating the RR Lyrae distance scale and bias problems that must be faced [101], or the recommendation to augment its science capability by devoting a modest fraction of its catalog to selected targets that are fainter than its magnitude limits [98]) raise issues that should be generally applicable to the next generation of astrometric missions. And *DIVA* (planned launch date 2004), *SIM* (planned launch date 2009), and *GAIA* (planned launch date 2010–2012) have the capability to resolve many outstanding problems.

DIVA, which in many ways is a precursor for the technology of the *GAIA* mission, will measure the parallaxes of 5×10^5 stars with a relative accuracy of better than 10%. It will calibrate the MS of about 30 open clusters, which will resolve the 'Pleiades problem' found by *Hipparcos*. It will also determine the luminosities of about 30 Cepheid and 200 Mira variables, which will greatly improve calibrations of local rungs of the distance ladder. *SIM* and *GAIA* will provide even more impressive distance determinations, including rotational parallaxes to nearby galaxies [79, 116] as well as parallaxes of a full range of Cepheids and RR Lyrae stars. If these missions are successful, then we will have many aspects of the cosmic distance scale on a firm footing within a decade. (There will also, of course, be many other important science benefits, unrelated to the distance scale.)

Until results from these astrometric missions are analysed, what progress can we expect to see in the elucidation of the distance scale?

The exciting discovery [43] of a luminous H_2O megamaser in the Sa galaxy NGC 2960, with features that are strikingly similar to NGC 4258, holds the promise that a precise geometric distance may be established to this galaxy. NGC 2960 is approximately ten times more distant than NGC 4258, so this galaxy could become an important target for establishing the cosmic distance scale on geometric foundations. One can only hope that, in the future, the distance to more galaxies will be determined with similar geometric techniques.

In the shorter term, the use of the eclipsing binary technique is likely to be more widely used as more target systems are identified. (Many EBs have been discovered in the LMC as a byproduct of microlensing surveys.) Furthermore, new methods of distance determination based upon EBs might be developed. The authors of [97], for example, suggest an empirical method based on the existence of a tight linear relationship between the V-band zero magnitude angular diameter and the Strömgren color index c_1 for B-stars; they claim that accurate Strömgren photometry – obtainable with 1.5 m-class telescopes – of HV 2274 and HV 982 will give the empirical LMC distance to an accuracy of 0.13 mag.

The difficulties of extending the SBF method to the infrared are being overcome. In [64], for example, the authors describe an infrared SBF survey of 19 early-type galaxies in the Fornax cluster, the results of which should strengthen the use of the IR SBF method as a distance indicator. This is important because with this IR method the SBF signal is 20–40 times larger, the seeing is usually better, and dust is less of a problem in the target galaxy.

The 'one-step' global measurements of H_0 should continue to improve in accuracy and robustness. For example, the *Planck Surveyor* satellite, due for launch in 2007, will produce SZ data on thousands of clusters. Even before then, with some modest improvements in SZ observations and with the addition of gravitational lensing observations, we may be able to correct for cluster asphericity and determine true distances to clusters with much better accuracy than at present [34]. In [37], it is shown how the discovery of lensed multiple images of high-z SNe, with future instruments like *SNAP* and *NGST*, will lead not only to constraints on H_0 but also other cosmological parameters.

Certainly there are outstanding questions regarding all aspects of the cosmic distance scale – from the details of the solar neighborhood through to the cosmological parameters that shape our universe. But just as certainly, these are exciting times to be studying such questions.

References

[1] Alves D R et al. (2002) K-band red clump distances to the Large Magellanic Cloud. Ap. J. **573** L51
[2] Arp H (2002) Arguments for a Hubble constant near $H_0 = 55$. Ap. J. **571** 615
[3] Arp H et al. (2002) NGC 3628: ejection activity associated with quasars. Astron. Astrophys. **391** 833
[4] Barnes T G and Evans D S (1976) Stellar angular diameters and visual surface brightness I. Late spectral types. MNRAS **174** 489

[5] Baryshev Yu V, Chernin A D and Teerikorpi P (2001) The cold local Hubble flow as a signature of dark energy. Astron. Astrophys. **378** 729

[6] Bell M B (2002) Further evidence for large intrinsic redshifts. Ap. J. **566** 705

[7] Bell M B (2002) On quasar distances and lifetimes in a local model. Ap. J. **567** 801

[8] Benedict G F et al (1999) Interferometric astrometry of Proxima Centauri and Barnard's star using the *Hubble Space Telescope* Fine Sensor Guidance 3: detection limits for substellar companions. Astron. J. **118** 1086

[9] Benedict G F et al (2002) Astrometry with the *Hubble Space Telescope*: a parallax of the fundamental distance calibrator RR Lyrae. Astron. J. **123** 473

[10] Benedict G F et al (2002) Astrometry with the *Hubble Space Telescope*: a parallax of the fundamental distance calibrator δ Cephei. Astron. J. **124** 1695

[11] Bernardi M et al. (2002) Redshift–distance survey of early-type galaxies. II. The D_n–σ relation. Astron. J. **123** 2159

[12] Birkinshaw M (1999) The Sunyaev–Zel'dovich effect. Phys. Rep. **310** 97

[13] Blakeslee J P et al. (2002) Early-type galaxy distances from the fundamental plane and surface brightness fluctuations. MNRAS **330** 443

[14] Bono G et al. (2002) On the pulsation parallax of the variable star RR Lyr. MNRAS **332** L78

[15] Bono G et al. (2002) On the distance of Magellanic Clouds: first overtone Cepheids. Ap. J. **574** L33

[16] Bradley A et al. (1991) The flight hardware and ground system for *Hubble Space Telescope* astrometry. PASP **103** 317

[17] Branch D (1998) Type Ia supernovae and the Hubble constant. Ann. Rev. Astron. Astrophys. **36** 15

[18] Brisken W F et al. (2002) Very long baseline array measurement of nine pulsar parallaxes. Ap. J. **571** 906

[19] Burud I et al. (2002) Time delay and lens redshift for the doubly imaged BAL quasar SBS 1520+530. Astron. Astrophys. **391** 481

[20] Caputo F, Marconi M and Musella I (2002) The Cepheid period–luminosity relation and the maser distance to NGC 4258. Ap. J. **566** 833

[21] Cardone V F et al. (2002) A new method for the estimate of H_0 from quadruply imaged gravitational lens systems. Astron. Astrophys. **382** 792

[22] Ciardullo R et al. (2002) Planetary nebulae as standard candles. XII. Connecting the population I and population II distance scales. Ap. J. **577** 31

[23] Clememti G et al. (2003) Distance to the Large Magellanic Cloud: the RR Lyrae stars. Astron. J. **125** 1309

[24] Cole A A (1998) Age, metallicity, and the distance to the Magellanic Clouds from red clump stars. Ap. J. **500** L137

[25] Cooke W J and Eichhorn H (1997) A new and comprehensive determination of the distance to member stars of the Hyades. MNRAS **288** 319

[26] Di Benedetto G P (2002) On the absolute calibration of the Cepheid distance scale using *Hipparcos* parallaxes. Astron. J. **124** 1213

[27] Di Fabrizio L et al. (2002) Anomalous RR Lyrae stars (?): CM Leonis. MNRAS **336** 841

[28] Feast M W et al. (1989) A period–luminosity–colour relation for Mira variables. MNRAS **241** 375
[29] Feast M W and Catchpole R M (1997) The Cepheid PL zero-point from *Hipparcos* trigonometrical parallaxes. MNRAS **286** L1
[30] Feast M W, Whitelock P and Menzies J (2002) Globular clusters and the Mira period–luminosity relation. MNRAS **329** L7
[31] Fernie J D (1969) The period–luminosity relation: a historical review. PASP **81** 707
[32] Fiorentino G et al. (2002) Theoretical models for classical Cepheids. VIII. Effects of helium and heavy-element abundance on the Cepheid distance scale. Ap. J. **576** 402
[33] Fitzpatrick E L et al. (2002) Fundamental properties and distances of the Large Magellanic Cloud from eclipsing binaries. II. HV 982. Ap. J. **564** 260
[34] Fox D C and Pen U-L (2002) The distance to clusters: correcting for asphericity. Ap. J. **574** 38
[35] Freedman W L et al. (2001) Final results from the *HST* Key Project to measure the Hubble constant. Ap. J. **553** 47
[36] Gieren W, Fouqué P and Gomez M (1998) Cepheid period–radius and period–luminosity relations and the distance to the LMC. Astrophys. J. **496** 17
[37] Goobar A et al. (2002) Cosmological parameters from lensed supernovae. Astron. Astrophys. **393** 25
[38] Gott J R III et al. (2001) Median statistics, H_0 and the accelerating universe. Ap. J. **549** 1
[39] Gould A and Uza O (1998) Upper limit to the distance to the Large Magellanic Cloud. Ap. J. **494** 118
[40] Grainge K et al (2002) Measuring the Hubble constant from Ryle Telescope and X-ray observations, with applications to Abell 1413. MNRAS **333** 318
[41] Guinan E F et al. (1998) The distance to the Large Magellanic Cloud from the eclipsing binary HV 2274. Ap. J. **509** L21
[42] Hamuy M and Pinto P A (2002) Type II SN as standardized candles. Ap. J. **566** L63
[43] Henkel C et al. (2002) Discovery of water vapor megamaser emission from Mrk 1419 (NGC 2960): an analogue of NGC 4258? Astron. Astrophys. **394** L23
[44] Herrnstein J R et al. (1999) A geometric distance to the galaxy NGC 4258 from orbital motions in a nuclear gas disk. Nature **400** 539
[45] Hjorth J et al (2002) The time delay of the quadrupole quasar RX J0911.4+0551. Ap. J. **572** L11
[46] Hu E M et al. (2002) A redshift $z = 6.56$ galaxy behind the cluster Abell 370. Ap. J. **568** L75
[47] Huchra J P (2002) Estimates of the Hubble constant. http://cfa-www.harvard.edu/~huchra/hubble.plot.dat
[48] Jacoby G H et al. (1992) A critical review of selected techniques for measuring extragalactic distances. PASP **104** 599
[49] Kannappan S J, Fabricant D G and Franx M (2002) Physical sources of scatter in the Tully–Fisher relation. Astron. J. **123** 2358
[50] Kaplan D L, van Kerkwijk M H and Anderson J (2002) The parallax and proper motion of RX J1856.5–3754 revisited. Ap. J. **571** 447

[51] Karachentsev I D et al. (2002) The M 81 group of galaxies: new distances, kinematics and structure. Astron. Astrophys. **383** 125

[52] Karachentsev I D et al. (2002) New distances to galaxies in the Centaurus A group. Astron. Astrophys. **385** 21

[53] Karachentsev I D et al. (2002) The very local Hubble flow. Astron. Astrophys. **389** 812

[54] Kim M et al. (2002) Determination of the distance to M 33 based on the tip of the red giant brach and the red clump. Astron. J. **123** 244

[55] Klebesadel R W, Strong I B and Olsen R A (1973) Observation of gamma-ray bursts of cosmic origin. Ap. J. **182** L85

[56] Kochanek C S (2002) What do gravitational lens time delays measure? Ap. J. **578** 25

[57] Kochanek C S (2002) Gravitational lenses, the distance ladder and the Hubble constant: a new dark matter problem. [astro-ph/0204043]

[58] Kochanek C S (2002) The gravitational lens time delays in CDM. [astro-ph/0206006]

[59] Koen C et al (2002) $UBV(RI)_C$ photometry of *Hipparcos* red stars. MNRAS **334** 20

[60] Lee M G, Freedman W L and Madore B F (1993) The tip of the red giant branch as a distance indicator for resolved galaxies. Ap. J. **417** 553

[61] Lee M G et al. (2002) Determination of the distance to M 33 based on single-epoch I-band observation of Cepheids. Ap. J. **565** 959

[62] Leonard D C et al. (2002) A study of the Type II-plateau supernova 1999gi and the distance to its host galaxy, NGC 3184. Astron. J. **124** 2490

[63] Li J, Harrington J P and Borkowski K J (2002) The angular expansion and distance of the planetary nebula BD $^+30°$ 3639. Astron. J. **123** 2676

[64] Liu C M, Graham J R and Charlot S (2002) Surface brightness fluctuations of Fornax cluster galaxies: calibration of infrared surface brightness fluctuations and evidence for recent star formation. Ap. J. **564** 216

[65] Luri X et al. (1998) The LMC distance modulus from *Hipparcos* RR Lyrae and classical Cepheid data. Astron. Astrophys. **335** L81

[66] Lutz T E and Kelker D H (1973) On the use of trigonometric parallaxes for the calibration of luminosity systems: theory. PASP **85** 573

[67] Madore B F and Freedman W L (1991) The Cepheid distance scale. PASP **103** 933

[68] Maíz-Apellániz J, Cieza R and MacKenty J W (2002) Tip of the red giant branch distances to NGC 4214, UGC 685, and UGC 5456. Astron. J. **123** 1307

[69] Makarov V V (2002) Computing the parallax of the Pleiades from the *Hipparcos* intermediate astrometry data: an alternative approach. Astron. J. **124** 3299

[70] Maoz E et al. (1999) A distance to the galaxy NGC 4258 from observations of Cepheid variable stars. Nature **401** 351

[71] Mathur S, Wilkes B J and Ghosh H (2002) *Chandra* detection of highest-redshift (~ 6) quasars in X rays. Ap. J. **570** L5

[72] Mei S et al. (2003) H_0 measurement from VLT deep I-band surface brightness fluctuations in NGC 564 und NGC 7619. Astron. Astrophys. **399** 441

[73] Méndez B et al. (2002) Deviations from the local Hubble flow. I. The tip of the red giant branch as a distance indicator. Astron. J. **124** 213

[74] Mermilliod J-C et al. (1997) The distance of the Pleiades and nearby clusters. In: Hipparcos Venice '97 ed. B Battrick (Noordwijk: ESA) p. 643
[75] Metzger M R et al. (1997) Spectral constraints on the redshift of the optical counterpart to the γ-ray burst of 8 May 1997. Nature **387** 878
[76] Miyoshi M et al (1995) Evidence for a black hole from high rotation velocities in a sub-parsec region of NGC 4258. Nature **373** L127
[77] Narayanan V K and Gould A (1999) Correlated errors in *Hipparcos* parallaxes towards the Pleiades and the Hyades. Ap. J. **523** 328
[78] Nordgren T E et al. (2002) Calibration of the Barnes–Evans relation using interferometric observations of Cepheids. Astron. J. **123** 3380
[79] Olling R P and Peterson D M (2000) Galaxy distances via rotational parallaxes. [astro-ph/0204043]
[80] Paczyński B and Stanek K Z (1998) Galactocentric distance with the OGLE and *Hipparcos* red clump stars. Ap. J. **494** L219
[81] Panagia N et al. (1991) Properties of the SN1987A circumstellar ring and the distance to the Large Magellanic Cloud. Ap. J. **380** L23
[82] Paturel G et al. (2002) Calibration of the distance scale from galactic Cepheids I. Calibration based on the GFG sample. Astron. Astrophys. **383** 398
[83] Paturel G et al. (2002) Calibration of the distance scale from galactic Cepheids II. Use of the *Hipparcos* calibration. Astron. Astrophys. **389** 19
[84] Paturel G et al. (2003) In preparation. (Data available at http://leda.univ-lyon.fr)
[85] Percival S M et al. (2002) Resolving the 47 Tucanae distance problem. Ap. J. **573** 174
[86] Perlmutter S (1999) Measurements of Ω and Λ from 42 high-redshift supernovae. Ap. J. **517** 565
[87] Perryman M A C et al. (1998) The Hyades: distance, structure, dynamics, and age. Astron. Astrophys. **331** 81
[88] Pietrzyński G and Gieren W (2002) The Araucaria project: deep near-infrared survey of nearby galaxies. I. The distance to the Large Magellanic Cloud from K-band photometry of red clump stars. Astron. J. **124** 2633
[89] Pinsonneault M H et al. (1998) The problem of *Hipparcos* distances to open clusters. I. Constraints from multicolor main-sequence fitting. Ap. J. **504** 170
[90] Pons J A (2002) Toward a mass and radius determination of the nearby isolated neutron star RX J1856.5–3754. Ap. J. **564** 981
[91] Refsdal S (1964) The gravitational lens effect. MNRAS **128** 295
[92] Refsdal S (1964) On the possibility of determining Hubble's parameter and the masses of galaxies from the gravitational lens effect. MNRAS **128** 307
[93] Ribas I et al. (2002) Fundamental properties and distances of Large Magellanic Cloud eclipsing binaries. III. Eros 1044. Ap. J. **574** 771
[94] Riess A G et al. (1998) Observational evidence from supernovae for an accelerating universe and a cosmological constant. Astron. J. **116** 1009
[95] Saha A et al. (2001) Cepheid calibration of the peak brightness of Type Ia supernova. XI. SN1998aq in NGC 3982. Ap. J. **562** 314
[96] Salaris M and Girardi L (2002) Population effects on the red giant clump absolute magnitude: the K band. MNRAS **337** 332

[97] Salaris M and Groenewegen M A T (2002) An empirical method to estimate the LMC distance using B-stars in eclipsing binary systems. Astron. Astrophys. **381** 440

[98] Salim S, Gould A and Olling R P (2002) Astrometry survey missions beyond the magnitude limit. Ap. J. **573** 631

[99] Sandage A R (1993) The Oosterhoff period effect and the age of the galactic globular cluster system. In: New Perspectives on Stellar Pulsation and Pulsating Variable Stars ed. J. M. Nemec and J. M. Matthews. (Cambridge: CUP)

[100] Sandage A R (2002) Bias properties of extragalactic distance indicators. X. The Teerikorpi cluster incompleteness bias for a modified Lemaître–Robertson–Hubble–Humason distance method that uses luminosity functions. Astron. J. **123** 1179

[101] Sandage A R and Saha A (2002) Bias properties of extragalactic distance indicators. XI. Methods to correct for observational selection bias for RR Lyrae absolute magnitudes from trigonometric parallaxes expected from the *Full-Sky Astrometric Mapping Explorer* satellite. Astron. J. **123** 2047

[102] Sarajedini A et al. (2000) *Hubble Space Telescope* WFPC2 photometry of M 33: properties of the halo star clusters and surrounding fields. Astron. J. **120** 2437

[103] Sarajedini A et al. (2002) K-band red clump distances to the Large Magellanic Cloud clusters Hodge 4 and NGC 1651. Astron. J. **124** 2625

[104] Solanes J M et al. (2002) The three-dimensional structure of the Virgo cluster region from Tully–Fisher and H I data. Astron. J. **124** 2440

[105] Spergel D N et al. (2003) First year Wilkinson Microwave Anisotropy Probe (WMAP) observation: determination of cosmological parameters. Ap. J. preprint doi: 10.1086/377226

[106] Strauss M A and Willick J A (1995) The density and peculiar velocity fields of nearby galaxies. Phys. Rep. **261** 271

[107] Tanvir N R (1997) Cepheids as distance indicators. In: The Extragalactic Distance Scale ed. M. Livio, M. Donahue and N. Panagia. (Cambridge University Press: Cambridge) pp 91

[108] Teerikorpi P (1997) Observational selection bias affecting the determination of the extragalactic distance scale. Ann. Rev. Astron. Astrophys. **35** 101

[109] Teerikorpi P and Paturel G (2002) Evidence for the extragalactic Cepheid distance bias from the kinematical distance scale. Astron. Astrophys. **381** L37

[110] Tonry J L (1997) In: The Extragalactic Distance Scale ed. M. Livio, M. Donahue and N. Panagio (Cambridge: CUP)

[111] Tonry J L and Schneider D P (1988) A new technique for measuring extragalactic distances. Astron. J. **96** 807

[112] Tonry J L et al. (2001) The SBF survey of galaxy distances. IV. SBF magnitudes, colors and distances. Ap. J. **546** 681

[113] Treu T and Koopmans L V E (2002) The internal structure of the lens PG115+080: breaking degeneracies in the value of the Hubble constant. MNRAS **337** L6

[114] Turner D G and Burke J F (2002) The distance scale for classical Cepheid variables. Astron. J. **124** 2931

[115] Udalski A et al. (1999) The optical gravitational lensing experiment. Cepheids in the Magellanic Clouds. III. Period–luminosity–color and period–luminosity relations of clasical Cepheids. Acta Astronomica **49** 201

[116] van der Marel R et al. (2002) New understanding of Large Magellanic Cloud structure, dynamics and orbit from carbon star dynamics. Astron. J. **124** 2639

[117] van Leeuwen F, Feast M W, Whitelock P A and Yudin B (1997) First results from *Hipparcos* trigonometric parallaxes of Mira-type variables. MNRAS **287** 955

[118] Walker A (1999) The distances of the Magellanic Clouds. In: Post-Hipparcos Cosmic Candles ed. A. Heck and F. Caputo (Dordrecht: Kluwer) p. 125

[119] Walter F M (2001) The proper motion, parallax and origin of the isolated neutron star RX J185635–3754. Ap. J. **549** 433

[120] Walter F M and Lattimer J M (2002) A revised parallax and its implications for RX J185635–3754. Ap. J. **576** L145

[121] Walter F M and Matthews L D (1997) The optical counterpart of the isolated neutron star RX J185635–3754. Nature **389** 358

[122] Walter F M, Wolk S J and Neuhäuser R (1996) Discovery of a nearby isolated neutron star. Nature **379** 233

[123] Webb S (1999) Measuring the Universe (Springer: London)

[124] Westerlund B E (1997) The Magellanic Clouds (Cambridge: CUP)

[125] Williams B A, Un Min S and Verdes-Montenegro L (2002) The VLA H I observations of Stephan's Quintet (HCG 92). Astron. J. **123** 2417

[126] Whitelock P A and Feast M W (2000) Dust enshrouded AGB variables in the LMC. Mem. Soc. Astron. It. **71** 601

[127] Whitelock P A and Feast M W (2000) Hipparcos parallaxes for Mira-like long period variables. MNRAS **319** 759

[128] Wyithe J S B and Wilson R E (2002) Photometric solutions for semidetached eclipsing binaries: selection of distance indicators in the Small Magellanic Cloud. Ap. J. **571** 293

5 The Cosmic Microwave Background

A.H. Jaffe

5.1 Introduction

The Cosmic Microwave Background (CMB) is a nearly perfect blackbody that traces the state of the Universe as the baryons and electrons underwent a transition from being an ionized plasma, opaque to photons, to a transparent neutral gas. At this time, the Universe was about 400,000 years old and had a root mean square density fluctuation of only one part in one hundred thousand. The photons that make up the CMB were at this time released from thermal contact with the baryons, and have been streaming freely through the Universe since. The characteristics of the resulting snapshot of the early Universe depend upon the cosmological parameters describing its contents and evolution; a careful statistical analysis of CMB data thereby allows us to measure these parameters.

The existence of what we now call the CMB was predicted in 1948 by work done by various combinations of Alpher, Gamow and Herman (with Bethe appearing in Alpher, Bethe and Gamow (1948) for the pleasing name). This work was primarily concerned with the abundances of the elements (what we now call Big-Bang Nucleosynthesis; the number of photons per baryons, needed for this calculation, is defined by the temperature of the CMB). It was finally discovered serendipitously in 1964 by Penzias and Wilson (1965) at AT&T Bell Labs in New Jersey; simultaneously a group at Princeton (also in New Jersey), led by Dicke, Peebles, Roll and Wilkinson (1964) (the latter commemorated in the renamed Wilkinson Microwave Anisotropy Probe (WMAP) satellite after his untimely death in late 2002) were designing an experiment to find it. Moreover, there had been data taken by McKellar (1940) looking at excitations of cosmic CN gas which, in retrospect, clearly shows the signature of absorption of CMB photons.

The spectrum of the CMB as observed by these pioneers was a featureless blackbody, isotropic over the sky. As experimenters looked with ever-more sensitive instruments, the blackbody spectrum continued to hold: the limits are now that the CMB is a blackbody with $T_0 = 2.725 \pm 0.0001$ K. Departures from a perfect black body are given in terms of the chemical potential, $|\mu| < 9 \times 10^{-5}$ and the Compton parameter $|y| < 1.5 \times 10^{-5}$ (at 95% confidence). These limits are from the FIRAS instrument on the COBE satellite, launched in 1989 (Wright et al. 1991; Fixsen et al. 1996).

The first detection of anisotropy in the temperature of the CMB came with the detection of a dipole pattern—hotter in a particular direction on the sky, cooler at the antipodal point—with an amplitude of $\Delta T/T \simeq 0.1\%$. This is interpreted as being due to our motion with respect to the frame in which the CMB is isotropic. It corresponds to a motion of the Local Group of galaxies of 600 km/sec (Smoot et al. 1977).

Fluctuations in the CMB on smaller scales and with smaller amplitude are, however, expected to reflect the same physics that has created the large-scale structure we observe in the Universe today. Today we observe structures in the Universe many megaparsecs in size. On still larger scales, we observe correlations in the distribution of matter out to the largest distances on which they can be measured (see, for example, Chap. 10 by A.P. Fairall in this volume).

The distance that a beam of light could travel since the Big Bang is known as the *particle horizon*. In the most naive Big Bang scenario, this is the maximum distance on which physics can act to build — or smooth out — structures at any time. The horizon scale grows approximately linearly with the age of the Universe. At the time when the CMB photons were freed from the primordial plasma, known as *last scattering*, the Universe was about 400,000 year old, corresponding to a horizon size of roughly 200 kiloparsecs. This corresponds to about one degree in angle on the sky.

These facts lead inexorably to a conclusion that at first seems bizarre and against our physical intuition: there must be some way to bring matter into contact when seemingly separated by distances larger than any causal physical process could act. It turns out that such an idea was proposed around 1980 by Guth (1980) in the USA and Starobinsky in the former USSR: inflation. Inflation posits a period of accelerated expansion in the early Universe. If this expansion proceeds for long enough (so that the scale factor grows by a factor greater than about e^{50}), inflation predicts the following:

- The Universe will be geometrically flat (defined by the cosmological parameter, $\Omega_{\text{tot}} = 1$), just as any small area of a balloon looks flat if you inflate it enough.
- The local Universe will appear to be homogeneous and isotropic, because what is now our observable Universe was once much smaller (and, by assumption, in "causal contact" at an early time). In particular, the CMB should have a nearly constant temperature.
- There will be small fluctuations on top of this homogeneous and isotropic background, generated from quantum fluctuations during inflation that have been blown up to macroscopic size. Inflation predicts not just the existence of these fluctuations, but that their spatial pattern is given by a so-called *scale-invariant* spectrum (note that this terminology is somewhat different than as used in other fields of mathematics and physics).

Indeed, our observations of the CMB itself provide the best evidence for each of these predictions. Conversely, we must realize that our observations of the

CMB test only these *repercussions* of the inflationary paradigm, rather than the underlying physics of inflation itself: inflation occurs at some time like 10^{-34} seconds; the CMB is formed at 400,000 years; and our observations are performed at 15 Gyrs. Any mechanism that makes these same predictions is of course also still allowed by the data. One new idea (arguably the only such) is the 'Cyclic Universe' (Turok et al. 2002), which thermalized the Universe by making it much older than it appears, while inflation makes the Universe (more precisely, the Horizon) much larger than it appears.

The anisotropy of the CMB was first detected by the DMR instrument on COBE (Bennett et al. 1996; see Table 5.1). DMR observed the CMB at frequencies of roughly 30, 60 and 90 GHz. The DMR instrument detected fluctuations to the CMB on angular scales from that of the dipole (180°) all the way down to the instrument's seven degree resolution. These fluctuations had the characteristic pattern of a scale invariant spectrum — the prediction of inflation.

Unfortunately, DMR's seven degree resolution is considerably larger than the roughly one degree size of the horizon at last scattering. Hence, these results were sensitive to the primordial distribution of matter on the largest scales, but not to the mechanisms by which matter on smaller scales was moved around to form today's galaxies and clusters of galaxies.

Conversely, these results implied that there must be some mechanism for forming small fluctuations in the matter on scales larger than this causal horizon, in addition to the aforementioned mechanism for smoothing the matter distribution. Again, inflation provided such a mechanism.

The COBE results spawned feverish theoretical and experimental activity in the study of the CMB which continues through today. Throughout the 1990s, CMB experiments probed angular scales down to roughly one degree. By 1999, a consistent picture was beginning to emerge. At one degree, the very smallest scales probed by most experiments, there did seem to be an excess of fluctuation power — as predicted within the paradigm of inflation. The few very ambitious experiments to probe with yet higher angular resolution

Table 5.1. Satellite CMB Anisotropy experiments, 1989-2007

	Beam FWHM	Frequency GHz	Technology
COBE/DMR (NASA) (Bennett et al. 1996)	7°	30-90 GHz	Differential Radiometers
WMAP (NASA) (Bennett et al. 2003)	10'	20-90 GHz	Differential Radiometers
Planck LFI (ESA) http://astro.estec.esa.nl/SA-general/Projects/Planck/	10'	30-100 GHz	Radiometers
Planck HFI (ESA) http://astro.estec.esa.nl/SA-general/Projects/Planck/	4'	100-800 GHz	Bolometers

also showed results consistent with the predictions (e.g., Bond & Jaffe 1998; Knox & Page 2000)

Starting in the year 2000, we have seen results from a second generation of CMB experiments specifically designed to probe the CMB at scales less than a degree over large areas of sky (see Table 5.2). To compare these observations with theoretical predictions, consider $T(\hat{\mathbf{x}})$, the temperature at some position $\hat{\mathbf{x}}$ (a unit vector) on the sky. We can decompose T into its spherical harmonic coefficients, $a_{\ell m}$ (analogous to a two-dimensional Fourier transform on a plane):

$$T(\hat{\mathbf{x}}) = \sum_{\ell m} a_{\ell m} Y_{\ell m}(\hat{\mathbf{x}}) , \qquad (5.1)$$

where the $Y_{\ell m}(\hat{\mathbf{x}})$ are the spherical harmonic functions. We can then define the CMB power spectrum, C_ℓ, from

$$\langle a_{\ell m} a_{\ell' m'} \rangle = \delta_{ll'} \delta_{mm'} , \qquad (5.2)$$

Table 5.2. Balloon-borne and ground-based CMB experiments, 2000-2003. The values given for the beam are typically the smallest used in the given experiment. The references cited are the most recent power spectrum results from each of the teams. See references therein for further information. 'LDB' stands for 'Long Duration Balloon' and further information about 'Spider Web Bolometers' can be found in the references for the individual experiments.

	Beam FWHM arcmin	Area deg^2	Frequency GHz	Technology
Balloon-borne experiments				
BOOMERANG (Ruhl et al. 2003)	10	3000	90–410	Spider-web bolo. LDB
MAXIMA (Lee et al. 2002)	10	120	150–410	Spider-web bolo.
Archeops (Benoit et al. 2003)	11	5000	143–545	Spider-web bolo. LDB
Ground-based experiments				
ACBAR (Kuo et al. 2003)	4	24	150–300	Spider-web bolo.
CBI (Pearson et al. 2003)	5	40	26–36	Interferometer
DASI (Halverson et al. 2002)	20	400	26–36	Interferometer
VSA (Grainge et al. 2002)	12	120	26–36	Interferometer

where the Kronecker delta factors and the fact that C_ℓ is only a function of ℓ enforce the statistical isotropy of the temperature field. The multipole ℓ corresponds roughly to a scale of $180°/\ell$ on the sky. Hence, a peak in C_ℓ at $\ell \simeq 180$ corresponds to structure at the vaunted one degree scale. In Fig. 5.1, we show the combined data from experiments through the end of 2002.

As this review was being prepared, the results from NASA's newest CMB anisotropy telescope, the Wilkinson Microwave Anisotropy Probe (WMAP) were released (Bennett et al. 2003 and references therein). The temperature anisotropy results from WMAP are shown in Fig. 5.2. For $\ell < 500$, these corroborate the features of the previous data: the high first peak at $\ell \simeq 200$, the further oscillations at higher ℓ, and the overall amplitude of the fluctuations. While it was still possible to believe that the 2002 data beyond the first peak were inconclusive, the WMAP data with their minuscule error bars (and, crucially, fine control and understanding of possible systematic errors) brook little disagreement with this small-scale structure, and thus with the underlying physical paradigm which we see predicts such structure: the hot Big Bang, a flat Universe, and small fluctuations generated somehow (e.g., inflation) on scales seemingly beyond the causal horizon. This conclusion is

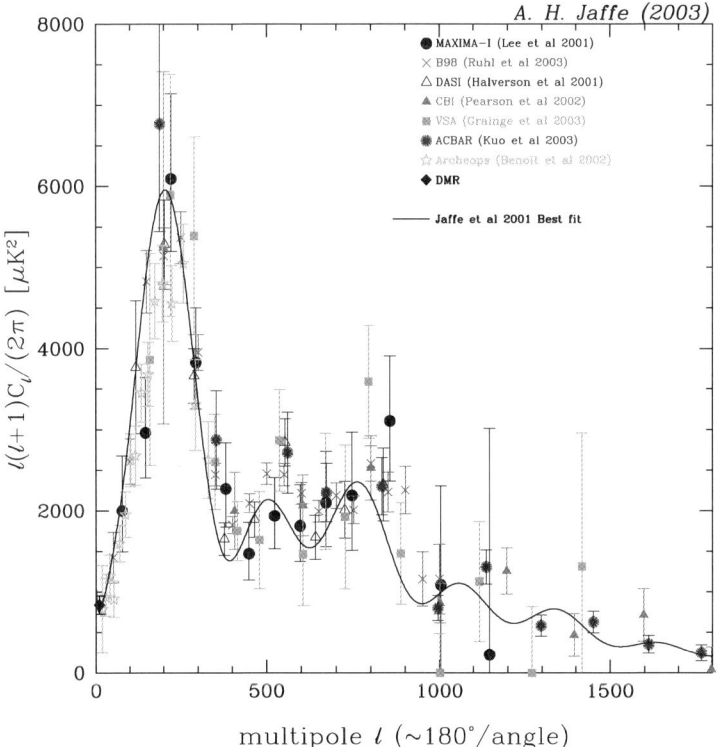

Fig. 5.1. The CMB power spectrum, C_ℓ, from experiments 1990-2002.

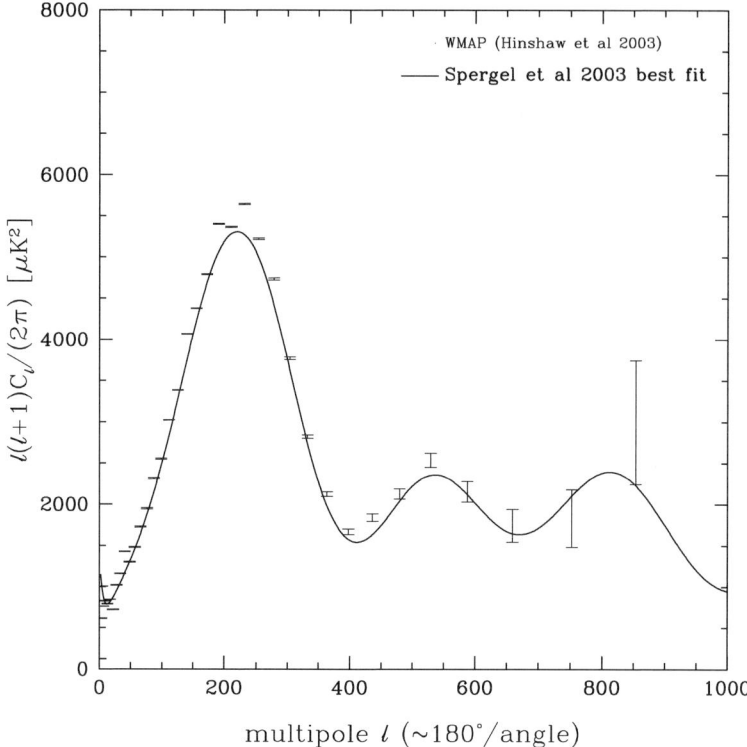

Fig. 5.2. The CMB power spectrum as measured by the WMAP satellite. Note that the error bars in this figure include only the variance due to the instrumental noise of the experiment, not the so-called cosmic or sample variance. The solid line is the WMAP team's best fit model.

further cemented by the detection by WMAP of a correlation between the temperature of the CMB and its polarization which, as we shall see below, is due to the streaming of matter in response to the gravitational field of density inhomogeneities.

5.2 The CMB and the Physics of the Early Universe

First, we review the physical processes that produce the CMB photons we observe today. We assume that the Universe began with a Hot Big Bang and is described by the Friedmann-Robertson-Walker (FRW) metric. The physical properties such a Universe are described in great detail in Cosmology textbooks such as Kolb & Turner (1991), Peebles (1993) and Peacock (1999). The scale factor, which gives the relative size of a spatial section of the cosmological manifold, is $a(z) = a_0/(1+z)$ at redshift z; here and throughout, a subscript, 0, gives the value at the present time. The distribution function of the photons (and other relativistic species, those with $mc^2 \ll kT$) has a temperature, $T(z) = T_0(1 + z)$.

The present day expansion rate of the Universe, the Hubble Constant, is $\dot{a}/a|_0 = H_0 = 100h\text{km/s/Mpc}$, with $h = 0.72 \pm 0.02$ (see Chap. 4 by Stephen Webb in this volume). The age of the Universe is $t_0 \simeq 14\text{Gyr}$. The constituents of the Universe are described by the density parameter, $\Omega_i = \rho_i/\rho_c$, where i denotes the constituent, such as matter, radiation, baryons, dark energy (such as a Cosmological Constant or Quintessence), etc. The critical density is $\rho_c = 8\pi G/(3H_0^2)$ (so we often quote densities as Ωh^2 to take into account our lack of perfect knowledge of the Hubble constant). The quantity Ω_{tot} gives the sum of all contributions: all forms of matter and radiation as well as any possible cosmological constant or "Quintessence" (Caldwell et al. 1998; the latter two also known as "Dark Energy"). Ω_{tot} also determines the curvature of the Universe: $\Omega_{\text{tot}} = 1$ is geometrically flat, $\Omega_{\text{tot}} > 1$ is closed (curved like a sphere) and $\Omega_{\text{tot}} < 1$ is open (a hyperboloid). Allowing for the presence of some form of dark energy means that Ω_{tot} does *not* determine the ultimate evolution of the Universe (i.e., whether it will expand forever or re-collapse).

To describe the Universe more fully we also need to know the way in which fluctuations in the constituents are initially distributed. The simplest possibility is that the ratio of the number densities of the different species are proportional to one another. This means, in particular, that the entropy per particle, proportional to the ratio of the number density of baryons to that of photons, is constant. Hence, these are known as *adiabatic* perturbations (although *isentropic* is a more technically correct term that is occasionally used). These are actual perturbations to the spacetime metric (spatial curvature) around the mean FRW value. Adiabatic perturbations are naturally expected well inside the cosmological horizon, where all species fall into gravitational wells. Outside the horizon, such perturbations are much more difficult to set up, as they seem to violate causality. Inflation is the most well-understood way for evading this bound.

Conversely, the small-scale curvature of the Universe could remain constant, with the perturbations to each of the components adjusted to cancel the net density perturbation; these are known as *isocurvature* perturbations. Finally, the most general perturbation (that is, arbitrary perturbations to each species) is a combination of adiabatic and isocurvature perturbations.

In the following, we will concentrate on adiabatic perturbations; these are, not incidentally, well-motivated within the inflationary picture and, as we will see, fit the CMB data, while a significant admixture of isocurvature perturbations does not.

These perturbations are thus arranged at a very early time, well before the Universe has aged even a second. As the Universe aged and cooled thereafter, the baryonic constituents of the Universe evolved from a plasma of quarks and gluons, condensing to neutrons and protons, and then undergoing Big-Bang Nucleosynthesis, as the free neutrons joined with protons to form heavier nuclei. All along, the baryons were accompanied by photons and leptons (i.e., electrons and neutrinos). Through this time, the photons were tightly coupled to the positively charged electrons, scattering off of them via Thomson scattering; the reaction rate was small compared to the expansion rate, $\dot{a}/a$. Eventually, the Universe cooled sufficiently that the negatively charged electrons could combine with the positively charged nuclei and form neutral atoms for the first time in the history of the Universe. When this epoch—recombination—occurred, the charged plasma of ions and electrons became a neutral hydrogen gas (with an admixture of helium and other heavier elements). Whereas the ionized plasma was opaque to photons, the neutral gas is transparent. Hence, the photons were no longer tightly coupled to the electrons and protons, but were able to stream freely through the Universe, redshifting over time, eventually becoming the 2.73K background we observe today.

What can affect the temperature that we see at some point on the sky? If we assume that the surface of last scattering is infinitesimally thin, and ignore scattering between then and now, the fractional temperature difference from the average at a position (defined by a unit vector $\hat{\mathbf{x}}$) is

$$\frac{\delta T}{T}(\hat{\mathbf{x}}) = \frac{1}{4}\frac{\delta\rho_\gamma}{\rho_\gamma}(\mathbf{x}) - [\mathbf{v}(\mathbf{x}) - \mathbf{v}_0]\cdot\hat{\mathbf{x}} + \int_{\eta_{\rm LSS}}^{\eta_0} d\eta\, h_{ij}\hat{x}_i\hat{x}_j \qquad (5.3)$$

where the appropriate terms on the right hand side are evaluated at $\mathbf{x} = (\eta_0 - \eta_{\rm LSS})\hat{\mathbf{x}}$, the position on the Last Scattering Surface pointed at by the unit vector. We measure distance (or, equivalently, the 'conformal time') as η, related to time by $d\eta = adt$, putting distance and time on the same footing. This equation splits the temperature fluctuations into three separate physical effects. First, the region itself can be hotter or colder than average. For small fluctuations, this just corresponds to a change in the density of photons, $\rho_\gamma \propto T^4$, at the surface of last scattering. Second, the region from which the photons are coming to us can be moving with respect to us, with velocity $\mathbf{v}$, and thus we see them as hotter or colder due to the Doppler effect. Finally, the photons may have to traverse relativistic potential wells, measured by the perturbation to the three-metric, h_{ij}, gaining or losing energy as they do so; these wells can be on the last scattering surface itself, or they can be along the line of sight toward us.

Of course, all of these quantities—the photon density, the bulk velocity, and the metric—are linked with one another through the initial conditions of the perturbations in the Universe, and their subsequent evolution.

This evolution can be described in more detail using kinetic theory and the Boltzmann Equation, which determines the evolution of the distribution functions for each of the species. There are a suite of freely-available computer codes [CMBFAST (Seljak & Zaldarriaga 1996); CAMB (Lewis and

Challinor 2000)], which solve the combined Boltzmann and linearized Einstein equations in an expanding Universe. These allow one to calculate the CMB temperature and polarization power spectra for a given model. A sample of spectra for various input cosmological parameters is shown in Fig. 5.3.

But we can understand much of this evolution without resorting to such detailed calculations. Rather than think of individual over- or underdense lumps in the Universe, consider instead Fourier modes of a given scale. These modes, labeled by a three-dimensional wavenumber **k**, evolve independently as long as their amplitude is small. As long as the wavelength $\lambda \sim 2\pi/|k|$ is much greater than the horizon size, the amplitude of the mode just increases linearly with time (this is a coordinate-dependent statement in general relativity, however). Inside the horizon, the interaction among the different species of matter and radiation affects the perturbation evolution. Pressure and gravitational potential gradients drive sound waves in the coupled plasma of baryons and photons. An important length scale for this dynamical process is the *sound horizon*, the distance a sound wave could have travelled since the Big Bang. Because the Universe has been dominated by radiation (with sound speed $c/\sqrt{3}$) for most of its his-

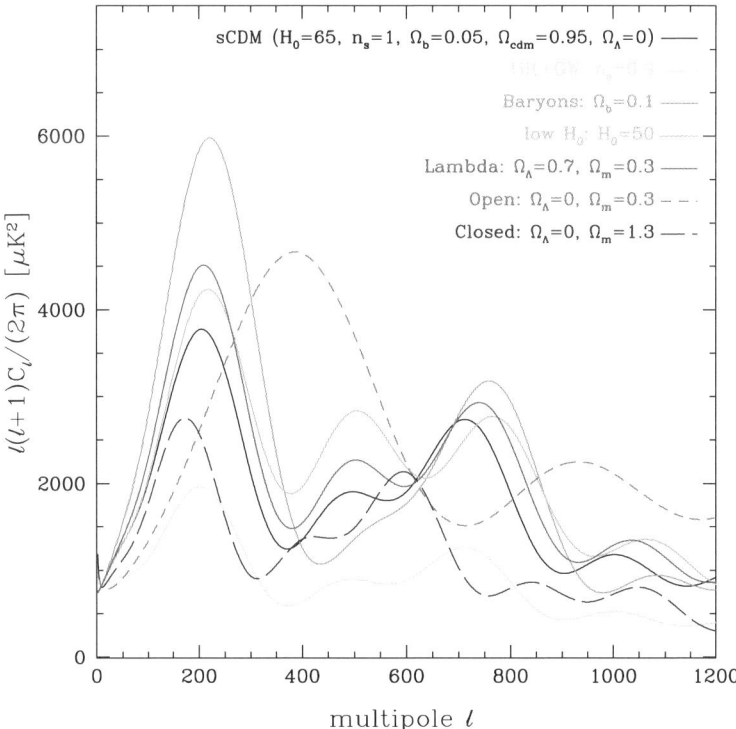

Fig. 5.3. A sample of theoretical power spectra for various cosmological parameters, as marked.

tory, the sound horizon is about $1/\sqrt{3}$ of the (classical Big Bang) particle horizon.

First, consider waves entering the (sound) horizon around the time of last scattering: these are the largest waves that could have formed a coherent structure at this time. Indeed, by determining the characteristic *angular* scales of the CMB fluctuation pattern, and matching this to the physical scale of the sound horizon at last scattering we can determine the *angular diameter distance* to the last scattering surface, which is mostly dependent on the geometry of the Universe (see Fig 5.4): in a flat Universe, angular and physical scales obey the usual Euclidean formulae; in a closed (positively curved) Universe, geodesics converge and a given physical scale corresponds to a larger angular scale (and hence smaller multipole ℓ); conversely, in a negatively curved Universe the same physical scale corresponds to a smaller angular scale and larger ℓ.

Consider now a wave that enters the horizon some time considerably before Last Scattering, when the density of the Universe is still dominated by radiation, and the baryons are tightly-coupled to the photons. Although the dark matter is pressureless, the dominant radiation has pressure $p \simeq \rho/3$, or somewhat less due to the baryons. Although the dark matter can continue to collapse, the radiation rebounds when the pressure and density become

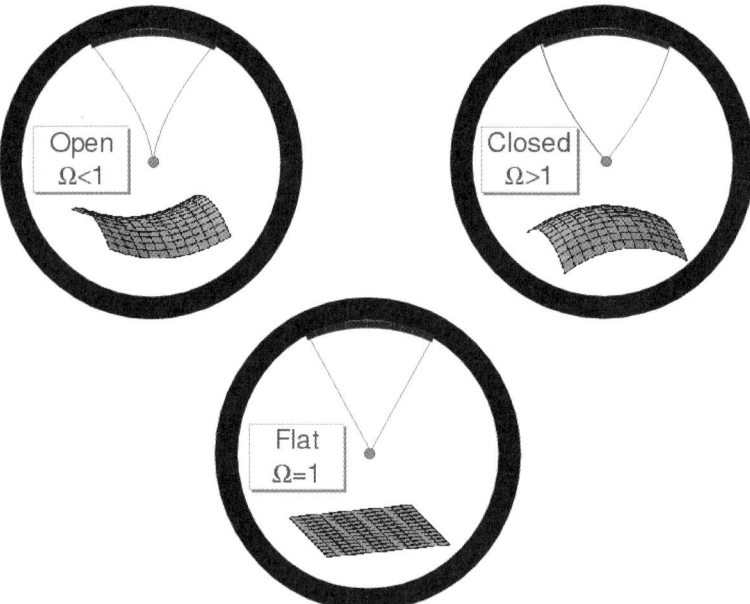

Fig. 5.4. Schematic of the effect of spatial curvature on the angular scale of the CMB. The outer circle for each geometry represents the last scattering surface; the thick arc represents a known physical scale along the surface; the thin lines represent the path of light rays from last scattering to the present.

sufficiently high. Eventually, gravity may take over again and cause the perturbation to collapse yet again, one or more times. Larger and larger scales, entering the horizon at later and later times, will thus experience fewer and fewer collapse and rebound cycles. Moreover, because of the effect of the baryons on the pressure, the strength of the rebound is decreased as we increase the baryon density. It is this cycle of collapse and rebound that we see as peaks in the CMB power spectrum, often called *acoustic peaks* after the acoustic waves responsible for them. We thus use their heights to measure the relative contributions of baryon and photons to the pressure, and their angular scale to determine the geometry, as well as the history of the sound speed in the baryon-photon plasma. (And of course other cosmological parameters also affect the spectrum in yet other ways)

There are yet other physical effects that affect the power spectrum. Although the photons and baryons are tightly bound to one another via scattering, the coupling is not perfect. Hence, there is a scale (known as the *Silk* damping scale) below which the photons can stream freely and wash out perturbations. This free-streaming damps perturbations on small scales, resulting in the damping of C_ℓ at large ℓ; there is very little power left by $\ell \sim 1500$.

Most generally, the formula for the CMB power spectrum can be expressed as

$$C_\ell = \int dk P_p(k) T_\ell^2(k) \tag{5.4}$$

where $P_p(k)$ is the *primordial* power spectrum of density fluctuations, and $T_\ell^2(k)$ is the *transfer function* which depends on the cosmological parameters, and encodes the physics we have discussed above; this equation merely at last scattering reflects the linearity of the perturbation evolution. The primordial power spectrum is usually modeled as a power law: $P_p(k) = Ak^{n_s}$. The scale-invariant spectrum has $n_s = 1$ (the so-called "scalar spectral tilt") but inflationary generically predicts a slight departure from scale-invariance; usually $n_s \simeq 0.9$–1.0 with the exact value of n_s depending on the physics of inflation. In fact, we don't expect exact power-law behavior; the detailed shape of the primordial spectrum depends on the shape of the inflationary potential as it 'rolls down' toward its minimum during inflation. Thus measuring the CMB power spectrum tells us about inflation, if we can disentangle the parameters encoded in the transfer function from the primordial spectrum.

This indicates that the CMB is not, however, a cosmological panacea: there are parameters that the CMB alone cannot measure. For example, we have seen above how the location of the peaks are determined by the overall curvature of the Universe, $\Omega_{\text{tot}} = \Omega_m + \Omega_\Lambda$.[1]. Conversely the CMB

---

[1] More precisely, the scale of the peaks are determined by angular size of the sound horizon at last scattering, itself determined by the ratio of the physical size of the sound horizon to the angular diameter distance to the last scattering surface (e.g., Stompor & Efstathiou 1999)

fluctuations are not strongly determined by the individual Ω_m and Ω_Λ. There are other such degeneracies that all together mean we must combine the CMB observations with other cosmological measurements, especially various more local measurements of large-scale structure in the Universe, in order to determine the parameters (as well as to check our measurements!).

So far we have been discussing fluctuations only to the intensity (i.e., temperature) of the CMB. The matter transport caused by the pressure and gravitational potential gradients means there are *velocity* perturbations as well. Hence, the photons can scatter off of moving electrons, which generates a net linear polarization of the photons (e.g., Jaffe et al. 2000). For the sound waves we are considering, the velocities are greatest when the density contrast is smallest, and vice versa: the velocity is out of phase with the density – and hence the polarization signal is out of phase with the temperature. Unfortunately, due to the relative inefficiency of scattering off of the moving electrons, the polarization fraction is only about 10%, and the polarization spectra are correspondingly suppressed.

The polarization pattern that results is 'curl-free'; that is, it can be represented as the 2-d divergence (gradient) of some scalar on the sky. This is easy to understand if we heuristically identify the polarization pattern with the flow of the plasma. In linear perturbation theory this flow is a potential flow; any vorticity is damped by expansion and not driven to grow by gravity.

Gravitational waves at last–scattering can produce a 'curl' pattern, since gravity waves can produce non–potential flows in the plasma. For full-sky maps with high sensitivity and resolution, the "gradient" and "curl" components (also known as 'Electric' and 'Magnetic' or E and B from the obvious analogy to electromagnetic fields) can be separated completely; in more realistic situations statistical techniques are necessary.

Density perturbations do not themselves produce B polarization. In the inflationary scenario the properties of the Gravitational-Wave (or Tensor) background spectrum are defined by the same physics that determines the density (or scalar) spectrum. Thus, observations of inflationary B polarization would be a strong indication of the occurrence of inflation (if, that is, we can separate the cosmological contribution from the astrophysical foregrounds).

5.2.1 Astrophysical Sources of Microwave Radiation

Of course, not all of the microwave photons we observe are of cosmological origin. Any intervening sources of microwave emission or absorption will contaminate our data. The frequency spectra of some possible astrophysical microwave sources are shown in Fig. 5.5. In principle, the foreground spectra are clearly distinguishable from the CMB. At low frequencies the major contaminant is from extragalactic point sources (synchrotron and so-called free-free emission or bremsstrahlung); at high frequencies it is galactic dust. In practice we have so far been quite lucky that foreground contamination

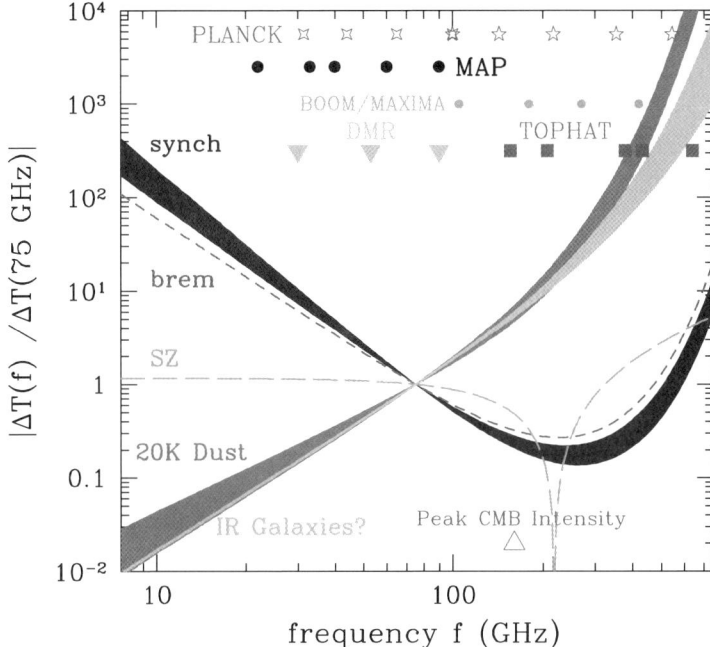

Fig. 5.5. The spectra of various astrophysical sources of microwave radiation, along with the observing bands of some experiments. The spectra are all normalized to be equal at 75 GHz. The units are thermodynamic temperature (the CMB itself is therefore a horizontal line). From Bond et al. (1999).

has been negligible over much of the sky at all frequencies. As our sensitivity increases it is likely that the residual foreground contamination will become a significant part of the error budget of future experiments, as it is already seems to be for the new WMAP results.

Most of these so-called foregrounds are objects of study in their own right. From a cosmological point of view, none have more potential than the Sunyaev-Zel'dovich (1981) effect, which is the upscattering of CMB photons by electrons in hot gas usually in the extended halos of galaxy clusters. This upscattering produces a specific spectra distortion (with some caveats due to relativistic effects at the highest temperatures) which can be distinguished from the CMB and other foregrounds. It has already been measured in dozens of clusters (e.g., Carlstrom et al. 2002), with observing campaigns for many hundreds more being planned by several groups. SZ observations have the advantage that their brightness is independent of distance and so with the SZ effect we can easily observe clusters at the highest redshifts. These observations allow us to map the internal properties of clusters (SZ observations measure, effectively, the integral of the pressure along the line of sight) allowing the use of the clusters as standard candles and rulers for use in cosmological tests.

For a fuller review of the issues involved in CMB foregrounds, see the collection by de Oliveira-Costa & Tegmark (2000).

5.3 CMB Experiments and CMB Data

After the detection of anisotropy by the DMR instrument on the COBE satellite which flew from 1989-1993, the CMB community en masse spent the 1990s refining the sensitivity and angular resolution of their detectors. By the end of the decade, there was a strong suggestion of a peak in C_ℓ at $\ell \sim 200$ corresponding to the predictions of a flat Universe (Knox & Page 2000). These results were from a wide variety of experiments using a variety of experimental techniques: balloons and ground-based platforms; bolometers and radiometers; scanning and differencing; interferometers and telescopy, with experiments covering almost all combinations of these. Since then, a second generation of ground- and balloon-based experiments has perfected these techniques and observed the CMB down to a resolution of less than an arcminute.

Most of these recent experiments can be broadly split into two classes. The first are the *balloon-borne-bolometers* and the second are the *ground-based interferometers*. This review can only scratch the surface of the beautiful and difficult work that has gone into the development of hardware capable of detecting temperature differences of one part in 10^5 on scales of a few arcminutes on the sky. For reviews, see, for example, the papers by the individual experimental teams, as listed in the accompanying tables.

A bolometer is essentially a thermometer: the detector heats up upon absorption of a CMB photon. Attempting to measure temperature differences of one part in 10^5, or about $30\mu K$, puts severe constraints on the thermal environment of the detector, and on the electronics required to convert these temperature differences into an electronic signal. These bolometers work best at photon frequencies above 100 GHz (corresponding to a wavelength of 1/3 cm). Air itself emits strongly at these frequencies; hence, balloons flying at 40,000 metres are used to escape the bulk of the atmosphere. The logistics of balloon flights then constrains the size of the telescope which can be flown, in turn finally limiting the beam size (the experimental resolution) at a given frequency.

Recently, balloon-borne experiments have been able to take advantage of Long-Duration Ballooning (LDB). The Arctic and Antarctic polar vortices are atmospheric phenomena enforcing smooth circulation of air around the poles for months at a time, allowing ballooning times of 1–2 weeks rather than the traditional single-night experiments possible from North America.

The other major class of experiment is that of the interferometers, which compare the detection of electromagnetic waves at different points to build up a picture of the source. With the CMB, these have the advantage of, in effect, directly measuring components of the Fourier transform of the temperature field on the sky – i.e., the multipole components $a_{\ell m}$. Instead of the incoherent heating measured by the bolometers, the interferometers usually use coherent receivers (which, not incidentally, make them automatically sensitive to polarization). Complementary to bolometers, these receivers are usually limited to frequencies below 100 GHz, and at this writing have lower sensitivity for fixed observing time. Unlike the single-mirror bolometer experiments, however, the resolution is given by the distance between individual receiver dishes; hence they are capable of higher resolution without considerable effort. Also unlike the bolometers, the main astrophysical contaminant is emission from synchrotron and free-free emission from extragalactic point sources, dominating over the CMB at high angular scales.

A few experiments do not fit into these broad classes. The ACBAR experiment is a ground-based bolometer array, combining the sensitivity of bolometers with the long experimental durations possible from the ground. Other groups are planning more ambitious hybrids, including the use of bolometers for interferometry, marrying the higher sensitivity of bolometers with the direct Fourier-domain observations of interferometers.

5.3.1 Data Analysis

Along with this increased sophistication in the technology of CMB detection have come parallel developments in the analysis of CMB data. This is a hard problem because we are trying to estimate the CMB power spectrum: a variance. Thus, we must have a very precise characterization of the sources of noise in our system. We cannot just increase our error bars to take into account any lack of understanding of the noise: getting the noise wrong is a bias. In addition, CMB data analysis is computationally intensive: brute-force algorithms to manipulate the likelihood function of CMB data scale as the cube of the number of pixels, which was difficult enough (at the time) for the few thousand pixels of COBE/DMR, but prohibitive even today for the megapixels of WMAP. Some of these problems are ameliorated with Monte Carlo methods, but more work remains necessary to understand the various approximations. A review of these and further data analysis issues can be found in Bond et al. (1999) and Jaffe (2003).

5.4 Current Observations of the CMB

We have already seen how some of the cosmological parameters affect the power spectrum of the CMB. Conversely, our measurements of C_ℓ allow us to determine the parameters. Since the release of the COBE/DMR data in the early 1990s, it has been realized that the CMB power spectrum is perhaps the cleanest route to many of the cosmological parameters: the physical processes are well-understood, and the perturbations are small (so a linearized treatment suffices). Indeed, it was the pioneering work of Jungman et al. (1996) that made the power of the CMB apparent. By the late 1990s, the results of the first degree-scale experiments were in, and with them the first evidence for a peak in C_ℓ at $\ell \sim 200$ (Knox & Page 2000; Bond & Jaffe 1997), corresponding to the flat Universe value of $\Omega_{\rm tot} = 1$. With MAXIMA-1 and BOOMERANG-98 in 2000, this interpretation was solidified. These experiments also gave the first continuous measurements from the peak scale of one degree down to tens of arcminutes, covering $100 < \ell < 800$ and allowing the first hints of structure *beyond* the first peak in C_ℓ. In fact, these experiments initially showed scant evidence for further peaks, with the data instead mildly preferring a flat spectrum after the first peak's rise and fall. As we saw above, the amplitude of these further peaks is an indicator of the baryon density. A low second peak would thus indicate a *high* baryon density ($\Omega_B h^2 \simeq 0.03$ instead of the value of 0.02 ± 0.02 inferred from Big-Bang Nucleosynthesis calculations combined with observations of light element abundances).

In 2001, along with a further analysis of both MAXIMA and BOOMERANG data, the first data from the interferometers DASI and CBI was released, confirming the overall picture, and beginning to 'resolve' the higher peaks. Results from the VSA interferometer followed shortly. In 2002, these experiments delivered two further coups: CBI made the first comprehensive measurements over the range $1000 < \ell < 3000$, and DASI made the first detection of CMB polarization. In late 2003, the French ARCHEOPS experiment released data probing larger angular than any experiment since COBE, while ACBAR examined the high-ℓ structure probed by CBI, but with its completely different bolometric techniques.

And then, in February, 2003, WMAP released its results: maps of the sky over the frequency range 20-90 GHz, with an angular resolution ranging from 20 arcminutes (at 90 GHz) to just over a degree at 20 GHz (see Table 5.1 on page 91). They also released power spectra of both temperature and polarization, as well as their own cosmological and astrophysical analyses (Bennett et al. 2003, references therein, and http://lambda.gsfc.nasa.gov/). The statistical weight of the WMAP data dominates for $\ell < 500$ or so. To gain the full value of these results, they must be combined with higher resolution CMB observations (e.g., CBI and ACBAR) as well as observations of large-scale structure, the accelerating Universe (see Chap. 6 by L. Bergström and A. Goobar in this volume) and direct measurements of parameters such as the Hubble Constant (Freedman et al. 2001).

Such an exercise largely confirms the prior wisdom so spectacularly that, despite decreasing the error bars by orders of magnitude, the WMAP results can be said to more revolutionary than evolutionary. When analyzing their data alone, the WMAP team restrict themselves to a flat Universe, but when combined with the full suite of cosmological data they confirm that the Universe is indeed flat: $\Omega_{\rm tot} = 1.02 \pm 0.02$, with a Hubble constant $h = 0.71^{+0.04}_{-0.03}$ and a baryon density $\Omega_B h^2 = 0.022 \pm 0.001$. Combined with evidence for an accelerating Universe on the one hand, or evidence for a low matter density ($\Omega_m \simeq 0.3$), this implies the existence of a "Dark Energy" component with $\Omega_\Lambda = 0.73 \pm 0.04$, perhaps the most puzzling cosmological data to come from this age of "precision cosmology." How do we explain a non-zero vacuum energy that is more than a hundred orders of magnitude below that expected if driven by Planck-scale physics? Just as an earlier period of acceleration — inflation — is postulated to be driven by a scalar field, such a field ("quintessence") is postulated to be responsible for the latter-day acceleration, as well (Caldwell et al. 1998), but no indication of its nature beyond acceleration has been found. Indeed, the data suggest that the acceleration is most consistent with that expected from a pure cosmological constant (with equation of state $p/\rho = -1$) rather than the richer phenomenology of quintessence.

When combined with large-scale structure data on small and medium scales, the WMAP team seem to find that the 'scalar tilt' parameter is not a constant. That is, it is near one at the scales probed by the CMB, but considerably lower at the small scales probed by large-scale structure. They interpret this as a marginal detection of the 'running' of n_s (as introduced by Kosowsky & Turner 1995). Although expected in some inflationary models, the large amplitude of the signal is a puzzle; some other authors have already disputed these results (Bridle et al. 2003; Mandelbaum et al. 2003).

The DASI detection of polarization was confirmed in spectacular fashion by WMAP. In addition, they reported a more intriguing and unexpected find: the temperature and polarization are much more correlated with one another on the largest scales (corresponding to $\ell < 10$) than expected in the simplest models. This is interpreted as a signal of the reionization of the Universe at redshift $z \simeq 17$–20 producing a second surface of last [sic] scattering, reflected at these scales rather than at the $\ell \sim 200$ of the first acoustic peak. These results contradict — or at least complicate — the interpretation of recent evidence for the reionization of the Universe at $z \simeq 6$ from the study of the absorption spectra of $z > 6$ QSOs (e.g., Becker et al. 2001). If both the CMB and QSO observations stand, then the epoch of astrophysical reionization must have occured over a long period of time ($z = 20$ corresponds to $t \simeq 0.1$ Gyr, $z = 6$ to $t \simeq 1$ Gyr. In addition, this scenario raises considerable theoretical problems: how do we form the first objects at such an early time, especially in the light of WMAP's purported observation of a tilt in the primordial power spectrum, lowering further the power at scales that could form objects sufficiently early.

5.5 Summary

The WMAP results were released on 11 February, 2003 at a NASA press conference and in thirteen papers submitted to *The Astrophysical Journal*, after which the data were immediately made available at the inauguration of the website http://lambda.gsfc.nasa.gov. It is a testament to the quality of the data, and the keen expectations with which they were awaited (and perhaps something more distressing about the level of competition in the field), that a check on the astro-ph preprint server shows thirty papers written in the four weeks following their release. In addition to the basic cosmological physics discussed here, these papers investigate neutrino physics in the early Universe; foreground contamination; the issue of the running of n_s; reionization at $z \simeq 20$; and the lack of large-scale power seen by WMAP. Still, the convergence of the community upon what was named the Concordance Model (Ostriker and Steinhardt 1995) even before the CMB data had solidified is either a triumph of good data and theory or a disturbing example of the triumph of power and bandwagon-jumping.

Some of these questions will undoubtedly be resolved by the WMAP data alone, as they continue to observe and release further results. Others will await observations by further instruments. ESA is planning to launch the Planck Surveyor satellite in 2007, which will observe over the range 30–800 GHz and have a resolution in its finest channel of 4 arcminutes (see Table 5.1 on page 91). Simultaneously, many groups are planning high-sensitivity observations of the polarization of the CMB, hoping to measure the B modes that are a possible signature of the inflationary epoch; these campaigns require the use of not the several or tens of detectors as in current observations, but hundreds or thousands. This opens up considerable technical challenges of detector fabrication and data analysis. Similar techniques will be used for the large-scale Sunyaev-Zel'dovich effect observations mentioned above.

Over the last decade, but especially since 2000, the Cosmic Microwave Background has proven itself to be one of the most fruitful areas of cosmological research. It has solidified, if not proven, the flatness of the Universe, and, when combined with other cosmological data, made nearly inevitable the idea that the Universe is dominated by some sort of accelerating "Dark Energy." It provides independent confirmation of measurements of the baryon density and the Hubble constant. Perhaps more importantly in the long run, the microwave band is becoming useful for more than just C_ℓ as the 'foregrounds' become an object of study in their own right, opening up new wavebands for observers.

The next few years will see no letup in the rate of information gathered from the CMB. WMAP will continue to take data, as will other ground-based experiments. New many-element instruments will begin to observe the cosmological signal and the Sunyaev-Zeldovich effect. By decade's end, we should reach levels of sensitivity necessary to begin to observe (or rule out) the B mode of polarization and thus primordial gravitational radiation. In 2007, the launch of the Planck satellite should mark the beginning of the end of a major phase of observational cosmology: Planck should see *all* the primary anisotropy visible from our part of the Universe (at least for the temperature and E polarization).

In addition to being a heroic experimental achievement, the measurement of CMB fluctuations at a level of one part in 10^5 is, in some sense, a triumph of theory. As the observations of Figs. 5.1–5.2 converge to the curves of Fig. 5.3, we are left with a Universe whose parameters are strange, but whose initial conditions lie still tantalizingly out of reach.

References

Alpher, R.A., Bethe, H., & Gamow, G. 1948, The Origin of Chemical Elements, Physical Review, **73**, 803

Alpher, R.A., Herman, R., & Gamow, G.A. 1948, Thermonuclear Reactions in the Expanding Universe, Physical Review, **74**, 1198

Becker, R.H. et al. 2001, Evidence for Reionization at $z \sim 6$: Detection of a Gunn-Peterson Trough in a $z = 6.28$ Quasar, Astron. J., **122**, 2850

Bennett, C.L. et al. 1996, Four-Year COBE DMR Cosmic Microwave Background Observations: Maps and Basic Results, The Astrophysical Journal Letters, **464**, L1

Bennett, C.L. et al. 2003, First Year Wilkinson Microwave Anisotropy Probe (WMAP) Observations: Preliminary Maps and Basic Results, astro-ph/0302207

Benoit, A. et al. 2002, The Cosmic Microwave Background Anisotropy Power Spectrum measured by Archeops, astro-ph/0210305

Bond, J.R. & Jaffe, A.H., "Cosmic Parameter Estimation Combining Sub-Degree CMB Experiments With COBE", in *Proceedings of the XVIth Moriond meeting, "Microwave Background Anisotropies,"* ed. F.R. Bouchet et al., Editions Frontieres, 1997

Bond, J.R., Crittenden, R., Jaffe, A.H. & L.E. Knox, "Computing Challenges for the Cosmic Microwave Background," *Computing in Science and Engineering*, invited article, March/April, Vol. 1, No. 2, p. 21, 1999. `astro-ph/9903166`

Bridle, S.L., Lewis, A.M., Weller, J., & Efstathiou, G. 2003, Reconstructing the primordial power spectrum, astro-ph/0302306

Caldwell, R.R., Dave, R., & Steinhardt, P.J. 1998, Cosmological Imprint of an Energy Component with General Equation of State, Physical Review Letters, **80**, 1582

Carlstrom, J.E., Holder, G.P., & Reese, E.D. 2002, Cosmology with the Sunyaev-Zel'dovich Effect, Ann. Rev. Astron. Astrophys., **40**, 643

Dicke, R.H., Peebles, P.J. E., Roll, P.G., & Wilkinson, D.T. 1965, Cosmic Black-Body Radiation., The Astrophysical Journal, **142**, 414

Fixsen, D.J., Cheng, E.S., Gales, J.M., Mather, J.C., Shafer, R.A., & Wright, E.L. 1996, The Cosmic Microwave Background Spectrum from the Full COBE FIRAS Data Set, The Astrophysical Journal, **473**, 576

Freedman, W.L. et al. 2001, Final Results from the Hubble Space Telescope Key Project to Measure the Hubble Constant, The Astrophysical Journal, **553**, 47

Gamow, G. 1948, The Origin of Elements and the Separation of Galaxies, Physical Review, **74**, 505

Gorski, K.M., Banday, A.J., Bennett, C.L., Hinshaw, G., Kogut, A., Smoot, G.F., & Wright, E.L. 1996, Power Spectrum of Primordial Inhomogeneity Determined from the Four-Year COBE DMR Sky Maps, The Astrophysical Journal Letters, **464**, L11

Grainge, K. et al. 2002, The CMB power spectrum out to l=1400 measured by the VSA, astro-ph/0212495

Guth, A.H. 1981, Inflationary Universe: A possible solution to the horizon and flatness problems, Physical Review D, **23**, 347

Halverson, N.W. et al. 2002, Degree Angular Scale Interferometer First Results: A Measurement of the Cosmic Microwave Background Angular Power Spectrum, The Astrophysical Journal, **568**, 38

Hinshaw, G., Banday, A.J., Bennett, C.L., Gorski, K.M., Kogut, A., Smoot, G.F., & Wright, E.L. 1996, Band Power Spectra in the COBE DMR Four-Year Anisotropy Maps, The Astrophysical Journal Letters, **464**, L17

Jaffe, A.H. 2003, Statistics and the Cosmic Microwave Background, in Feigelson & Babu, *Statistical Challenges in Modern Astronomy*, 197-211, Springer-Verlag, New York, 2003

Jaffe, A.H., Kamionkowski, M. & L. Wang, "A Polarization Pursuers' Guide," *Phys. Rev. D*, 61, 083501, 2000. (`astro-ph/9909281`)

Jungman, G., Kamionkowski, M., Kosowsky, A., & Spergel, D.N. 1996, Cosmological-parameter determination with microwave background maps, Physical Review D, **54**, 1332

Knox, L. & Page, L. 2000, Characterizing the Peak in the Cosmic Microwave Background Angular Power Spectrum, Physical Review Letters, **85**, 1366

Kolb, E.W. & Turner, M.S. 1990, *The early Universe*, Frontiers in Physics, Reading, MA: Addison-Wesley, 1988, 1990

Kosowsky, A. & Turner, M.S. 1995, CBR anisotropy and the running of the scalar spectral index, Physical Review D, **52**, 1739

Kovac, J.M., Leitch, E.M., Pryke, C., Carlstrom, J.E., Halverson, N.W., & Holzapfel, W.L. 2002, Detection of polarization in the cosmic microwave background using DASI, Nature, **420**, 772

Kuo, C.L. et al. 2002, High Resolution Observations of the CMB Power Spectrum with ACBAR, astro-ph/0212289

Lee, A.T. et al. 2001, A High Spatial Resolution Analysis of the MAXIMA-1 Cosmic Microwave Background Anisotropy Data, The Astrophysical Journal Letters, **561**, L1

Lewis, A., Challinor, A., & Lasenby, A. 2000, Efficient Computation of Cosmic Microwave Background Anisotropies in Closed Friedmann-Robertson-Walker Models, The Astrophysical Journal, **538**, 473; http://camb.info/

Mandelbaum, R., McDonald, P., Seljak, U., & Cen, R. 2003, Precision Cosmology from the Lyman-alpha Forest: Power Spectrum and Bispectrum, astro-ph/0302112

McKellar, A. 1940, Evidence for the Molecular Origin of Some Hitherto Unidentified Interstellar Lines, Pub. Astron. Soc. Pacific, **52**, 187

de Oliveira-Costa, A. & Tegmark, M. 1999, Microwave Foregrounds, ASP Conf. Ser. 181: Microwave Foregrounds

Ostriker, J.P. & Steinhardt, P.J. 1995, The Observational Case for a Low Density Universe with a Non-Zero Cosmological Constant, Nature, **377**, 600

Peacock, J.A. 1999, *Cosmological physics*. Publisher: Cambridge, UK: Cambridge University Press, 1999. ISBN: 0521422701

Pearson, T.J. et al. 2002, The Anisotropy of the Microwave Background to l = 3500: Mosaic Observations with the Cosmic Background Imager, astro-ph/0205388

Peebles, P.J. E. 1993, *Principles of physical cosmology*, Princeton Series in Physics, Princeton, NJ: Princeton University Press, —c1993

Penzias, A.A. & Wilson, R.W. 1965, A Measurement of Excess Antenna Temperature at 4080 Mc/s., The Astrophysical Journal, **142**, 419

Ruhl, J.E. et al. 2002, Improved Measurement of the Angular Power Spectrum of Temperature Anisotropy in the CMB from Two New Analyses of BOOMERANG Observations, astro-ph/0212229

Seljak, U. & Zaldarriaga, M., http://physics.nyu.edu/matiasz/CMBFAST/cmbfast.html

Smoot, G.F., Gorenstein, M.V., & Muller, R.A. 1977, Detection of anisotropy in the cosmic blackbody radiation, Physical Review Letters, **39**, 898

Steinhardt, P.J. & Turok, N. 2002, Cosmic evolution in a cyclic Universe, Physical Review D, **65**, 126003

Stompor, R. & Efstathiou, G. 1999, Gravitational lensing of cosmic microwave background anisotropies and cosmological parameter estimation, Mon. Not. R. Astron. Soc., **302**, 735

Sunyaev, R.A. & Zeldovich, I.B. 1981, Intergalactic gas in clusters of galaxies, the microwave background, and cosmology, Soviet Scientific Reviews Section E Astrophysics and Space Physics Reviews, **1**, 1

Wright, E.L. et al. 1991, Preliminary spectral observations of the Galaxy with a 7 deg beam by the Cosmic Background Explorer (COBE), The Astrophysical Journal, **381**, 200

6 Particle Astrophysics and the Dark Sector of the Universe

L. Bergström and A. Goobar

6.1 Introduction

At present, there are a great variety of experiments checking various cosmological parameters. We are therefore in a situation of gathering data also from the furthermost regions and are already puzzled by some of these results.

The first, and plausibly most exciting result, is the amount of dark energy in the Universe. This is what we will start with here, then we will investigate what the cosmic matter density tells us, and finally we discuss a couple of topics which have to do with neutrino physics.

In the Standard Model of cosmology the Universe started with a Big Bang. The expansion of an isotropic and homogeneous Universe is described by the Friedmann-Lemaitre-Robertson-Walker model (or FLRW model, for short).

The free parameters of the FLRW model are the energy contributions from radiation, matter and vacuum fluctuations. At the present epoch, the energy density in the form of radiation ρ_r can be neglected in comparison with the matter density ρ_m, and the Friedmann equation for the Hubble parameter (H) becomes:

$$H^2 \equiv \left(\frac{\dot{a}}{a}\right)^2 = \frac{8\pi G}{3}\rho_m + \frac{\Lambda}{3} - \frac{k}{a^2}, \qquad (6.1)$$

where a is the growing scale factor of the Universe and k=-1,0 or 1 represent the three possible geometries for the Universe: open, flat or closed.

Thus the expansion rate of the Universe depends on the matter density, the cosmological constant ($\Lambda = 8\pi G \rho_{vac}$) and the geometry of the Universe. It is also customary to rewrite (6.1) so that it instead contains the fractional energy density contributions at the present epoch ($z = 0$). We thus introduce the definition:

$$\begin{cases} \Omega_M \equiv \frac{8\pi G \rho_m^0}{3H_0^2} \\ \Omega_\Lambda \equiv \frac{\Lambda}{3H_0^2} \\ \Omega_K \equiv \frac{-k}{a_0^2 H_0^2} \end{cases} \qquad (6.2)$$

There are only two independent contributions to the energy density since in the FLRW model (remember that we are generally at an epoch when $\Omega_r \sim 0$):

$$\Omega_M + \Omega_\Lambda + \Omega_K = 1 \tag{6.3}$$

The time evolution of the scale factor a and thus the *fate* of the Universe as determined by the two independent cosmological parameters is shown in Fig. 6.1. A large cosmological constant, for example, leads to rapid "inflation" of the Universe.

The *deceleration* parameter (at z=0), q_0, is defined as:

$$q_0 = \frac{\Omega_M}{2} - \Omega_\Lambda, \tag{6.4}$$

thus, a *negative* value of q_0 implies that the rate of expansion of the Universe is increasing, i.e. the expansion is accelerating.

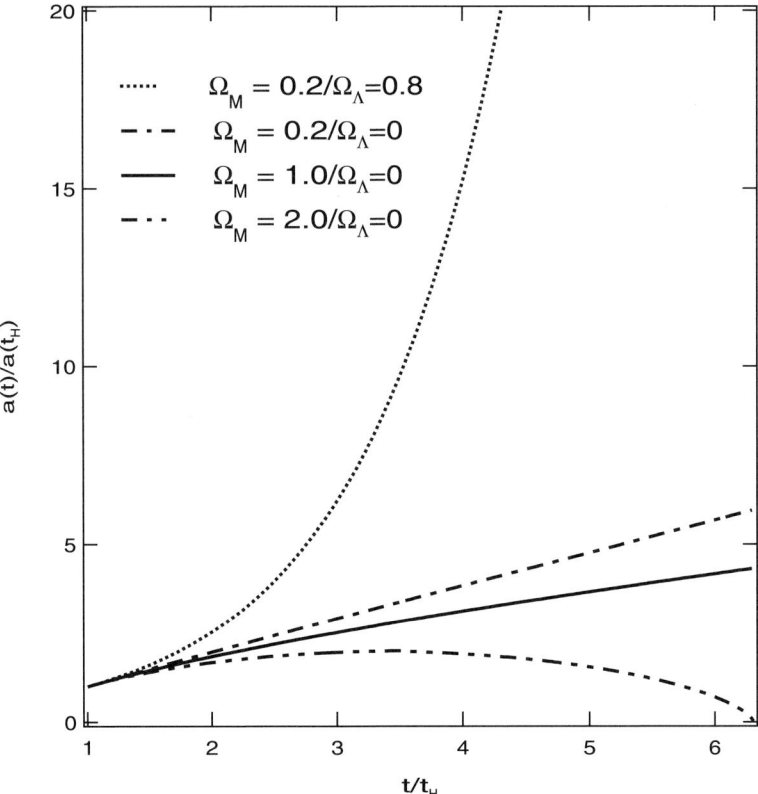

Fig. 6.1. Time evolution of the scale factor of the Universe, normalized to its current value, where t_H is the age of the Universe today. Various combinations of cosmological parameters are shown. The Universe with non-vanishing cosmological constant (dotted line) will go through an exponential expansion. Figure from [1].

In the next section we generalize the discussion as to also include the contribution from any arbitrary energy form characterized by the the relation between its pressure and density.

6.2 Cosmological Parameters from 'Standard Candles'

One of the most direct ways to measure the expansion history of the Universe is through the magnitude-redshift relation of a source of known strength, a 'standard candle'. Relative distances of objects as a function of their redshift provide information on the cosmological parameters [2]. Ω_M and Ω_X denote the present-day energy density parameters of "ordinary" matter $\Omega_m(z)$ (including also dark matter) and a "Dark Energy" component $\Omega_x(z)$, respectively. The "Dark Energy" is characterized by the equation of state parameter, $w(z)$, where $p_X = w \cdot \rho_X$. For the specific case of the cosmological constant, $w = -1$, i.e. $p_\Lambda = -\rho_\Lambda$. On the other hand, as usual $\Omega_m(z)$ corresponds to $p_M = 0$.

Supernovae (SNe) are the brightest astrophysical objects used for accurate distance estimations. (See also Chap. 4 by S. Webb in this volume). Here, we concentrate on the use of most homogeneous class: Type Ia SNe. These are believed to result from the ignition of an accreting white dwarf in a binary system, as the mass exceed the Chandrasekhar limit. The apparent magnitude m of a supernova at redshift z is then given by

$$m(z) = M + 5 \log_{10}\left[d'_L(z)\right], \tag{6.5}$$
$$\mathcal{M} = 25 + M + 5 \log_{10}(c/H_0), \tag{6.6}$$

where M is the absolute magnitude of the supernova, and $d'_L \equiv H_0 \, d_L$ is the H_0-'independent' luminosity distance, where H_0 is the Hubble parameter[1]. Hence, the intercept $\mathcal{M}$ contains the 'nuisance' parameters M and H_0 that apply equally to all magnitude measurements (in this section we do not consider possible evolutionary effects $M = M(z)$). The H_0-independent luminosity distance d'_L is given by

$$d'_L = \begin{cases} (1+z)\frac{1}{\sqrt{-\Omega_K}} \sin(\sqrt{-\Omega_K}\, I), & \Omega_K < 0 \\ (1+z)\, I, & \Omega_K = 0 \\ (1+z)\frac{1}{\sqrt{\Omega_K}} \sinh(\sqrt{\Omega_K}\, I), & \Omega_K > 0 \end{cases} \tag{6.7}$$

$$\Omega_K = 1 - \Omega_M - \Omega_X, \tag{6.8}$$

$$I = \int_0^z \frac{dz'}{H'(z')}, \tag{6.9}$$

---

[1] In the expression for $\mathcal{M}$, the units of c and H_0 are km s^{-1} and km s^{-1} Mpc^{-1}, respectively.

$$H'(z) = H(z)/H_0 =$$
$$= \sqrt{(1+z)^3 \, \Omega_M + f(z) \, \Omega_X + (1+z)^2 \, \Omega_K}, \quad (6.10)$$
$$f(z) = \exp\left[3 \int_0^z dz' \, \frac{1+w(z')}{1+z'}\right]. \quad (6.11)$$

As the measurements are performed through broad-band filters one has to correct for the fact that different parts of the supernova spectrum are detected depending on the redshift z of the source. For example, at a redshift $z \sim 0.5$ the light captured with a red (R) filter at a telescope at Earth originates from the blue (B) part of the spectrum. This so called 'K-correction' is preferentially done using blue (B) absolute magnitudes in the rest-frame and V,R,I filters for the observation of supernovae with increasing redshift [3].

6.3 Current Results

Two collaborations, the SCP [4, 5] and the High-Z team [6–8], have been searching for high-redshift Type Ia supernovae with the aim to measure cosmological parameters. Both groups find that the data is consistent with the existence of some "Dark Energy" form that is accelerating the rate of expansion of the Universe at present, i.e. $q_0 < 0$, as shown in Fig. 6.2. The Hubble diagram for 42 supernovae found by the SCP along with 18 low-z supernovae from the Calán/Tololo Supernova Survey [9] indicates that supernovae at $z \sim 0.5$ are 0.2 – 0.5 magnitudes too faint to be consistent with an open or flat Universe with $\Lambda = 0$.

The error bars in Fig. 6.2 represent the photometric uncertainty with 0.17 magnitudes of intrinsic dispersion of SN Ia magnitudes that remain after applying the width-luminosity correction add in quadrature. The theoretical curves for a Universe with no cosmological constant are shown as red (open) and green (flat) lines. The blue line shows the best fit-cosmology for which the total mass-energy density $\Omega_M + \Omega_\Lambda = 1$. The best fit value for the mass density in a flat Universe is ($\Omega_\Lambda = 1 - \Omega_M^{\text{flat}}$):

$$\Omega_M^{\text{flat}} = 0.28^{+0.08 \; +0.05}_{-0.08 \; -0.04},$$

where the first uncertainty is statistical and the second due to known systematics. The details of the estimation of systematic errors such as from extinction, Mamlquist bias and brightness evolution of type Ia supernovae can be found in [4].

6.4 Cross-Cutting Measurements

Because of the potentially revolutionary nature of the supernovae results, it is appropriate to look for alternative explanations for the dimming of supernovae at $z \sim 0.5$. Two important sources of error to consider are brightness

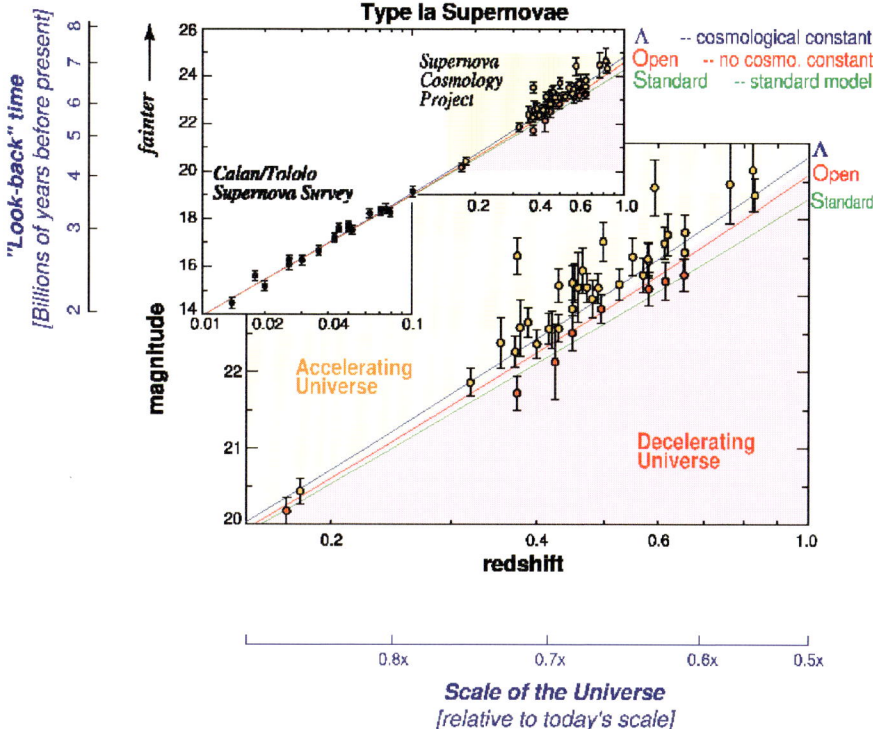

Fig. 6.2. Hubble diagram for 42 high-redshift Type Ia supernovae from the Supernova Cosmology Project [4], and 18 low-redshift Type Ia supernovae from the Calán/Tololo Supernova Survey [9], plotted along with 3 combinations of cosmological parameters, two of which have $\Omega_\Lambda = 0$. The best-fit flat cosmology is $(\Omega_M, \Omega_\Lambda) = (0.28, 0.72)$. Yellow data points indicate that the central value is in the $q_0 < 0$ region (acceleration) while the red points are in the $q_0 > 0$ region (deceleration). Clearly, the data favours an accelerating Universe. Courtesy of S. Perlmutter, [10].

evolution of Type Ia supernovae and extinction of the supernova light by dust grains along the line of sight. In Sect. 6.6 we will discuss these systematic uncertainties further. Next, however, we examine the constraints in the $\Omega_M - \Omega_\Lambda$ parameter space from a combination of cosmological methods. Several of these will be reviewed in coming chapters of this book. Figure 6.3 shows the the allowed regions derived from two completely different techniques, in addition to the SCP results. The CMBR anisotropies at scales 1° or smaller give a firm constraint on the geometry of the Universe indicating that the sum of all energy densities, i.e. $\Omega_M + \Omega_X$ must be unity with only 5 % uncertainty [11–13]. Constraints on the matter density Ω_M from cluster abundances [14, 15] and large-scale structure [16] has left cosmology with a concordance model with $\Omega_M \approx 0.3$ and $\Omega_\Lambda \approx 0.7$. Further, Fig. 6.4 shows the isochrones corresponding to the age of the Universe as derived from the

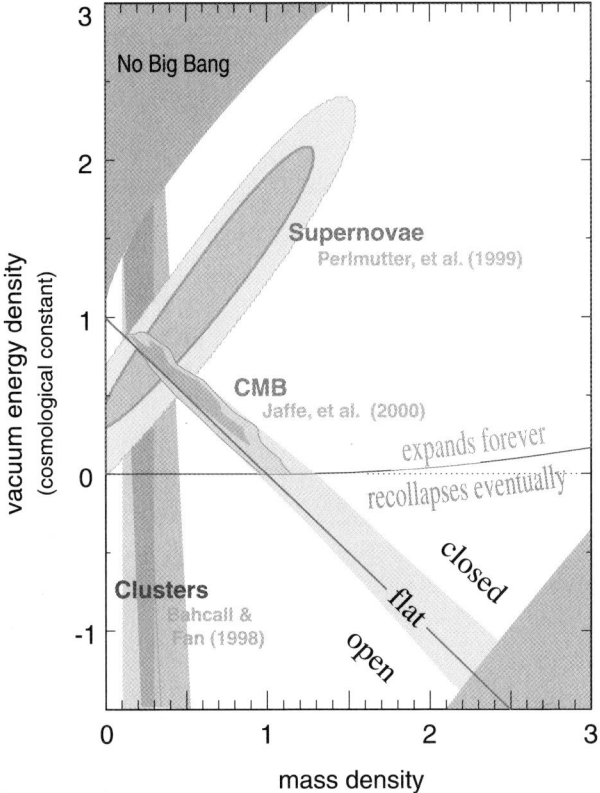

Fig. 6.3. Constraints from the Supernova Cosmology Project [4]. Top: Best-fit confidence regions in the Ω_M–Ω_Λ plane. In cosmologies above this near-horizontal line the Universe will expand forever, while below this line the expansion of the Universe will eventually come to a halt and collapse.

different combinations of Ω_M and Ω_Λ assuming a Hubble constant value of 63 km s^{-1}Mpc^{-1}. The allowed region as derived from the SCP data is thus also consistent with the age of the oldest stars in our galaxy.

6.5 How Much Better Can We Do?

Figure 6.5 shows the degeneracy in the CL-region of the $\Omega_M - \Omega_\Lambda$ parameter space defined by observations at *single* redshifts, ranging from z=0.2 to 1.8, assuming an accuracy of $\Delta m = 0.02$ mag in the measured mean. In Fig. 6.6, a hypothetical data-set including supernovae at z=0.2–1.8 is used to demonstrate how the major axis of the confidence region could be dramatically shrunk. Clearly, enlarging the redshift range of the followed supernovae has the potential of refining our understanding of the cosmological parameters.

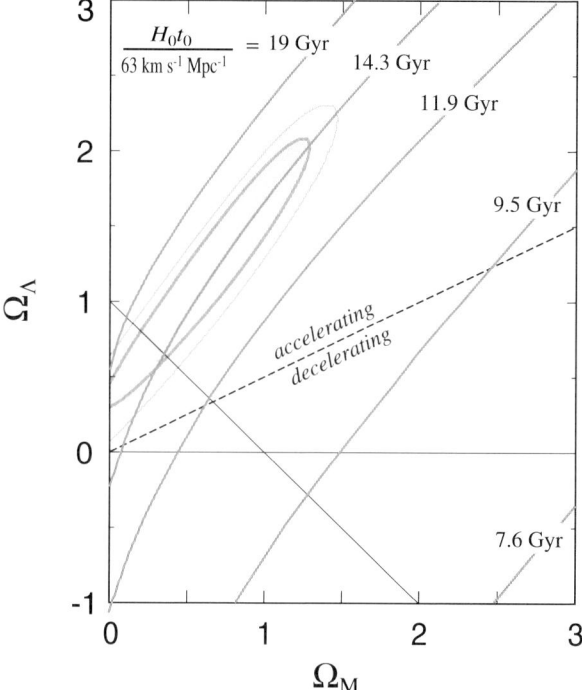

Fig. 6.4. Isochrones of constant $H_0 t_0$, the age of the Universe relative to the Hubble time, H_0^{-1}, with the best-fit 68% and 90% confidence regions in the Ω_M–Ω_Λ plane for the primary analysis, Fit C. The isochrones are labeled for the case of $H_0 = 63$ km s^{-1} Mpc^{-1}, representing a typical value found from studies of SNe Ia [9, 17, 18].

6.6 The Highest Redshift Supernova: SN1997ff

In Fig. 6.6 we demonstrated how the accuracy in the magnitude–redshift method increases as supernovae at higher redshifts are added to the sample. In particular, at redshifts above $z \sim 1$ one can study the transition from acceleration to deceleration as the mass density term contribution, enhanced by the the shrinking volume as $(1+z)^3$, overtakes the effect of Ω_Λ, as shown in Fig. 6.7.

The discovery of a likely Type Ia supernova at a remarkable redshift of $z \sim 1.7$ [19, 20] generated a great deal of excitement in the cosmology community. The measurement the rest-frame B-band brightness of this object turned out to be very bright but with a considerable uncertainty. Riess et al concluded that if the faintness of Type Ia supernovae at $z \sim 0.5$ was due to either the fact that high-z supernovae are intrinsically fainter than nearby ones or that the light had been absorbed/scattered away by intergalactic dust, then a $z \sim 1.7$ supernova had to be very dim, much dimmer than what was measured in SN1997ff. The qualitative argument is that whatever effect is going on, it should be monotonic, i.e. increasing with distance.

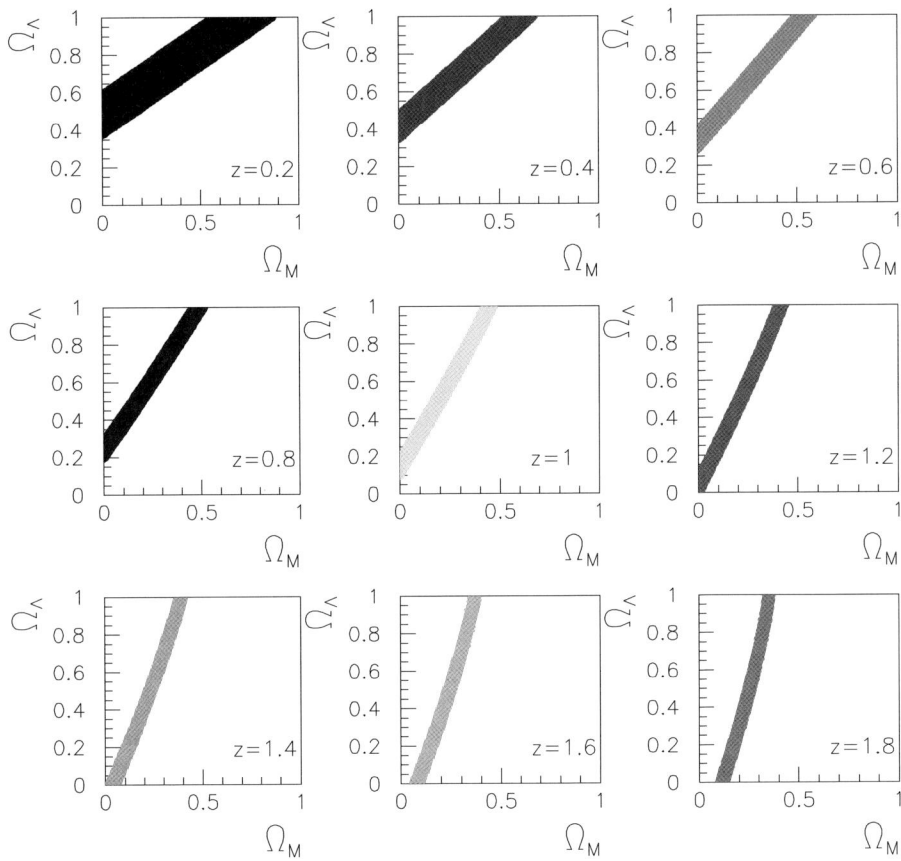

Fig. 6.5. 68 % CL-regions in the $\Omega_M - \Omega_\Lambda$ parameter space defined by each redshift bin ($\Delta z = 0.2$) assuming a total uncertainty in the mean brightness of $\Delta m = 0.02$ /bin.

Unfortunately, there are considerable caveats to the results of [20]. Benitez et al. [21] and Mörtsell, Gunnarsson & Goobar [22] showed that, because of the presence several foreground galaxies at $z \sim 0.5$ very close to the line of sight of SN1997ff, one cannot exclude a significant magnification of the brightness of the supernova by gravitational lensing.

Even disregarding the gravitational lensing uncertainties, one single supernova cannot exclude the presence of intergalactic gray dust. While Riess et al. [20] consider an extremely simple redshift distribution of putative dust in the Universe, nature might be more complex. In [23] we showed that viable models of intergalactic dust can be constructed that are consistent with the data at $z \sim 0.5$ and SN1997ff.

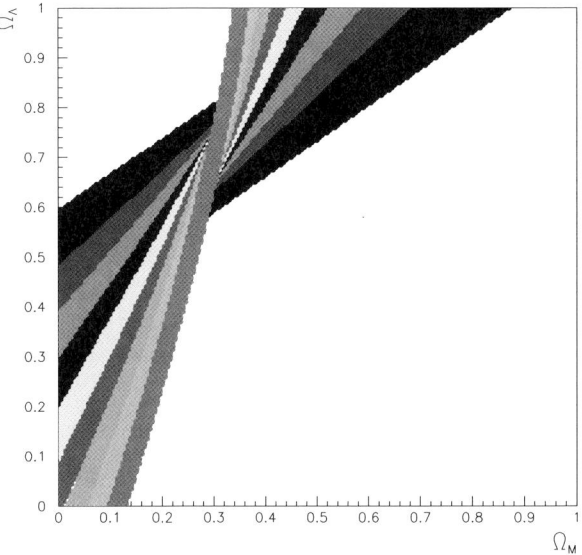

Fig. 6.6. The bands in 6.5 are superimposed. The resulting CL region is defined by the common area.

It is clear that more than one supernova is needed to exclude systematic effects and at the same time improve our knowledge of the "Dark Energy".

6.7 The Next Generation of SN Experiments

Several projects with the aim of discovering thousands of high-z supernovae are being proposed. One of the most interesting ones is the SNAP satellite [24], a 2-m telescope equipped with an optical and NIR mosaic camera with a field of view of 1 square degree.

In addition of having the capability of discovering about 2500 SNe a year up to a redshift $z = 1.7$, the design of the SNAP satellite also includes an integral field spectrograph. This will allow for detailed spectroscopic studies of the supernovae and their host galaxies. Thus, systematic uncertainties on the measured supernova brightnesses are supposed to stay below 0.02 mag in which case one can expect to measure Ω_M and Ω_Λ simultaneously to about 2% and 5% respectively, as shown in Fig. 6.8.

6.8 The Quintessence Alternative

The exciting results from the SCP and High-Z teams suggest that the method can be used to further improve our knowledge of cosmological parameters

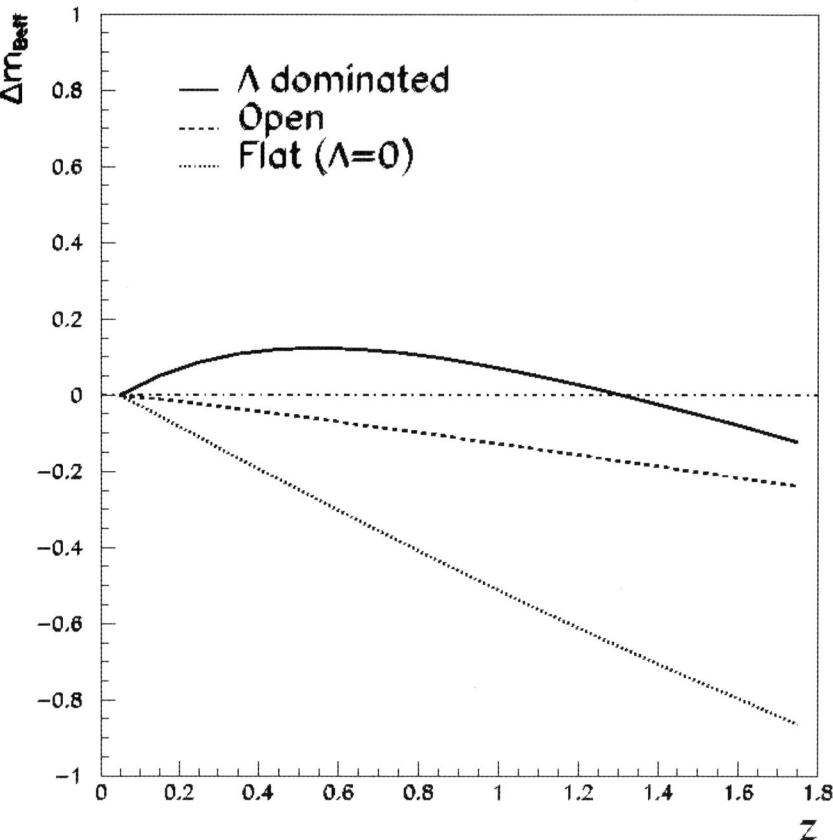

Fig. 6.7. Differential magnitude for three cosmologies, $\Omega_M, \Omega_\Lambda =$ (0.3,0.7) (solid line), (0.2,0) (dashed line) and (1,0) (dotted line), compared with an empty Universe, $(\Omega_M, \Omega_\Lambda)=$(0,0) (horizontal, dash-dotted line).

with Type Ia supernovae. While the existence of an energy form with negative pressure is strongly supported by the present data, it is not clear that the "Dark Energy" really is identical with the cosmological constant. Alternative solutions have been proposed. Steinhardt [25] suggests that the effect might be caused by a different type of matter characterized by an equation of state $p = w(z)\rho$, where $w > -1$, as shown in (6.11). The "quintessence" models do not suffer from the two fundamental problems of the cosmological constant: a) a value of $\Omega_\Lambda \sim 0.7$ is about 122 orders of magnitude from the naive theoretical calculation(!) b) It seems somewhat unnatural that we happen to live in a time when $\frac{\Omega_\Lambda}{\Omega_M} \approx 2$ since this ratio depends on the third power of the redshift. For instance, at the epoch of radiation decoupling $\frac{\Omega_\Lambda}{\Omega_M} \sim 10^{-9}$. In "quintessence" models, the "Dark Energy" density tracks the development of the leading energy term making both comparable.

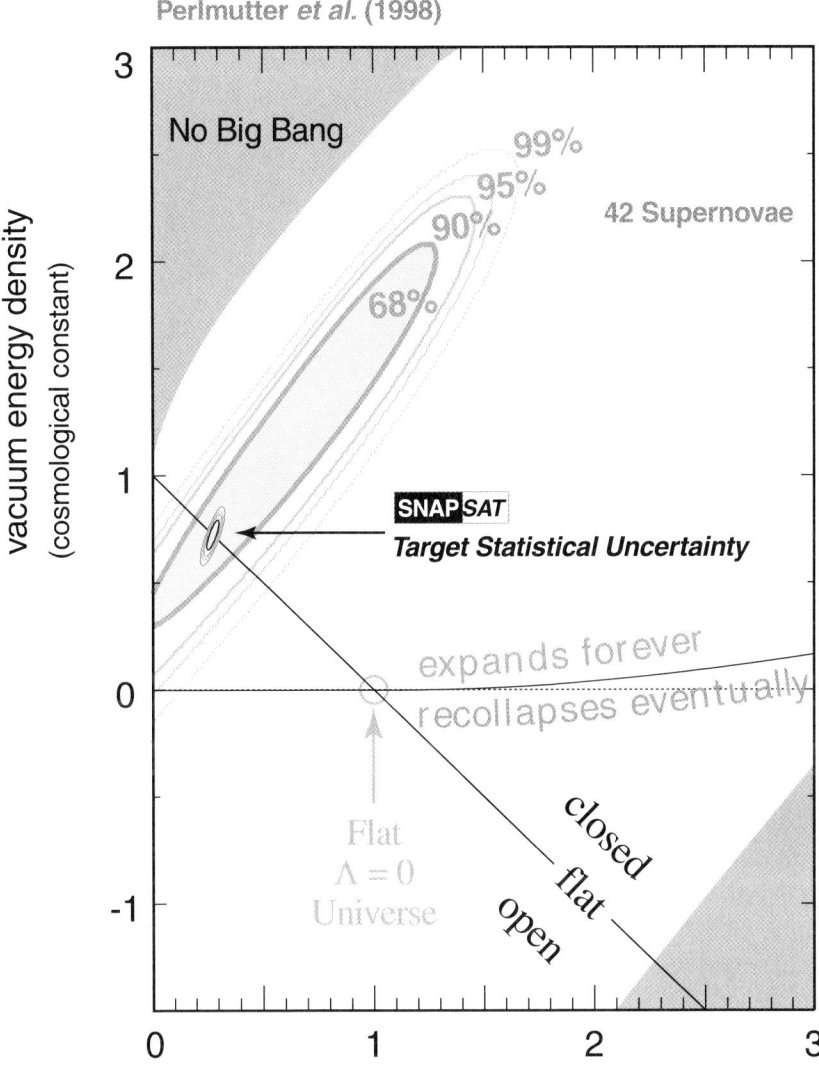

Fig. 6.8. Target uncertainty for the SNAP satellite experiment (small ellipses) compared to the published results in [4].

Figures 6.9 and 6.10 indicate the accuracy to which the effective equation of-state parameter w can be measured in a flat Universe using supernovae ranging from z=0.2 to 1.8, assuming an accuracy of $\Delta m = 0.02$ mag in the measured mean.

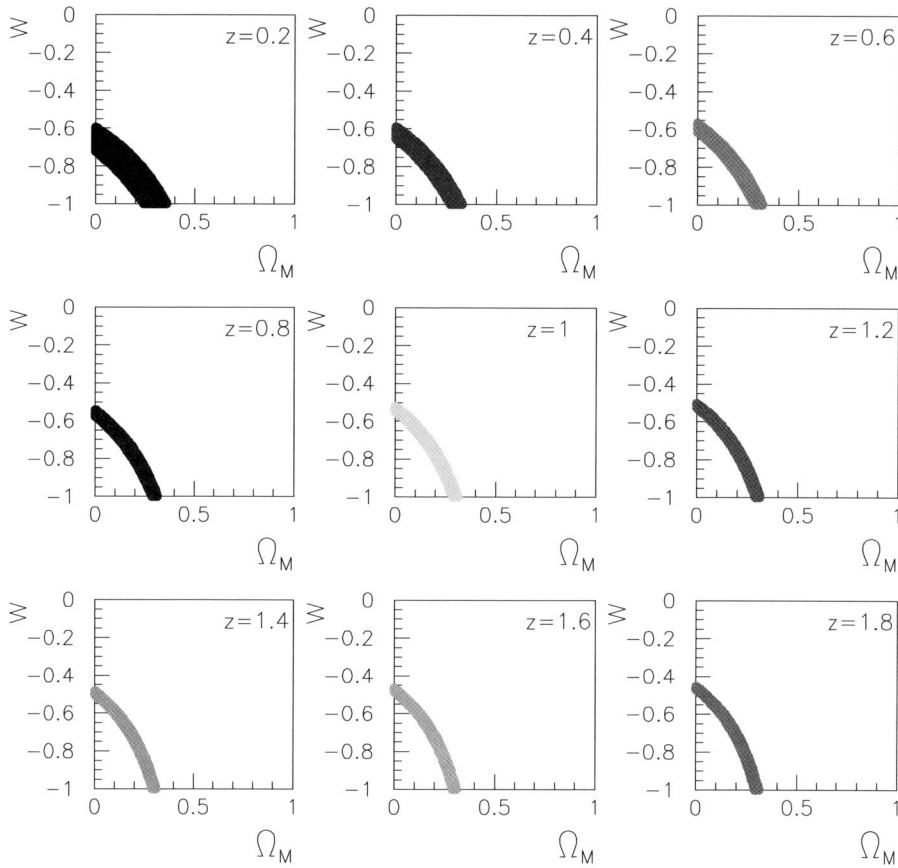

Fig. 6.9. 68 % CL-regions in the $w - \Omega_M$ parameter space defined by each redshift bin ($\Delta z = 0.2$) assuming a total uncertainty in the mean brightness of $\Delta m = 0.02$ /bin and a flat Universe.

The situation becomes more complicated once we try to measure the time evolution of the equation of state parameter. Assuming a linear expansion, $w(z) = w_0 + w_1 \cdot z$, is sufficient for the small redshift range $z < 2$, one additional parameter has to be considered. Figure 6.11 (from [26]) shows what the fit of simulated data corresponding to one year of the SNAP satellite. The accuracy on the estimate of the nature of the "Dark Energy" will depend on independent knowledge, especially, of the Ω_M from e.g. weak lensing measurements. The SNAP satellite, with is large field of view, will also provide extremely accurate measurements of cosmic shear. In addition, dedicated low-z supernova searches will be required in order to bound the intercept of the Hubble diagram, $\mathcal{M}$.

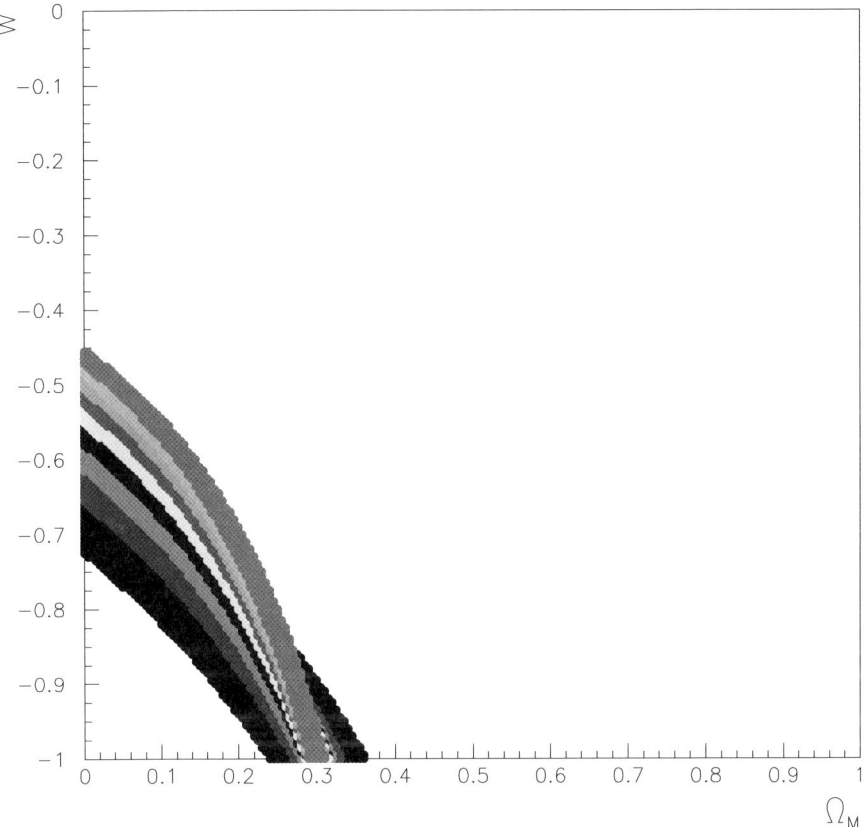

Fig. 6.10. The bands in Fig. 6.9 are superimposed. The resulting CL region is defined by the common area.

6.9 The Nature of Dark Matter

With Type Ia supernovae it may be also possible to shed light on the nature of Dark Matter. Gravitational lensing in the inhomogeneous path that the beam of high-z supernovae follow from the source to us, affects the dispersion of the data points in the Hubble diagram. Thus, with a large sample of high-z supernovae, it is possible to measure the fraction of compact objects in the Universe from the residuals of the Hubble diagram. While the compact objects are likely to be of astrophysical nature, e.g. faint stars or black holes, a smooth dark matter component would indicate that the missing mass is in the form of particles, such as the lightest stable supersymmetric particles. In [27] we used Monte-Carlo simulations to show that with one year of SNAP data, the fraction of compact objects can be measured with 5% absolute precision.

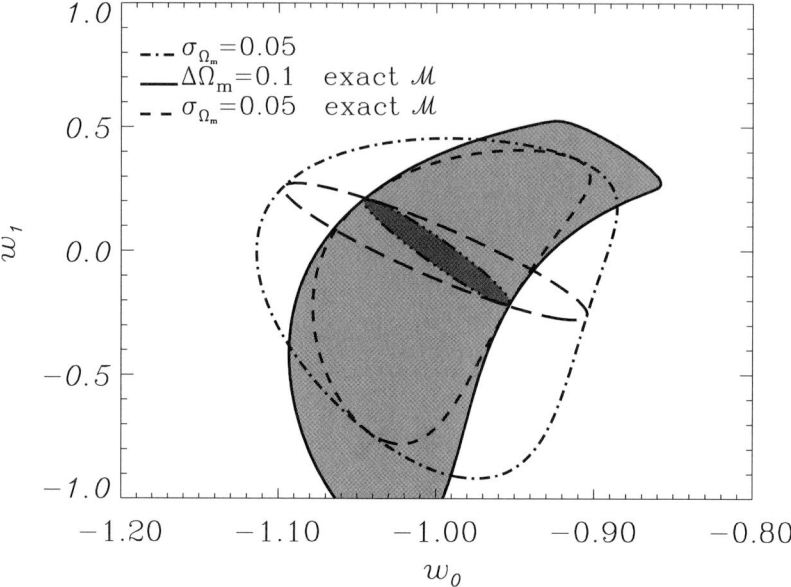

Fig. 6.11. 68.3 % confidence regions for (w_0, w_1) in the one-year SNAP scenario. The elongated ellipses correspond to the assumption of exact knowledge of Ω_m: the dash-dot-dot-dotted line is with exact $\mathcal{M}$ and the long-dashed line corresponds to no knowledge of $\mathcal{M}$. The larger, non-elliptic regions assume prior knowledge of Ω_m: the dash-dotted line assumes that Ω_m is known with a Gaussian prior for which $\sigma_{\Omega_m-\text{prior}} = 0.05$; the short-dashed line assumes the same prior and exact knowledge of $\mathcal{M}$; finally, the solid line is with Ω_m confined to the interval $\Omega_m \pm 0.1$ and exact knowledge of $\mathcal{M}$. From [26].

6.10 More on the Dark Matter

Since Big Bang nucleosynthesis (BBN) puts an upper limit to the baryonic contribution Ω_B to Ω_M of [28]

$$\Omega_B h^2 \leq 0.022, \tag{6.12}$$

non-baryonic dark matter is required beyond any doubt also in the ΛCDM model, which has $\Omega_M \sim 0.3$. In fact, is has yet turned out to be impossible to explain the CMBR data and the large scale distribution of galaxies in models with only baryons.

The non-baryonic dark matter candidates we will discuss here, in particular weakly interacting particles (WIMPs) such as neutralinos, have the virtue of lending themselves to experimental investigations at a level that is already starting to probe relevant regions of the parameter space which defines the particle physics properties of such models.

However, there are still large uncertainties related to the way the dark matter is distributed in present-day galactic haloes. On large scales like that of clusters of galaxies, gravitational lensing indicates that the dark matter is smoothly distributed, on the average. When it comes to the question of how the dark matter is distributed on the smallest, galactic and sub-galactic, scales the situation is much less clear, however (for a review, see, e.g. [29]). After being subject to an extensive debate, with both theoretical and observational controversies, it seems that the Cold Dark Matter model, with dark matter made of, e.g., weakly interacting massive particles, is in good agreement with current observations, so that drastic modifications like strong self-interaction are not urgently called for (see, e.g. [30]).

6.11 Dark Matter Candidates

In principle, one could imagine having a sterile neutrino as dark matter, if it is non-thermally produced, e.g., generated through mixing with the active neutrinos. Generally, this candidate will have a mass in the keV to MeV range and would act as something in between cold and hot dark matter (sometimes named "warm dark matter", WDM). An unpleasant feature of these models is a necessary, delicate fine tuning of the mixing angle versus mass to get the right abundance, but models of this kind have been constructed which so far evade experimental constraints [31, 32].

The right-handed neutrino, needed to give mass to the three known neutrino species, is in most models in the GUT mass range and cannot have been produced by thermal processes. Non-thermal production is again possible, but involves elements of fine-tuning. Recently, a version of the Zee model has been proposed [33], where a right-handed Majorana neutrino N_R has a TeV-scale mass. As we will see, this would be a favourable candidate for detection in gamma-rays.

One of the prime candidates for the non-baryonic component is otherwise provided by the lightest supersymmetric particle, plausibly the lightest neutralino χ.

If the scale of supersymmetry breaking is related to that of electroweak breaking, Ω_χ comes out in the right order of magnitude to explain the non-baryonic dark matter. This may be a numerical coincidence, or a sign of a deep connection between dark matter and whatever causes the breaking of electroweak symmetry.

The lightest neutralino χ is a mixture of the supersymmetric partners of the photon, the Z and the two neutral CP-even Higgs bosons present in the minimal extension of the supersymmetric standard model (for reviews see, e.g. Bergström [34], Jungman, Kamionkowski & Griest [35]). The attractiveness of this dark matter candidate stems from the fact that its generic couplings and mass range naturally gives a relic density in the required range to explain halo dark matter. Besides, its motivation from particle physics, which was originally based on solving the so-called hierarchy problem (the puzzling discrepancy between the mass scales of electroweak interactions and gravity), has become stronger due to the apparent need for 100 GeV - 10 TeV scale supersymmetry to achieve unification of the gauge couplings in view of LEP results [36], and the prediction that the lightest Higgs boson should be below 135 GeV, as seems also favoured by LEP data [37].

Supersymmetry is a mathematically beautiful theory, and would give rise to a very predictive scenario, if it were not broken in an unknown way which unfortunately introduces a large number of unknown parameters.

When using the minimal supersymmetric standard model in calculations of relic dark matter density, one should make sure that all accelerator constraints on supersymmetric particles and couplings are imposed. In addition to significant restrictions on parameters given by LEP (see, e.g. [38]), the measurement of the b → sγ process is providing important bounds.

Recently, there has been much discussion (see, e.g. [39]) about the constraints on the MSSM which follow from the measurements of $(g-2)_\mu$, the anomalous magnetic moment of the muon [40]. The first set of data indicated a large discrepancy with theoretical calculations within the standard (non-supersymmetric) model. The requirement that the discrepancy be explained by MSSM contributions led to the identification of a region in supersymmetric parameter space where neutralinos couple relatively strongly to ordinary matter and therefore have large cross section for various detection methods [39]. However, it has subsequently appeared [41–43] that the original calculations of the standard model contributions contained errors. On the other hand, the new set of data which has recently been released still shows a discrepancy at the $2 - 3$ σ level, which can in principle be due to supersymmetry [44]. However, the case is not compelling due to the large uncertainties in the calculation of the hadronic part of the standard model contribution.

The relic density calculation in the MSSM for a given set of parameters is nowadays accurate to 10 % or so. A recent important improvement is the inclusion of co-annihilations, which can change the relic abundance by a large factor in some instances [45]. Much of the effort that has gone into this field has resulted in publicly available computer program packages, for instance DarkSUSY [46], which is used in the examples below.

We start with a discussion of the direct detection limits. The experiment giving the most important limit, but also a claimed detection, is the DAMA experiment [47]. This experiment is still the largest, but is now being challenged by other detectors. For instance, combining CDMS and Edelweiss results [48, 49], it seems that essentially all of the parameter space singled out by the DAMA measurements can be excluded. The new results envisaged from CDMSII (after their removal to a deep site in the Soudan mine) may well be the final test of the DAMA results. The larger sensitivity of the scattering rate will be one of the first tests of rates which appear well into the region of realistic supersymmetric dark matter parameters.

It is essential to notice that the limit signal for direct detection also puts in a limit for the neutrino signal form the Earth that can be measured using present-day neutrino telescopes. The signal from the Sun can still be high, mainly because it is the spin-dependent cross section which is probed, and as yet there are no good direct detection devices with a high spin-dependent detection rate.

For detection of gamma-rays in Air Cerenkov Telescopes (ACTs), it is important to note that there are supersymmetric models with masses up to 10 TeV (or even higher, if co-annihilations are considered) which give the correct relic density and satisfy all other experimental constraints. For these high masses, it may be that ACTs are the only instruments capable of detecting a signal from dark matter annihilation. This occurs through the annihilation process when two neutralinos meet in the galactic halo. (The neutralino is a Majorana fermion and therefore its own antiparticle.) We now discuss this process in some detail.

6.12 Indirect Detection Through Gamma-Rays

When neutralinos collide and annihilate, the primary annihilation products are fermion-antifermion pairs (quark and leptons), or W^+W^-, ZZ, WH, ZH, or HH states. (Which states are kinematically allowed depends only on the mass of the neutralino, since galactic velocities $v/c \sim 10^{-3}$ means that the annihilations take place essentially at rest.) The gamma ray spectrum arising from the fragmentation of fermion and gauge boson final states is quite featureless and gives the bulk of the gammas at low energy where the cosmic gamma ray background is severe. However, the signal should be correlated with the mass distribution of the dark matter, which may be used to discriminate against more diffusely distributed backgrounds. In particular, there should be a noticeable enhancement toward the Galactic Center, as the annihilation rate grows with the square of the dark matter number density distribution (squared because two particles have to be at the same place for the annihilations to take place).

Since annihilations take place almost at rest, sharp (almost monoenergetic) high-energy gamma rays may result from the loop-induced annihilations $\chi\chi \to \gamma\gamma$ [50–52] or $\chi\chi \to Z\gamma$ [53, 54].

The rates of these processes are difficult to estimate because of uncertainties in the supersymmetric parameters, cross sections and halo density profile. However, in contrast to other proposed detection methods they have the virtue of giving very distinct, 'smoking gun' signals: monoenergetic photons with $E_\gamma = m_\chi$ or $E_\gamma = m_\chi(1 - m_Z^2/4m_\chi^2)$ from the halo.

Unfortunately, it is difficult to give reliable quantitative estimates of the line rates expected from these processes, since the detection probability of the gamma line signal depends, as does the continuum signal, on the very poorly known density profile of the dark matter halo.

To illustrate this point, let us consider the characteristic angular dependence of the gamma-ray intensity from neutralino annihilation in the galactic halo. Annihilation of neutralinos in an isothermal halo leads to a gamma-ray flux of

$$\frac{d\mathcal{F}}{d\Omega} \simeq (2 \times 10^{-15} \mathrm{cm}^{-2} \mathrm{s}^{-1} \mathrm{sr}^{-1}) \times \frac{(\sigma_{\gamma\gamma}v)_{29}(\rho_\chi^{0.3})^2}{(m_\chi/1\,\mathrm{TeV})^2} \left(\frac{R}{8.5\,\mathrm{kpc}}\right) J(\Psi) \quad (6.13)$$

where $(\sigma_{\gamma\gamma}v)_{29}$ is the annihilation rate in units of $10^{-29}\,\mathrm{cm}^3\,\mathrm{s}^{-1}$, $\rho_\chi^{0.3}$ is the local neutralino halo density in units of $0.3\,\mathrm{GeV\,cm^{-3}}$ and R is the distance to the galactic center. The integral $J(\Psi)$ is given by

$$J(\Psi) = \frac{1}{R\rho_0^2} \int_{\mathrm{line-of-sight}} \rho^2(\ell) d\ell(\Psi), \quad (6.14)$$

and is evidently very sensitive to local density variations along the line-of-sight path of integration.

We remind of the fact that since the neutralino velocities in the halo are of the order of 10^{-3} of the velocity of light, the annihilation can be considered to be at rest. The resulting gamma ray spectrum is a line at $E_\gamma = m_\chi$ of relative linewidth 10^{-3} which in favourable cases will stand out against background. The process $\chi\chi \to Z\gamma$ is treated analogously and has a similar rate [53, 54].

To compute $J(\Psi)$, a model of the dark matter halo has to be chosen. Recently, N-body simulations have given a clue to the final halo profile obtained by hierarchical clustering in a CDM scenario [55]. It turns out that the universal halo profile found in these simulations has a rather significant enhancement $\propto 1/r$ near the halo center. If applicable to the Milky Way, this would lead to a much enhanced annihilation rate toward the Galactic Center, and also to a very characteristic angular dependence of the line signal. This would be very beneficial when discriminating against the galactic and extragalactic γ ray background, and ACTs would be eminently suited to look for these signals, if the energy resolution is at the $10-20\,\%$ level. However, both the N-body simulations and the observations of rotation curves of galaxies are controversial at the present time [30], so it is not possible to give solid predictions for the expected fluxes. Besides the steep profiles, or 'cusps', seen in the N-body simulations, there is also a noticeable tendency for substructure (dark matter 'clumps') to be formed. This would of course increase the expected signals even further, but again an exact quantitative treatment is lacking at present.

In Fig. 6.12, we show the gamma-ray line flux given in a scan of supersymmetric models consistent with all experimental bounds, assuming an effective value of 10^3 for the average of $J(\Psi)$ over the 10^{-3} steradians that typically an Air Cerenkov Telescope would cover.

It can be seen that the models which give the highest rates should be within reach of the new generation of ACTs presently being constructed. These will have an effective area of almost $10^5\,\mathrm{m}^2$, a threshold of some tens

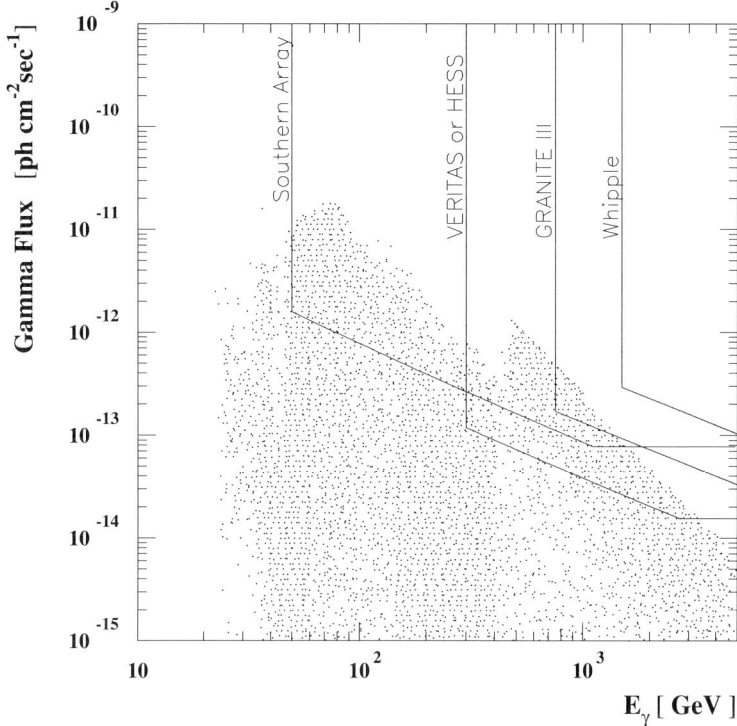

Fig. 6.12. Results for the gamma-ray line flux in an extensive scan of supersymmetric parameter space in the MSSM [56]. Shown is the number of events versus photon energy in an Air Cerenkov Telescope of area $5 \cdot 10^4$ m^2 viewing the galactic center for one year. The halo profile of [55] for the dark matter has been assumed.

of GeV and an energy resolution around 10 %. For low-mass models, the space-borne telescope GLAST may have a better sensitivity. (See [56] for details.)

Another possibility to detect dark matter in gamma-rays has recently been investigated [57, 58]. If N-body simulations of structure formation are taken seriously, it appears that the average enhancement of the integrated signal from all cosmic structure in the Universe would be several orders of magnitude compared to the case when the dark matter density only scales with the cosmic dilution factor $(1+z)^3$. The signature would be a continuum from neutralino annihilations plus a characteristic redshift-smeared line with a very rapid fall-off beyond the energy corresponding to the neutralino mass. As an example, in Fig. 6.13 from [58], the expected diffuse gamma-ray signal predicted for GLAST in a couple of the high-rate MSSM models is shown.

Of course, the detection method we have been focusing on here, indirect detection through annihilation into gamma-rays, is only one of a number of different possibilities. Also, annihilation in the the halo giving antiprotons or

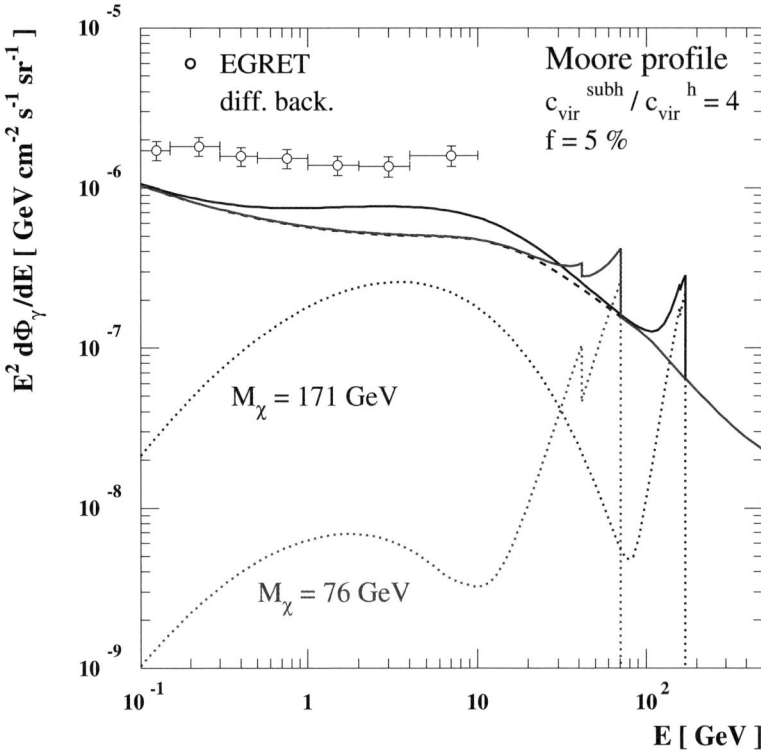

Fig. 6.13. Extragalactic gamma-ray flux (multiplied by E^2) for two sample thermal relic neutralinos in the MSSM (dotted curves), summed to the blazar background expected for GLAST (dashed curve). See [58] for details.

positrons may yield a signature if the rate is above that expected from other sources, such as cosmic ray collisions with interstellar material.

Generally, all these rates depend in different ways on the supersymmetric parameters, so the best strategy seems to be to probe them all, with the hope that at least one method may eventually give a signal. Despite some preliminary indications of possible signals in some experiments, there is not yet consensus of any detection.

6.13 Non-Supersymmetric Candidates

The phenomenology of supersymmetric dark matter (neutralinos) may be very similar for other types of weakly interacting massive particles (WIMPs). However, one can also imagine models where the WIMP only couples to leptons. These leptonic WIMPs, or LIMPs, may at first seem essentially undetectable in present-day experiments. It may be shown, however, that in most cases they necessarily give energetic gamma rays in their annihilations, due to higher-order processes [59].

In Fig. 6.14 (a) we show the flux predicted for the continuum and line gamma-ray fluxes together with the estimated background toward the Galactic Center for a 100 GeV LIMP, and 110 GeV charged scalar S_2, Navarro-Frenk-White profile and angular acceptance $\Delta\Omega = 10^{-3}$, in the model explained in [33, 59]. For this energy range, we have used an energy resolution of 3% (GLAST). It may be difficult to push the N_R mass much below 100 GeV without fine tuning the parameters of this model (and the S_2 mass is also bounded by LEP results to be larger than around 100 GeV).

The natural mass range for the LIMP is around 1 - 10 TeV, where GLAST runs out of sensitivity but where ground-based arrays of Air Čerenkov Telescopes with large collecting area can detect a signal. Indeed, there are such arrays of telescopes planned or in operation such as CANGAROO, HESS, VERITAS and MAGIC. In particular, CANGAROO and HESS are well located to observe the Galactic Center for a sizable fraction of their observing time. As can be seen from the figure, the signal with these assumptions would stand out from the gamma-ray background. (We do not enter here into the more technical issue of rejecting other types of background, such as from hadrons and electrons, where there is a steady improvement in the techniques employed.)

In Fig. 6.14 (b) results are shown for a LIMP of mass $m_N = 8$ TeV and $m_{S_2} = 8.8$ TeV. These should be clearly observable with a very conspicuous

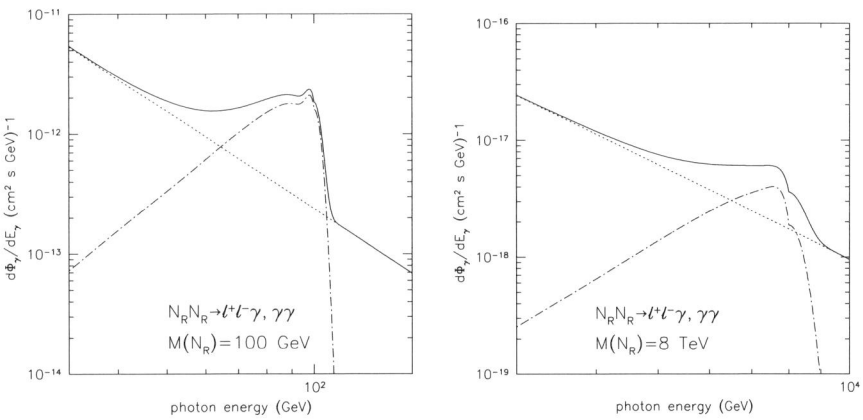

Fig. 6.14. (a) The total gamma-ray flux expected from a $\Delta\Omega = 10^{-3}$ sr cone around the Galactic Center (solid line). The flux is composed by a power-law background extrapolated from EGRET data (dotted line) and a 100 GeV LIMP annihilating with a cusped (NFW) density profile through a 110 GeV scalar S_2, giving both a continuous spectrum and a 2γ line. An energy resolution of 3% has been assumed for the line signal. (b) Same as (a) for an 8 TeV LIMP, $m_{S_2} = 8.8$ TeV. Here the line has been smeared by an assumed energy resolution of 5%.

'bump' in the spectrum, for the halo parameters chosen. We note with interest that preliminary results from the CANGAROO collaboration indeed show an excess flux of TeV gamma-rays from the Galactic Center [60]. The absolute flux level for this possible signal seems higher than that predicted here, so an enhancement beyond that provided by the NFW profile would then be indicated.

6.14 Fluxes of Neutrinos in Standard Models

The year 2002 was an amazing one for particle astrophysics. The Nobel prize in physics was awarded for two key discoveries which have changed our knowledge of and our outlook on the Universe. The first half of the Nobel prize went to R. Davis and M. Koshiba for pioneering contributions to astrophysics, in particular for the detection of cosmic neutrinos, and the second to R. Giacconi for pioneering contributions to astrophysics, which have led to the discovery of cosmic X-ray sources.

The detection of a neutrino flux from the Sun, and corroborated both by a continuous direct signal from the direction of the Sun and a remarkable few seconds of neutrinos pointing right back to the centre of a supernova (1987A) were striking pieces of evidence that the neutrino plays an important role in the Universe. These discoveries were of prime importance proving that neutrinos behaved the way they should do, and that theories of supernova explosions work more or less like they should.

Earlier this year there have in principle been two important measurements which shows that the three neutrinos behave as if they are three-fold, mixing with each other and making up a close to unitary triangle of mixing angles, just like quarks. However, the mixing matrix that one gets from the experiment at SNO [61] is of an interesting type in that it seems to constrain the measurements to be of large mixing angle. Similarly, the new evidence from the KamLAND experiment [62], which uses antineutrinos produced at nuclear rectors to propagate to the detector, has confirmed both that the neutrinos seems to validate CPT symmetry and that the large mixing angle solution is the one preferred by data.

We are thus in a situation at present, where we start to have a mass matrix also for the neutrinos. Of course it will be important to see, e.g., the outcome of the upgraded experiment where the hitherto unchecked 'LSND' result will be tested, i.e., if there is room for one more independent mass splitting. Also when looking for neutrinos as dark matter (the left-handed ones should be around at the per cent level or so) we have seen above that right-handed neutrinos in some instances will have TeV mass scale and perhaps be observable in annihilation into leptons and photons.

6.15 Particles Above the GZK Cutoff?

As a last example of an effect that has been rumored to be a prominent feature among the highest energy cosmic rays (protons), we can have a look at the incident arrival direction as a function of energy. It is well-known from textbooks [1] that the Greisen Zatsepin Kuzmin (GZK) limit is $E_{GZK} > 10^{20}$ eV. The meaning of this limit is that for an energy above this, a nucleon and a pion is produced on the Cosmic Microwave Background, meaning that there should be a cutoff of order this energy for the differential energy distribution. Now it has been proposed during some time that the AGASA data indeed show a surplus of events with energy higher than GZK. However, recently there has been additions from other data groups which do not show the same surplus [63]. On the one hand, the statistical significance is quite difficult to extract from these measurements. Perhaps the best situation at the moment is to wait until the Auger array becomes fully operational and tells us exactly what is the spectrum when it operates from within Argentina.

6.16 Summary and Conclusions

Observational cosmology is arguably one of the most exciting fields in physics at the moment. Techniques developed during the last years have provided new and unexpected results: the energy density of the Universe seems to be dominated by the Einstein's cosmological constant (Λ), or possibly some even more exotic form of "Dark Energy". Within the next decade, several measurement techniques are likely to provide conclusive evidence for the nature of the energy form that is currently causing the Universe to expand at an accelerated rate, and for the nature of the dark matter, opening a new era of precision observational cosmology.

In fact, non-baryonic dark matter seems to be needed more than ever to explain new cosmological data, in particular the recent high-precision measurements of the microwave background. The fact that the favoured value of Ω_M has gone down from near 1 to around 0.3 is good news for detection, since larger cross sections generally means lower relic density. In particular this is true for the main particle physics candidate, the neutralino, which we have presented in some detail here. Indirect detection methods have the potential to be very useful complements to direct detection of supersymmetric dark matter candidates. In particular, new gamma-ray telescopes may have the sensitivity to rule out or confirm the supersymmetry solution of the dark matter problem. If the dark matter particle is leptonic in nature, gamma-rays may provide the only window of opportunity for detecting them.

Finally, we have taken a look at a couple of specific models, namely neutrinos with standard couplings and standard relic abundance and looked at the way they have been mixed before being detected. We have also taken a critical path to the eventual overabundance of particles with momentum above the GZK cutoff. There, we find that the whole excess lies with one experiment. The Auger project, well under way to completion, will have a key role as the final deviser of this and other questions on the way to knowing most of our Universe.

Acknowledgements

A. G. is a Royal Swedish Academy Research Fellow supported by a grant from the Knut and Alice Wallenberg Foundation. This work has been supported in part by the Swedish Research Council (VR).

References

[1] Bergström, L. and Goobar, A. (1999) Cosmology and Particle Astrophysics, Wiley-Praxis Series in Astronomy & Astrophysics, John Wiley & Sons, Ltd., Chichester, UK

[2] Goobar, A. and Perlmutter, S., (1995), Feasibility of Measuring the Cosmological Constant Lambda and Mass Density Omega Using Type IA Supernovae, ApJ, **450**, 14

[3] Kim, A., Goobar, A. and Perlmutter, S. (1996), A Generalized K Correction for Type IA Supernovae: Comparing R-band Photometry beyond z=0.2 with B, V, and R-band Nearby Photometry, PASP, **108**, 190

[4] Perlmutter, S. et al. (1999), Measurements of Omega and Lambda from 42 High-Redshift Supernovae, ApJ. **517**, 565

[5] Sullivan. M. et al., (2002), The Hubble Diagram of Type Ia Supernovae as a Function of Host Galaxy Morphology, MNRAS in press, [arXiv:astro-ph/0211444]

[6] Garnavich, P., et al. (1998), Supernova Limits on the Cosmic Equation of State, ApJ, **493**, L53

[7] Riess, A. G., et al. (1998), Observational Evidence from Supernovae for an Accelerating Universe and a Cosmological Constant, **AJ**, 116, 1009

[8] Schmidt, B. P. et al (1998), The High-Z Supernova Search: Measuring Cosmic Deceleration and Global Curvature of the Universe Using Type IA Supernovae, ApJ, **507**, 46.

[9] Hamuy, M., Phillips, M.M., Maza, J., Suntzeff, N.B., Schommer, R.A. and Aviles, R. (1996), The Absolute Luminosities of the Calan/Tololo Type IA Supernovae, AJ, **112**, 2391

[10] Bahcall, N. A, Ostriker, J. P., Perlmutter, S. and Steinhardt, P. J. (1999), The Cosmic Triangle: Revealing the State of the Universe, Science, **284**, 1481

[11] Balbi, A. et al. (2000), Constraints on Cosmological Parameters from MAXIMA-1, ApJ, **545**,1; (2001) erratum: ApJ, **558**, 145

[12] de Bernardis, P. et al. (2000), A flat Universe from high-resolution maps of the cosmic microwave background radiation, Nature **404**, 955

[13] Pryke, C. et al. (2002), Cosmological Parameter Extraction from the First Season of Observations with the Degree Angular Scale Interferometer, ApJ, **568**, 46

[14] Carlberg, R. G. et al. (1999), The Ω_M-Ω_Λ Dependence of the Apparent Cluster Omega, ApJ. **516**, 552
[15] Bahcall, N. A and Fan, X. (1998), The Most Massive Distant Clusters: Determining Omega and delta 8, ApJ. **504**, 1
[16] Peacock, J. A. et al. (2001), A measurement of the cosmological mass density from clustering in the 2dF Galaxy Redshift Survey, Nature, **410**, 169
[17] Riess, A. G., Press, W. H. and Kirshner, R. P. (1996), A Precise Distance Indicator: Type IA Supernova Multicolor Light-Curve Shapes, ApJ, **473**, 88
[18] Saha, A. et al. (1997), Cepheid Calibration of the Peak Brightness of Type IA Supernovae. VIII. SN 1990N in NGC 4639, ApJ, **486**, 1
[19] Gilliand, R. L., Nugent, P. E., and Phillips, M. M. (1999), High-Redshift Supernovae in the Hubble Deep Field. ApJ, **521**, 30
[20] Riess, A. G. et al. (2001), The Farthest Known Supernova: Support for an Accelerating Universe and a Glimpse of the Epoch of Deceleration, ApJ, **560**, 49
[21] Benitez, N. et al. (2002), The Magnification of SN 1997ff, the Farthest Known Supernova, ApJL, **577**, 1
[22] Mörtsell, E., Gunnarsson, C., and Goobar, A. (2001), Gravitational Lensing of the Farthest Known Supernova SN 1997ff, ApJ, **2**, 53
[23] Goobar, A., Bergström, L. and Mörtsell, E., (2002), Measuring the properties of extragalactic dust and implications for the Hubble diagram, A&A, **384**, 1
[24] Perlmutter, S. et al. (2000), The SNAP Science Proposal, http://snap.lbl.gov
[25] Steinhardt, P. J. (2000), Quintessence and the Missing Energy Problem, Proc. of the Nobel Symposium "Particle Physics and the Universe", L. Bergström, P. Carlson and C. Fransson (eds.), T85, 177
[26] Goliath, G., Amanullah, R., Astier, P., Goobar, A., and Pain, R. (2001), Supernovae and the nature of the dark energy, A&A, **380**, 6
[27] Mörtsell, E., Goobar, A. and Bergström, L. (2001), Determining the Fraction of Compact Objects in the Universe Using Supernova Observations, ApJ, **559**, 53
[28] Tytler, D., Fan, X.-M. and Burles, S. (1996), Cosmological baryon density derived from the deuterium abundance at redshift Z = 3.57, Nature, **381**, 207
[29] Moore, B. (2001), The dark matter crisis, plenary talk at 20th Texas Symposium, [arXiv:astro-ph/0103100]
[30] Primack, J., (2001), The Nature of Dark Matter, [arXiv:astro-ph/0112255]
[31] Abazajian, K. N. and Fuller, G. M. (2002), Bulk QCD thermodynamics and sterile neutrino dark matter, Phys. Rev. D, **66**, 023526. [arXiv:astro-ph/0204293]
[32] Dolgov, A. D. and Hansen, S. H. (2001), Producing massive sterile neutrinos as warm dark matter, [arXiv:hep-ph/0103118.]
[33] Krauss, L. M., Nasri, S. and Trodden, M. (2002), A model for neutrino masses and dark matter, [arXiv:hep-ph/0210389.]
[34] Bergström, L. (2000), Non-baryonic dark matter: Observational evidence and detection methods, Rept. Prog. Phys. 63, 793. [arXiv:hep-ph/0002126].
[35] Jungman, G., Kamionkowski, M. and Griest, K. (1996), Phys. Rep., **267**, 195

[36] Amaldi, U., de Boer, W. and Furstenau, H. (1991), Comparison of Grand Unified Theories with electroweak and strong coupling constants measured at LEP, Phys. Lett., **B260**, 447
[37] Quadt, A. (2002), Higgs searches at LEP, [arXiv:hep-ex/0207050]
[38] Abreu, P. et al. (DELPHI Collaboration) (1998), Search for Charginos, Neutralinos and Gravitinos at LEP, Eur. Phys. Jour. **C1**, 1
[39] Baltz, E. A. and Gondolo, P. (2001), Implications of muon anomalous magnetic moment for supersymmetric dark matter, Phys. Rev. Lett. **86**, 5004. [arXiv:hep-ph/0102147]
[40] Bennett, G. W. et al. [Muon g-2 Collaboration] (2002), Measurement of the positive muon anomalous magnetic moment to 0.7 ppm, Phys. Rev. Lett. **89**, 101804, [Erratum-ibid. 89, 129903] [arXiv:hep-ex/0208001]
[41] Bijnens, J., Pallante, E. and Prades, J. (2002), Comment on the pion pole part of the light-by-light contribution to the muon g-2, Nucl. Phys. B, **626**, 410. [arXiv:hep-ph/0112255]
[42] Marciano, W. J. and Roberts, B. L. (2001), Status of the hadronic contribution to the muon (g-2) value, [arXiv:hep-ph/0105056]
[43] Nyffeler, A. (2002) Hadronic light-by-light scattering contribution to the muon g-2, [arXiv:hep-ph/0209329]
[44] Baltz, E. A. and Gondolo, P. (2002), Improved constraints on supersymmetric dark matter from muon g-2, [arXiv:astro-ph/0207673]
[45] Edsjö, J. and Gondolo, P. (1997), Neutralino relic density including coannihilations, Phys. Rev. **D56**, 1879
[46] Gondolo, P., Edsjö, J., Ullio, P, Bergström, L., Schelke, M and Baltz, E. A. (2002), DarkSUSY: A numerical package for supersymmetric dark matter calculations, [arXiv:astro-ph/0211238]
[47] Bernabei, R. et al. (2002), Search for WIMP annual modulation signature: results from DAMA/NaI-3 and DAMA/NaI-4 and the global combined analysis, Phys. Lett. **B480**, 23
[48] Abusaidi, R. et al. (2000), Exclusion Limits on the WIMP-Nucleon Cross Section from the Cryogenic Dark Matter Search, Phys. Rev. Lett., **84**, 5699
[49] Benoit, A. et al. (2002), Improved exclusion limits from the EDELWEISS WIMP search, Phys. Lett. **B545**, 43
[50] Bergström, L. and Snellman, H. (1988), Observable monochromatic photons from photino annihilation, Phys. Rev. **D37**, 3737
[51] Bouquet, A., Salati, P., and Silk, J., (1989) gamma-ray lines as a probe for a cold-dark-matter halo, Phys. Rev. **D40**, 3168
[52] Jungman, G. and Kamionkowski, M. (1995), gamma-rays from neutralino annihilation Phys. Rev. **D51**, 3121
[53] Bergström, L. and Kaplan, J. (1994), Gamma ray lines from TeV dark matter, Astropart, Phys., **2**, 261
[54] Ullio, P. and Bergström, L. (1998), Neutralino annihilation into a photon and a Z boson, Phys. Rev. **D57**, 1962
[55] Navarro, J. F., Frenk, C. S., and White, S. D. M. (1996), The Structure of Cold Dark Matter Halos, ApJ. **462**, 563
[56] Bergström, L., Ullio, P. and Buckley, J. H. (1998), Observability of gamma rays from dark matter neutralino annihilations in the Milky Way halo, Astropart. Phys. **9**, 137. [arXiv:astro-ph/9712318]

[57] Bergström, L., Edsjö, J. and Ullio, P. (2001), Spectral gamma-ray signatures of cosmological dark matter annihilations, Phys. Rev. Lett. **87**, 251301. [arXiv:astro-ph/0105048]
[58] Ullio, P., Bergström, L., Edsjö, J. and Lacey, C. (2002), Cosmological dark matter annihilations into gamma-rays: A closer look, Phys. Rev. **D66**, 123502
[59] Baltz, E. A. and Bergström, L. (2002), Detection of leptonic dark matter, [arXiv:hep-ph/0211325]
[60] Tsuchiya, K. et al. (CANGAROO Collaboration) (2002), to appear in the Proceedings of "The Universe Viewed in Gamma-rays", Kashiwa, Japan, September 2002
[61] Ahmad, Q. R. et al. (the SNO collaboration) (2002), Direct Evidence for Neutrino Flavor Transformation from Neutral-Current Interactions in the Sudbury Neutrino Observatory, Phys. Rev. Lett. **89**, 011301
[62] Eguchi, K. et al. (The KamLAND collaboration) (2002), First Results from KamLAND: Evidence for Reactor Anti-Neutrino Disappearance, [arXiv:hep-ex/0212021]
[63] Bahcall, J. N. and Waxman, E. A. (2002), Has the GZK cutoff been discovered?, [arXiv:hep-ph/0206217]

7 The Early Universe: From Recombination to Reionization

W.C. Keel

7.1 Introduction

Recent observations have led to remarkable progress in understanding the early history of galaxies. The 'initial conditions' for galaxy formation are largely encoded in the Cosmic Microwave Background, responding to density fluctuations of a few parts in 10^5 in baryonic matter at the start of its gravitational clumping. Prior to the epoch of recombination of which the microwave background is the escaping relic, ordinary matter was so strongly coupled to the general radiation field that it was kept highly uniform in temperature and density. Only after most of the gas became neutral and hence effectively transparent could it begin to fall together under the gravity of denser regions, where the density and therefore collapse history were initially set completely by the distribution of dark matter (which could have its own history of clumping before the time of recombination, not necessarily coupling directly to the radiation). The next things we can see are galaxies and quasars at large redshift, $z > 6$, about 800 million years later. This means that much of the important action in the formation of galaxies, and in the initial chemical enrichment required to set the stage for 'normal' star formation, took place in what are often called the 'Dark Ages'. This means, for a theorist, the period between recombination and the appearance of the first stars or active nuclei; and for an observer, the era between recombination and the earliest objects that we can now observe. At the moment, the observer's Dark Ages coincide closely with the cosmic nonaligned epoch, when most of the baryons were in neutral form, between recombination and reionization. Processes in this interval are of intense interest, holding keys to galaxy formation, the appearance of massive black holes, and the first production of heavy elements. New techniques, observational and numeric, are starting to show us some of these processes. This article reviews the epoch of reionization, whem luminous objects radiated enough energy to ionize most of the intergalactic gas which had been neutral since recombination; the expected properties of the first stars formed during the neutral epoch, with their impact on all subsequent generations; and evidence that the growth of black holes in galactic nuclei was prompt, rapid, and universal.

In assessing timescales, I will follow the remarkable consensus which has emerged in cosmology, taking a Hubble constant $H_0 = 70$ km s^{-1} Mpc^{-1}, overall matter density $\Omega_m = 0.3$, and a flat geometry.

7.2 Reionization

Our understanding of the state of the tenuous intergalactic medium, let alone its composition and history, has only recently reached a level that makes it a subject of detailed inquiry. However, the fact that most of it is ionized, with a very small neutral fraction, was established long ago for the portions of it dense enough to show up as Lyman α absorbers against QSO light. Since we see radiation from the initial recombination as the Cosmic Microwave Background, and the intergalactic medium is highly ionized today, it must have been reionized sometime later. This reionization epoch is important for the history of galaxy assembly and our ability to observe high-redshift objects; its progress likewise carries clues to the early history of galaxies and QSOs.

Numerical modelling of the recombination era tracks the same features seen in modelling H II regions - expanding ionization fronts, surrounding partially ionized zones - now set against the backdrop of cosmic expansion. The results have been reviewed in detail by Loeb & Barkana (2001). As the ionized volumes ('Strömgren spheres') expand around individual ionizing sources, it is useful to follow Gnedin (2000) in distinguishing phases before, during, and after the overlap of these volumes. Initially, each ionized bubble expands independently, fully surrounded by neutral material. The rate of expansion depends on the ionizing luminosity, shape of the ionizing spectrum, and density distribution of the neutral gas. The ionizing sources are likely to be in the highest-density regions, being surrounded by gas which recombines fastest. The ionization structure will trace in reverse the gas density in the 'cosmic web', with denser filaments becoming ionized later as the flux finally overcomes their (falling) recombination rate. Late in recombination, this has the somewhat paradoxical result that regions of neutral gas will still exist quite close to the ionizing sources while much more distant low-density regions are fully ionizing and remain so.

Hydrogen naturally dominates the physical state during recombination. Helium still plays a role, since the ionization potential for He II is 54 eV, making its appearance a still later event. In low-density regions, reionization of He II took place near $z = 2.7$ (Kriss et al. 2001).

On the observational side, there are a series of effects to show that we are seeing into and before the epoch of reionization. First we expect the classic signature of a neutral intergalactic medium, Gunn-Peterson (1965) absorption. At low redshifts, this takes the form of the Lyman α forest, discrete absorption features from the densest regions of the IGM, where the neutral fraction is highest (although in absolute terms still quite small). The density of features in the Lyman α forest increases with redshift, both as a result of genuine evolution in the density (and thus recombination rate) of the gas, and from cosmology, due to the nonlinear mapping between redshift and distance along the incoming ray. The appearance of genuinely diffuse, space-filling neutral gas will then be as excess absorption beyond the extrapolated behavior of the narrow individual lines, or, at high enough spectral resolution, a complete 'trough' of absorption beyond what one would expect from the total of individual lines. This has been observed for QSOs at $z > 6$, found from the Sloan Digital Sky Survey (SDSS) as reported by Becker et al. (2001). The rapid increase in H I opacity across the range $z = 5.7 - 6.4$ is strong evidence that we are seeing the end of reionization at these redshifts (see Fig. 7.1).

The Gunn-Peterson effect has been observed at much lower redshift in He II, a major goal of deep-ultraviolet observations for many years. A major part of its detection was in finding the right target - a high-redshift QSO which is bright, and where the line of sight does not intersect a foreground gas-rich system producing Lyman-limit absorption which blocks the spectral range shortward of 228 Å $\times (1+z)$. A measurement at lower redshifts is more secure in having less confusion with Lyman α forst lines from hydrogen.

The Gunn-Peterson troughs that are seen imply only the end of reionization. A neutral-hydrogen density of only a small fraction of the overall baryon density will produce stronger absorption than we can reliably measure, since all optical depths $\tau > 5$ are observationally indistinguishable. A fully neutral IGM should produce $\tau > 100$, so other signatures are needed to measure its neutral density. As pointed out by Miralda-Escude (1998), at such high column densities, the damping wings on the Lyman α absorption line become very strong and broad, so that the absorption signature is a broad absorption

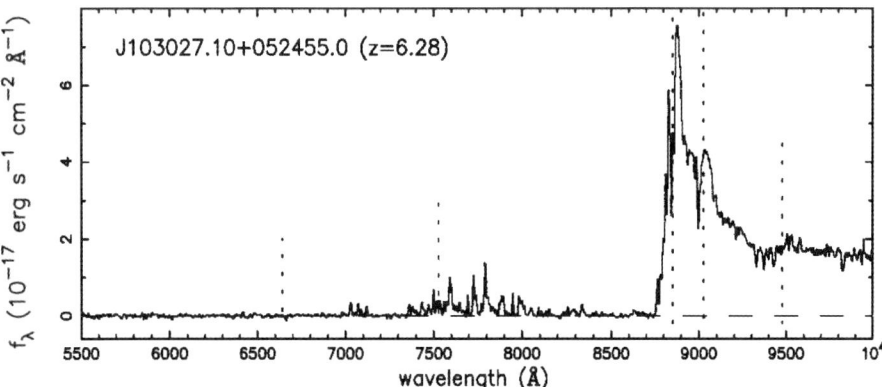

Fig. 7.1. Spectrum of the QSO 10327.0+052455.0 at $z = 6.28$, reproduced from Becker et al. 2001 by permission of the American Astronomical Society. Dashed lines marked the wavelengths of typical QSO emission lines. Gunn-Peterson absorption is manifested as the complete lack of flux just shortward of redshifted Lyman α, from about 8100–8750 Å.

edge of predicted shape, starting ~ 100 Å redward of the QSO's Lyman α wavelength. Finally, scattering of microwave-background photons by ionized material will imprint a signature of reionization on the polarization structure of fluctuations in the microwave background as detected by the Wilkinson Microwave Anisotropy Probe (WMAP) suggesting reionization at $z = 11-30$ (Bennett et al. 2003).

Once a significant volume of the ISM has become ionized, we expect ionized (and thus transparent) volumes around individual radiation sources. These would appear as holes in the absorption when a line of sight passes through such a volume, in a more powerful version of the proximity effect. This effect appears as a decrease in the number of Lyman α forest features near the redshifts of QSOs, whether the background QSO whose spectrum is being measured or fortuitously located foreground objects within a Mpc or so of the line of sight. As long as the differential Hubble expansion across an ionized volume is greater than the span of absorption wings from adjacent material, an ionized region will appear as a local deficit in absorption (as depicted by Loeb & Barkana 2001).

An emission signature of the final stages of reionization is expected, which could help specify its redshift. Broad and asymmetric Lyman α emission is predicted from blank sky areas in all directions (Baltz et al. 1998), whose intensity depends largely on baryon density, with a wavelength determined by the recombination redshift. A sensitive search for this feature has used a large number of HST STIS spectra covering blank-sky regions (Windhorst et al. 2001), yielding to date an intensity limit for this feature anywhere in the redshift range $z = 6 - 9$ somewhat higher than typical predictions.

Emission features may also be observable from the expanding H II regions around individual sources earlier in reionization. While the optical depth in the Lyman α transition will be enormous, some photons can be scattered far enough to the red wing of the line to escape. The situation is analogous to escape of Lyman α from a galactic wind, with the Hubble expansion playing the role of the wind's velocity field. Only the red wing matters, since photons scattered to the blue wing are promptly absorbed by foreground gas at a redshift for which the photons are close to the line center. Loeb & Rybicki (1999) have shown that we may hope to detect the haloes of scattered light around very luminous objects at $z \sim 10$, particularly with the James Webb Space Telescope. We might be able to see the H II regions themselves around some objects in dense environments at smaller redshifts, if the falling neutral fraction does not outweight the much smaller absorption optical depth to Lyman α.

Since we now find quasars into the epoch of reionization, strong radio sources from active nuclei may exist at significantly higher redshift. In principle, ionization structure at such epochs may be traced by highly redshifted 21-cm absorption (Carilli, Gnedin, & Owen 2002). While probably an endeavor for the next generation of radio telescopes, this would offer unique probes of these early epochs free of losses due to dust and Lyman-continuum absorption.

Further pursuit of most of these signatures to redshifts much beyond $z = 7$ requires achieving high sensitivity in the near-infrared bands, which faces formidable obstacles from the atmosphere's own molecular and thermal emission for ground-based instruments. Progress can be made in spectroscopy of compact objects using the new generation of 8-10m telescopes, particularly those optimised for infrared performance, and techniques to limit the impact of night-sky emission. These include spectrographs with airglow suppressors, which disperse the light spectrally and mask out wavelengths with strong OH airglow, and spectrographs with high dispersion so that these wavelengths can be ignored while retaining nearby wavelength bins with source information. For diffuse emission, such as scattered Lyman α halos around early QSOs or intense starbursts, the advantage of getting above the atmosphere is probably required. Such observations are primary goals of the James Webb Space Telescope (JWST), currently set to have 6.5m equivalent aperture in a deployable primary mirror, and operate in the L2 region, which will minimize heating from the Earth's own thermal radiation. Its useful wavelength range, 0.6-28 μm, was set by the scientific imperatives of understanding the formation of planets, stars, and galaxies. All three can best benefit from deep mid-infrared capability. The performance of this instrument is impressive, even in comparison to the current results from the Hubble Space Telescope. JWST promises breathrough results in probing events during photoionization, the buildup of early galaxies, and, if we are fortunate, the first generation of stars.

A key question must be answered from observation - what was the source of reionizing photons? One could imagine that either stars or active nuclei could provide the energy. Current surveys suggest that QSOs fall well short of the space density required to be important, although they will certainly dominate in their immediate neighborhoods. From metallicity arguments, first-generation stars were not numerous enough to ionize more than their immediate cloud vicinities. Further into galaxy buildup, we can count bright galaxies at redshifts $z \sim 6$ by Lyman-break selection, and find that there are too few luminous galaxies for their hot stellar populations to power reionization (Yan et al. 2003, Lehnert & Bremer 2003). This makes the most likely culprits lower-luminosity galaxies, which it is tempting to compare with the small, low-metallicity objects found in significant numbers at $z = 2 - 3$ and sometimes termed 'subgalactic'. If most of the star formation at these epochs was in low-mass systems, there is a good fit to predictions from structure-formation simulations for a Universe whose matter density is dominated by cold dark matter (CDM), which gives bottom-up or hierarchical histories for mass clumping and hence galaxy building.

Because of the Hubble expansion, it requires vastly less ionizing radiation to maintain the ionization of the intergalactic medium than to ionize it in the early Universe. As a two-body process, the timescale for recombination varies as the product of densities $n_p n_e$, which is to say n_p^2 for fully ionized hydrogen. Ignoring for the moment the dependence of cross-section on electron temperature, the basic redshift behavior for the recombination timescale will be $(1+z)^{-6}$ simply from density considerations. This has increased by more than 10^5 since the end of recombination for gas outside galaxies, at $z \sim 6.4$.

The mean ionizing intensity from star-forming galaxies and AGN is ample to maintain an ionized intergalactic medium today. In fact, there may be evidence that the intensity of this intergalactic ionizing radiation sets the cutoff in column density for H I envelopes of galaxies (Maloney 1993). The ionization balance in low-redshift QSO absorption-line systems suggests that the harder spectra of AGN are most important in the present ionizing intensity, in contrast to the situation at high redshifts, as shown by the recent analysis of Scott et al. (2002). They show that the density of ionizing radiation needed to match the extent of the 'proximity effect' in QSO absorption spectra matches that estimated for the current QSO population, and that this quantity has been dropping rapidly with cosmic time, by a factor of nearly seven for redshifts $z < 1$ and $z > 1$.

7.3 The First Stars

With the recognition that Baade's two stellar populations show a time sequence, and that the metal abundance in stars increases with time as successive generations enrich the interstellar medium, a puzzle became evident which is, in a way, an extension of the 'G Dwarf Problem'. Where is Population III? There must have been a first set of stars, to produce even the small fractions of heavy elements found in the most extreme Population II objects. Strenuous efforts have pushed the lower bound for stellar metallicity in the Milky Way to quite low levels - recently reported at [Fe/H]=-5.3 (Christlieb et al. 2002) - but hardly zero. Big Bang nucleosynthesis and the low abundances of gas in relatively pristine environments such as the Lyman α forest indicate that the first stars must have formed at very nearly zero metallicity, but we have yet to find any such star which has lasted to the present epoch.

Simulations of the behavior of pure H/He gas clouds indicate that the cooling, and thus gravitational collapse, of such clouds is quite different from what will happen even at metal fractions of order 10^{-8}. Cooling can proceed only through radiation from lines of molecular hydrogen (including its deuterated form HD), which will form at very low concentrations in the pregalactic Universe. It is important that the 'neutral' Universe still had an ionized fraction of order 10^{-4}, since this facilitated formation of molecules through a charge-exchange reaction (recently detailed by Abel et al. 1997) starting with
$$H^0 + e^- \to H^- + \gamma.$$

This weakly-bound ion then forms H$_2$ via

$$\mathrm{H}^- + \mathrm{H}^0 \to \mathrm{H}_2 + e^-$$

so that it is proper to speak of the free electrons as catalysts. The H$_2$ and HD molecules can cool through radiation in rotational and vibrational transitions lying in the near-infrared, down to temperature around 200 K (where the cooling is slowest). This temperature defines the mass scales on which objects can collapse if only this molecular cooling operates; the temperature is several times higher than the ultimate limit from equilibrium with the 'microwave' background temperature seen by an object at redshift z, of $2.7 \times (1+z)$ K.

Several recent numerical studies (Abel, Bryan, & Norman 2002; Bromm, Coppi, & Larson 2002) have converged on the results of such a collapse. The first stars were massive, hot, and efficient at enriching their surroundings in processed elements. The restricted cooling possibilities mean that only masses in the rough range 80-300 solar masses formed these stars, without further fragmentation as we see today. The lack of opacity from heavy elements made them substantially hotter than even comparably massive present-day stars, and copious emitters of Lyman-continuum photons. Not only did they ionize their surroundings, albeit briefly, but this radiation dissociated the H$_2$ and HD throughout their ancestral clouds, so that these stars formed one to a 'galaxy'.

A first-generation star of given mass was hotter than any later star of the same mass. With no elements to engage in the CNO cycle on the 'main sequence', core fusion proceeded by the $p-p$ chain, driving the core temperature to $\sim 10^8$ K to balance the pressure of the large masses (Tumlinson & Shull 2000; Tumlinson, Giroux, & Shull 2001). For chemically homogeneous stars, a hotter core also means a hotter surface.

There is less agreement among models about the exact nucleosynthetic yield and abundance pattern of the material returned to the intergalactic gas by the final explosions of these stars, largely driven by the considerable differences in supernova mechanism with mass and the strong sensitivity of the outcome to details of envelope ejection (the so-called mass cut, the point in a stellar envelope beyond which all the material is taken to escape the compact remnant, which probably has to be properly treated as more than a single cutoff value). Broadly, tens of solar masses of ejected material can be released from such an explosion. Details depend strongly on the mass distribution of the stars, both in the nucleosynthesis and because they may undergo different kinds of supernova explosion at different masses. Such exotic mechanisms as pair-production instability or reverse nucleosynthesis can drive core collapse in various mass ranges. The most massive stars, above about 260 solar masses, may swallow their entire masses into black holes rather than produce supernovae. In contrast, the pair-production collapse occurs for masses 140-260 solar masses, resulting in intense nuclear burning and completely disrupting the star (Heger and Woosley 2002). The relative amounts of heavy elements they produce, and expel to the surrounding medium, can depend strongly on the stars' mass distribution.

Since they occurred singly rather than in clusters, and are individually short-lived, detection of these stars during their normal lifespans will be beyond our means for some time. However, their supernova outbursts may be an order of magnitude more luminous than the type Ia supernovae that can now be discovered to at least $z = 1.7$, and for the redshift range $z = 10 - 30$ where we expect to see such objects, the peak flux is almost constant with redshift. They should reach a peak AB magnitude near 26, with the redshift determining what wavelengths we could detect them in (through absorption shortward of Lyman α from the neutral IGM). Time dilation improves the odds of detecting such explosions, since the observed outburst may last for several years. If, for example, 10^{-6} of the baryons were incorporated into these first stars, we would see a supernova explosion from this population every 6 seconds somewhere on the sky (Heger et al. 2002). This flux level in the K band, corresponding to supernova redshifts $z < 17$, is in principle detectable now with such instruments as the Keck telescopes, but not over wide enough areas to find these explosions by blind search. For larger redshifts, the thermal background for ground-based instruments makes their detection even more difficult, and a prime task for JWST. Searching for these supernovae (or hypernovae) suggests that deep fields, such as those to be observed by JWST, are most effectively built up by observations spread in time across a couple of years.

Ironically, we may be already observing some of these supernovae, while remaining ignorant of the fact. The connection between gamma-ray bursts and supernovae is enticing (some might say 'seductive') based on the properties of the long-wavelength afterglows, composition of hot gas in the afterglows from X-ray spectroscopy, location with respect to host galaxies, and calculations of stellar implosions which can produce a temporary accretion structure around a young black hole (reviewed by Mészáros 2002). If these more massive stars had similarly asymmetric explosions, which would include temporary formation of disks around young black holes, they should produce gamma-ray bursts as well over an appropriate solid angle. The striking lack of a fluence-redshift relation for bursts with identified host galaxies might mean that some fraction of the bursts already being observed, and for which we cannot identify a host galaxy, may lie at substantially higher redshifts.

These earliest stars played a crucial role in setting the stage for the 'normal' star formation to follow. Even a tiny salting of heavy elements changes the cooling of interstellar clouds and the properties of the resultant stars dramatically. As soon as the mean metallicity of cooling gas rose above $\sim 10^{-8}$ by mass, the nature of star formation switched to nearly what we see today, reflected in the extreme Population II stars still extant in our neighborhood.

The strongest constraint on the number and metal yield of these stars comes from the minimum metallicity encountered anywhere. It is not clear yet just what the distribution of stellar metallicity in the Milky Way is at the low end. There are several stars known with [Fe/H] − 4, and one at −5.3, found from winnowing thousands of candidates starting from broad-band colors or slitless-spectrum line strengths. The value of 10^{-4} solar is broadly consistent with the carbon abundance seen from co-adding large numbers of systems in the Lyman α forest (Cowie and Songaila 1998, Ellison et al. 2000) although the oxygen abundance may be an order of magnitude greater (Telfer et al. 2002). The abundances in low-density intergalactic regions are especially important, since these are unlikely to have been enriched by supernova ejecta in galactic winds as we see them in starburst systems today (Ostriker and Gnedin 1996). The recent detection of widespread O VI absorption from the cosmic web shows that enrichment took place over large regions, although the very uncertain ionization corrections, and easy detection of O VI in shocked regions where matter is falling into denser concentrations, make quantitative abundance estimates difficult (Tripp, Savage, & Jenkins 2000).

Finally, in connecting the products of the first stars to what we see in old populations today, we need to establish how homogeneous the mixing of their products was, and on what scales it occurred. This enters into what their overall nucleosynthentic signature is on extreme Population II stars. For example, should we consider their output to account only for the most metal-poor stars, as in the example at [Fe/H]=-5.3, or could our Galaxy have formed from gaseous regions that experienced such differing levels of early enrichment that the stars near [Fe/H]=-4 also bear their imprint? A hierarchical buildup of the Milky Way could allow it to contain stars originally in rather different chemical environments, if the dispersal of metals from primordial stars was uneven enough. Some of the complexities of applying this idea to the galactic abundance distribution are considered by Oey (2002).

7.4 Where Do Black Holes Come From?

Evidence at both very low and very high redshifts shows that massive (or indeed supermassive) black holes are common in luminous galaxies, and that they had already reached substantial masses at early times. This can be seen from the very existence of high-redshift quasars, currently known at $z = 6.4$. These QSOs have emission-line properties almost identical with those at lower redshifts, suggesting super-solar metal abundances in the gas of their broad-line regions. Such abundances can be easily understood only if these QSOs mark the locations of early and intense star formation, so that stellar products were copious, and in deep potential wells so that they were bound into small regions. Furthermore, the luminosities of these QSOs suggest that they had already achieved substantial central masses at these early times. Some galaxies collapsed so early that a rich cycle of star formation and supernova explosions had taken place, giving an interstellar medium strongly enriched in their nucleosynthetic products, in the same environment which fostered production and growth of massive black holes, all within about 0.8 Gyr. QSOs found at yet higher redshifts may well push this time limit farther, or perhaps show us how these processes take place.

Dynamical studies of nearby galaxy nuclei have shown an equally important, complementary conclusion - essentially every galaxy bulge contains a central black hole with about 0.5% of its stellar mass (sometimes known as the Magorrian relation; Magorrian et al. 1998). The relative tightness of this relation across more than three orders of magnitude in mass (see Fig. 7.2) suggests that the black holes and stellar bulges are closely related. This could, in principle, happen because either one came first and regulated the other. There are compelling models for the behavior of stellar orbits in the presence of a central point mass which suggest that the bulges began formation, feeding a low-mass black hole until it reached a limiting mass. At that point, the central mass changed the local potential sufficiently to alter the characteristic shapes of stellar orbits, effectively limiting its own growth (Merritt and Quinlan 1998). The limiting mass they derive is 3–4 times higher than we see empirically, but it is encouraging to see a result even this close. The required speed of growth for objects seen at high redshift remains somewhat unsettling, since rapid accretion may require the ability to tidally disrupt passing stars, easily accomplished only above $\sim 10^5$ solar masses. How could the black hole start off this massive?

Ordinary stellar evolution predicts that black holes up to a few tens of solar masses should result from the core collapse of massive stars, depending on their previous mass-loss history and details of the ensuing supernova explosion. Such a black hole grows only very slowly, through accretion of surrounding interstellar material and through very rare chance passages of stars close enough to be tidally disrupted, from which some material would be accreted. For a massive enough black hole, the tidal disruption radius becomes large enough to disrupt stars frequently (or indeed constantly). Once this happens, the black hole may grow until its contribution changes the overall potential and forces most stellar orbits toward circular shapes. A key issue then becomes how to get the black hole to the $\sim 10^5$ solar-mass value where such disruption becomes effective in the dense environment of a galactic nucleus. Even the 100-solar-mass remnants of modelled Population III stars aren't much help here, especially since they may well end up outside today's galactic nuclei.

Considerable interest in this connection was created by reports of black holes at the cores of the globular clusters M15 and G1 (one of the brightest such clusters associated with the Andromeda galaxy) by van der Marel et al. (2002), Gerssen et al. (2002), and Gebhardt, Rich, & Ho (2002). The enthusiasm was dampened by the demonstration that the evidence in M15 was more equivocal than first appeared (Gerssen et al. 2003). Still, black holes formed in dense stellar clusters would offer an attractive way to seed the growth of more massive ones. Relativistic clusters – configurations involving compact

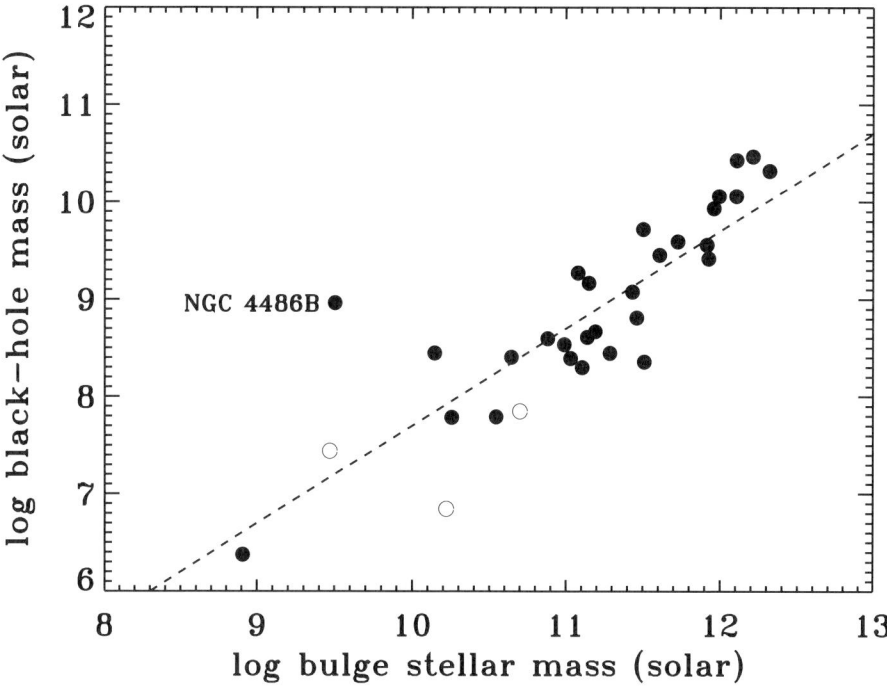

Fig. 7.2. The relation between the masses of black holes in galactic nuclei and the stellar mass in the surrounding bulge, derived from kinematic measures with the Hubble Space Telescope. Both are plotted logarithmically, with solar masses as the implied unit. The dashed line marks a constant fraction 0.5% of the bulge mass, which is a fair representation over three orders of magnitude here. The outlier is NGC 4486B, a companion to M87 in the Virgo Cluster which shows evidence that many of its stars have been tidally stripped during encounters with M87 itself. The three open symbols represent upper limits in black-hole mass. Data taken from Magorrian et al.(1998)

stellar remnants within very small volumes – might lead naturally to runaway collapse of many of them into a single object, depending on the initial conditions (recently discussed by Shapiro 2003). Alternatively, there is still exploration of the idea that primordial black holes of large mass could have been in place before galaxy formation. This seems less plausible, however, since it provides no natural explanation for the apparently close proportionality between masses of stellar bulges and black holes, or for the products of stellar evolution which are universally seen in the spectra of even the earliest quasars we observe.

Looking to yet higher redshifts will tell us much about the early growth of massive black holes and how they couple to surrounding star formation. In general, the luminosity of the active nucleus tells us about its growth rate through accretion, and dynamical clues (emission line widths during the active phases, and eventually stellar dynamics as we aproach the present epoch) give the total mass. Whether we eventually find low-metallicity quasars at the highest redshifts should reveal the relative timescales for massive star formation and the growth of the central mass.

7.5 Prospects

As rapid as progress has been within the last few years, our understanding of the early Universe is poised to grow even further. The Sloan survey should find significant additional numbers of quasars at $z > 6$, providing more lines of sight to probe the final stages of reionization using spectrposcopy from 8-10m ground-based telescopes. Additional deep imaging surveys reaching (such as the Ultra-Deep Field recently announced for observations by the Advanced Camera for Surveys on the Hubble Space Telescope) will generate large samples of galaxies at similar redshift based on their Lyman spectral breaks, reaching deep enough to survey the luminosity function of galaxies at the end of reionization. This will enter not only into our understanding of the energetics of the ionizing radiation, but the overall history of galaxy growth. The James Webb Space Telescope will probe into the crucial mid-infrared region, chasing the Lyman limit to higher redshifts still, perhaps finding the explosive ends of primordial massive stars.

Closer to home, the history of the first stars is written in the low-metallicity distribution of stars in our own Galaxy and in the enrichment of the intergalactic medium. Significant progress is going on in both these areas. The Far-Ultraviolet Spectroscopic Explorer continues to observe O VI absorption in various environments against the radiation from background quasars, and the Chandra X-ray observatory can in some cases complement these data by measurements of yet more highly-ionized forms of oxygen. The ionization balance of this element is a significant source of uncertainty in the heavy-element abundances in the IGM, so any common measurements will improve the situation. Multi-object spectrographs on the largest ground-based telescopes have vastly increased the survey speed with which candidate low-metallicity stars can be examined, which will bring us closer to the real 'floor' in pre-Milky-Way composition and its variation within the stars that came to comprise our galaxy. Both the local fossil record and the telescopic time machine applied to the distant Universe are set to make the Dark Ages a little less dark.

References

Abel, T., Anninos, P., Zhang, Y., & Norman, M.L. (1997) Modeling primordial gas in numerical cosmology. New Astron. **2**, 181–207.

Abel, T., Bryan, G.L. & Norman, M. L. (2002) The formation of the first star in the Universe, Science **295**, 93–98.

Baltz, E.A., Gnedin, N.Y., & Silk, J. (1998) Spectral features from the reionization epoch. Astrophys. J. Lett. **493**, L1–4.

Becker, R.H., Fan, X., White, R.L., Strauss, M.A., Narayanan, V.K., Lupton, R.H., Gunn, J.E., Annis, J., Bahcall, N.A., Brinkmann, J., Connolly, A.J., Csabai, I., Czarapata, P.C., Doi, M., Heckman, T.M., Hennessy, G.S., Ivezic, Z., Knapp, G.R., Lamb, D.Q., McKay, T.A., Munn, J.A., Nash, T., Nichol, R., Pier, J.R., Richards, G.T., Schneider, D.P., Stoughton, C., Szalay, A., Thakar, A., & York, D.G (2001) Evidence for reionization at $z \sim 6$: detection of a Gunn-Peterson trough in a $z = 6.28$ quasar. Astron. J. **122**, 2850–7.

Bennett, C.L. et al. (2003), First Year Wilkinson Microwave Anisotropy Probe (WMAP) Observations: Preliminary Maps and Basic Results, Astrophys. J., submitted (astro-ph/0302207).

Bromm, V., Coppi, P.S., & Larson, R.B. (2002) The formation of the first stars. I. The primordial star-forming cloud. Astrophys. J. **564**, 23–51.

Carilli, C.L., Gnedin, N.Y., & Owen, F.N. (2002) H I 21 centimeter absorption beyond the epoch of reionization. Astrophys. J. **577**, 22–30.

Christlieb, N., Bessell, M. S., Beers, T. C., Gustafsson, B., Korn, A., Barklem, P. S., Karlsson, T., Mizuno-Wiedner, M., & Rossi, S. (2002). A stellar relic from the early Milky Way. Nature **419**, 904–6.

Cowie, L.L. & Songaila, A. (1998) Heavy-element enrichment in low-density regions of the intergalactic medium. Nature **394**, 44–46.

Ellison, S.L., Songaila, A., Schaye, J., & Pettini, M. (2000) The enrichment history of the intergalactic medium - measuring the C IV/H I ratio in the Ly α Forest. Astron. J. **120**, 1175–1191.

Gebhardt, K., Rich, R.M.,& Ho, L.C. (2002) A 20,000 M_{solar} black hole in the stellar cluster G1. Astrophys. J. Lett. **578**, L41–L45.

Gerssen, J., van der Marel, R.P., Gebhardt, K., Guhathakurta, P., Peterson, R.C., & Pryor, C. (2002) Hubble Space Telescope evidence for an intermediate-mass black hole in the globular cluster M15. II. Kinematic analysis and dynamical modeling. Astron. J. **124**, 3207–3288.

Gerssen, J., van der Marel, R.P., Gebhardt, K., Guhathakurta, P., Peterson, R.C., & Pryor, C. (2003) Addendum: Hubble Space Telescope evidence for an intermediate-mass black hole in the globular cluster M15. II. Kinematic analysis and dynamical modeling. Astron. J. **125**, 376–377.

Gnedin, N.Y. (2000) Cosmological reionization by stellar sources. Astrophys. J. **535**, 530–554.

Gunn, J.E., and Peterson, B.A. (1965) On the density of neutral hydrogen in intergalactic space. Astrophys. J. **142**, 1633–6.

Heger, A. & Woosley, S.E. (2002) The nucleosynthetic signature of Population III. Astrophys. J. **567**, 532–543.

Heger, A., Woosley, S.E., Baraffe, I., & Abel, T. (2002) Evolution and explosion of very massive primordial stars. In MPA/ESO/MPE/USM Joint Astronomy

Conference "Lighthouses of the Universe: The Most Luminous Celestial Objects and their use for Cosmology, 369–375.

Kriss, G.A., Shull, J.M., Oegerle, W., Zheng, W., Davidsen, A.F., Songaila, A., Tumlinson, J., Cowie, L.L., Deharveng, J.-M., Friedman, S.D., Giroux, M.L., Green, R.F., Hutchings, J.B., Jenkins, E.B., Kruk, J.W., Moos, H.W., Morton, D.C., Sembach, K.R., & Tripp, T.M. (2001) Resolving the structure of ionized helium in the intergalactic medium with the Far Ultraviolet Spectroscopic Explorer. Science **293**, 1112–16.

Lehnert, M.D. & Bremer, M. (2003). Luminous Lyman break galaxies at $z > 5$ and the source of reionization. Astrophys. J. Lett., submitted (astro-ph/0212431).

Loeb, A. & Barkana, R. (2001) The reionization of the Universe by the first stars and quasars. Ann. Rev. Astr. Ap. **39**, 19–66.

Loeb, A., & Rybicki, G.B. (1999) Scattered Ly α radiation around sources before cosmological reionization. Astrophys. J. **524**, 527–535.

Magorrian, J., Tremaine, S., Richstone, D., Bender, R., Bower, G., Dressler, A., Faber, S.M., Gebhardt, K., Green, R., Grillmair, C., Kormendy, J., & Lauer, T. (1998). The demography of massive dark objects in galaxy centers. Astron. J. **115**, 2285–2305.

Maloney, P. (1993). Sharp edges to neutral hydrogen disks in galaxies and the extragalactic radiation field. Astrophys. J. **414**, 41–56.

Merritt, D. & Quinlan, G.D. (1998) Dynamical evolution of elliptical galaxies with central singularities. Astrophys. J. **498**, 625–639.

Mészáros, P. (2002) Theories of gamma-ray bursts. Ann. Rev. Astron. Astrophys. **40**, 137–169.

Miralda-Escudé, J. (1998) Reionization of the intergalactic medium and the damping wing of the Gunn-Peterson trough. Astrophys. J., **501**, 15–22.

Oey, M.S. (2002) Metal dispersal and the number of Population III stars. Astrophys. Space Sci. **281**, 483–486.

Ostriker, J.P. & Gnedin, N.Y. (1996) Reheating of the Universe and Population III. Astrophys. J. Lett. **472**, L63–L67.

Scott, J., Bechtold, J., Morita, M, Dobrzycki, A., & Kulkarni, V.P. (2002) A uniform analysis of the Ly α forest at $z = 0-5$. V. The extragalactic ionizing background at low redshift. Astrophys. J. **571**, 665–692.

Shapiro, S.L. (2003) Formation of supermassive black holes: simulations in general relativity, in *Carnegie Observatories Astrophyiscs Series, vol. 1: Coevolution of Black Holes and Galaxies*, ed. L. Ho, in press (astro-ph/0304202).

Telfer, R.C., Kriss, G.A., Zheng, W., Davidsen, A.F., & Tytler, D. (2002) Extreme-ultraviolet absorption lines in Ly α forest absorbers and the oxygen abundance in the intergalactic medium. Astrophys. J. **579**, 500–516.

Tripp, T.M., Davage, B.D., & Jenkins, E.B. (2000) Intervening O VI quasar absorption systems at low redshift: a significant baryon reservoir. Astrophys. J. Lett. **534**, L1–L5.

Tumlinson, J., Giroux, M.L., & Shull, J.M. (2001) Probing the first stars with hydrogen and helium recombination emission. Astrophys. J. Lett. **550**, L1–5.

Tumlinson, J. & Shull, J.M. (2000) Zero-metallicity stars and the effects of the first stars on reionization. Astrophys. J. Lett. **528**, L65–68.

van der Marel, R.P., Gerssen, J., Guhathakurta, P., Peterson, R.C., & Gebhardt, K. (2002) Hubble Space Telescope evidence for an intermediate-mass black hole in the globular cluster M15. I. STIS spectroscopy and WFPC2 photometry. Astron. J. **124**, 3255–3269.

Windhorst, R., Bernstein, R., Collins, N., Plait, P., Woodgate, B., Mather, J., Madau, P., & Shaver, P. (2001) Closing in on the hydrogen reionization edge at $z < 7.2$ with deep STIS/CCD parallels. *Deep Fields*, Proceedings of the ESO/ECF/STScI Workshop, eds. S. Cristiani, A. Renzini, & R.E. Williams, Springer, 357–361.

Yan, H., Windhorst, R.A., & Cohen, S.D. (2003) Searching for $z \simeq 6$ Objects with the HST Advanced Camera for Surveys: prelininary analysis of a deep parallel field. Astrophys. J. Lett., **585**, L93–96.

8 The Most Distant Galaxies

H. Spinrad

8.1 Introduction, Motivations and Questions

The study of distant galaxies is empirically demanding – not surprisingly, as these galaxies are very faint.

Of course there are a variety of motivations to observe and perhaps understand distant 'units of the Universe'. We would like to detail the present-day 'lumpiness' of the Cosmos and its evolution from a very smooth 'sea' at decoupling. At the nominal redshift of the Cosmic Microwave Background the key fluctuations on reasonable scales are only of order 10^{-5}. Of course at $z \sim 0$ we have a very inhomogenous distribution of baryons we call galaxies and the Intergalactic Medium (IGM hereafter).

Noting the obvious, studying distant galaxies is synonymous with travelling far back in cosmic time towards the birth of massive sub-structures and large galaxies. Can we now see directly the development of single galaxies of Milky Way dimensions?

We now believe that most galaxies form and accumulate either (1) by the infall of gas (and dark matter) as 'monolithic' entities, self-gravitating by the time we can observe them, or (2) by a series of major and/or minor mergers. This is the now-popular 'bottom-up' scenario. Here it is presumably difficult to catch the small and immature systems in the act of merging, depending perhaps on the appropriate dynamical time scales. Thus for scenario (2) we would anticipate young galaxies to illustrate complex morphologies, quite different from those of the mature galactic systems we study readily here and now, at zero redshift. There is indeed some evidence for 'recent' mergers from the fine images of distant galaxies observed with the Hubble Space Telescope (HST) - see Stern & Spinrad (1999) for some plausible early merger examples (Fig. 8.1). And we'd like to push these examples back in cosmic time to even 'younger' galaxy growth - but the first problem is, quite naturally, the location of small and dismally faint candidates for galaxies in formation.

Another important contemporary research area emerging is the study of intergalactic (gaseous) matter usually seen in silhouette against a bright background source like a QSO or an unusually bright and distant galaxy. And now, new observational techniques are beginning to tell us about the interaction history of galaxies and the IGM (*cf.* Adelberger et al., 2003).

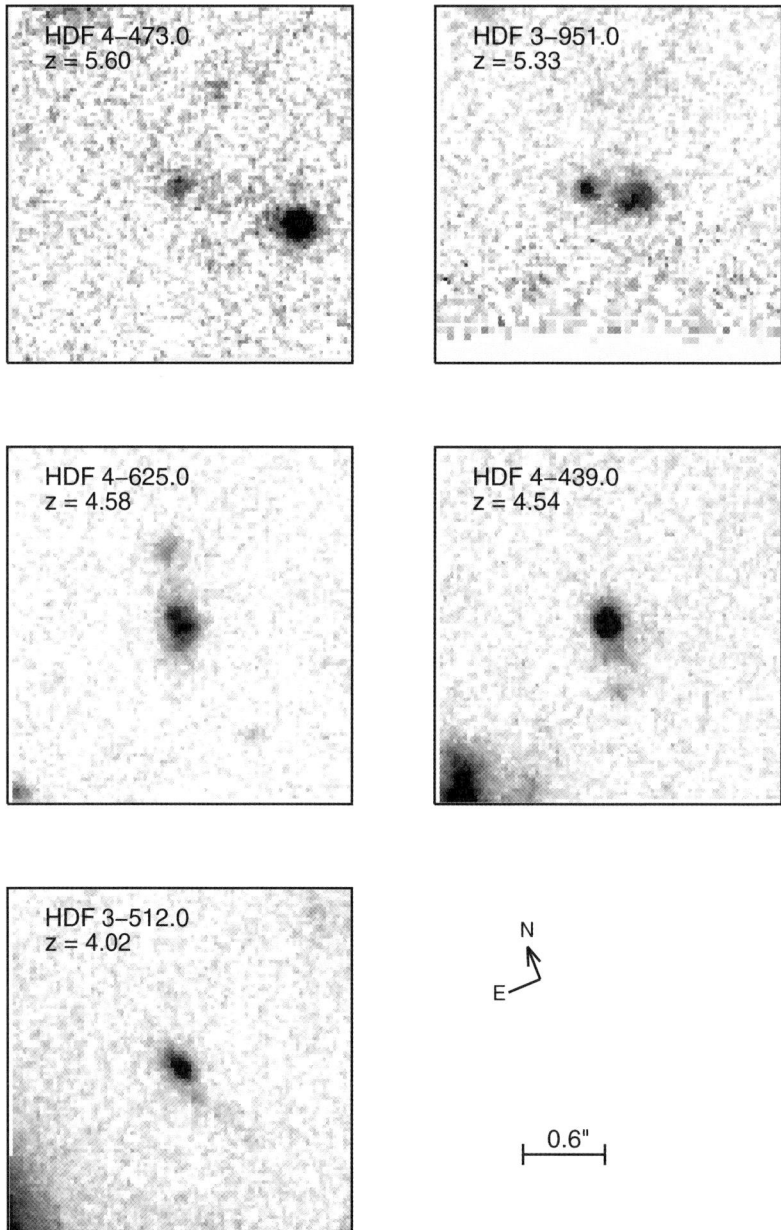

Fig. 8.1. HST images of five spectroscopically confirmed galaxies located in the HDF(N). Note the distortions, small tails, and multiple central components - presumably due to mergers. Overall the galaxies are obviously quite small at this stage of their evolution. From Stern & Spinrad (1999)

One of this review's topics, directly or indirectly stated, is just how early in cosmic epoch (parameterized by redshift) we can study individual galaxies or their 'pre-galactic' fragments. There is only a short time interval between the early epochs beyond $z = 3$ (see Fig. 8.2). How can the galaxies evolve so quickly?

The historical view of our empirical and theoretical march outward toward higher redshift has shown a fairly rapid expansion. By 1976 a few radio galaxies had been located and studied at $z > 0.5$. The $z = 1.0$ threshold (for galaxies) was crossed in 1981. Of course quasars and QSOs had been actively observed and known earlier at large distances - redshifts in the 1960s and 1970s taking us to $z = 2.01$ (Schmidt 1965) and then 2.88, and then to $z = 3.5$ (OQ 172; Baldwin et al., 1974). Finally, $z = 4$ for QSOs was surpassed by the Palomar two-color-based searches (Schneider, Schmidt & Gunn 1991),

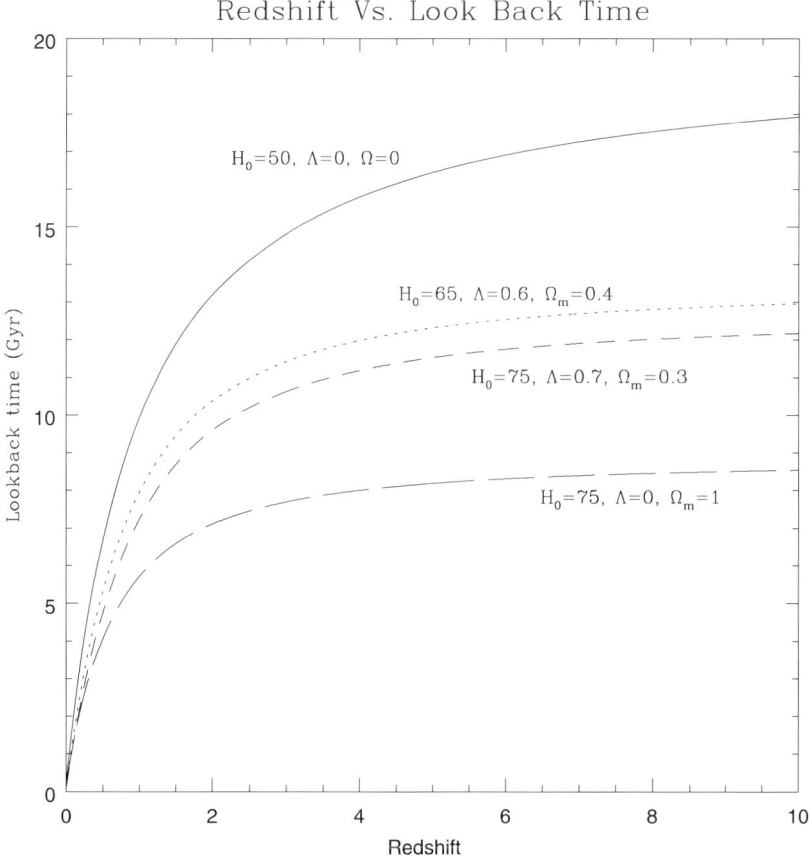

Fig. 8.2. A plot of look back time (in Gyrs) versus redshift for three cosmological models. Most might now prefer the short-dashed curve. Note that at high z ($z \gtrsim 3$) the time intervals become quite short. Figure by Curtis Manning

and searches for Lyα on low-resolution grism spectra (Osmer 1999) were equally successful. Almost all the recent stages of the 'QSO-z race' have emphasized red-IR photometry and unusual colors, since the $z \sim 5$ QSOs are heavily depressed by the Lyα forest of the IGM (see Fan et al. 2001). The largest published QSO redshift to date is $z = 6.28$ (Fan et al., 2002; Pentericci et al., 2002a).

Now we are witness to the era of a friendly race toward higher and record-breaking galaxy redshifts. The current limit for galaxies, which we shall detail later in this publication, is near $z = 6.5$. Is this redshift close to the end of the 'dark ages', where re-ionization by massive stars and/or early QSOs play as vital sources of ionizing radiation? We return to this topic, with empirical evidence, toward the conclusion of this review.

8.2 Some Issues in the Contemporary Theory of Early Galaxy Evolution

Over the past three or four years, the thoughts of theorists have narrowed on the birth and evolution of galaxies - including dark matter halos, plus the baryons we observe more directly. These adventurous researchers have bi-modally attacked the problems with a pair of model types. Most contemporary modeling assumes, *ab initio*, the Lambda Cold Dark Matter cosmology (LCDM).

Following Weinberg et al., (1999), we note that the current (broad) theory of galaxy formation and early evolution follows White & Rees (1978) and their 'successors' - gravitational collapse of a dark matter halo, gas falling into the potential well so defined, and then gas astrophysics (cooling, contracting, and eventually forming stars in a dense baryonic core). Now we often add inflationary cosmological parameters and thus demand $\Omega_m + \Omega_\Lambda = 1$.

The 'technology' for modeling often takes one of two paths. The first is hierarchical numerical simulations (with a realistic treatment of the collapse) including additional gas-phase physics and plenty of computational effort to cover the wide size range of non-spherical assemblies (the 'roots' of the assembly 'tree') that appear.

The second tool, deemed the semi-analytical approach, assumes again LCDM haloes. The proto-galaxies contract within, and then we find small sub-galactic systems (or fragments?) with the specific physically-motivated 'stories' given by the strengths of their starburst mergers. The mergers obviously increase the model masses, and also modify the relative numbers of luminous stars and the amount of residual gas. Even before that step, the semi-analytic models utilize the Press-Schecter (Press & Schechter 1974) formalism to describe the number of haloes as a function of their mass. This approach allows conventional and mature use of population synthesis and even chemical evolution schemes in conjunction with the mergers demanded to build up galaxies of reasonable mass with moderate star-formation rates.

One of the strengths of the direct numerical simulations is to utilize the non-spherical distribution of dark matter and baryons to produce a more realistic treatment of the model's gravitation. Then the more 'astrophysical' computations can proceed; Weinberg et al. (1999) predict the surface densities of galaxies as a function of their star-formation rate (SFR) over a relevant range of redshifts.

The semi-analytic models (*cf.* Baugh et al.1999; Somerville & Primak 1999) have now been amplified to include a range of interesting physical processes, hopefully relevant to early galaxy evolution. For example, the central baryons and the outer dark matter (DM) halo interact to change the halo structure and foster further contraction of the model galaxy. Baugh et al. (1998) mention that the main constraining property of local galaxies they favor for comparison with semi-analytic modeling is the field galaxy luminosity function. The agreement is good; one can then easily visualize the effect of omitting or including various individual physical processes, like star-formation (SF) feedback.

The SFR in the early Universe (say, to $z = 3$ or 4) of these models is also well-matched by observations of the SFR per unit volume. (Madau et al. 1999).

Weinberg et al. (1999) also show the numerical simulation's cumulative distribution of galaxies (with the parameter = surface density/$\Box'$/unit z) as a function of their SF rate from $z = 10$ to $z = 0.5$. At the moderately large galaxian SFR = 10 $M_\odot$ yr^{-1} for $z = 5$, the predicted surface density of galaxies is nearly $5/\Box'$. This surface density is rather higher (by a factor of ~ 3) than observed by Spinrad and collaborators (although some of this observational statistic is derived from the Lyα - SFR correlation, which may be suspect). The best unpublished observational estimate for the SFR surface density at $z \sim 5$ is now $2 \pm 1/\Box'$. However, this surface density for Lyα emitters is uncertain because their continua are often very weak and thus not necessarily sampled consistently in terms of galaxy luminosity. The theoretical simulations and follow-up astrophysical scaling may, of course, be systematically over-efficient in, for example, converting cooling gas to massive star births.

The numerical simulations with LCDM may have one flaw: they overpredict the number of small galaxies near larger ones (which are countable) and thus the number of stars at low redshifts. We are not positive that a real problem exists; it may be that dark haloes with coupled non-stellar baryons (*e.g.*, high velocity clouds (Klypin et al., 1999) are being 'counted' as observable systems.

The potential problems of early galaxy evolution from the theoretical side may well change, increasing or decreasing as their confrontations with empirical 'facts' or new understandings go forward. The general outlines of the theory and relevant observations are probably fairly firm.

8.3 A Race for the Maximum Redshift

It is a very human tendency to climb a celestial mountain. So it stands for any race, including that of finding individual objects at greater and greater distances, abbreviated usually as at larger redshift, or 'bigger z' (where $1+z = \lambda_{observed}/\lambda_{emitted}$).

As Stern & Spinrad (1999) pointed out in their Table 1, there has been a fairly rapid increase in 'zmax' for galaxies; we went from $z = 0.20$ in 1956 to $z = 1$ by 1982, but then to $z = 5.3$ in 1998 and $z = 5.7$ in 1999. The record-breaking progress since 1998 has been due to observations of the strong Lyα (from rest $\lambda 1216$ Å) emission line, shifted to the visible and red by the Universal expansion. Over the past year the 'LALA' (Large Area Lyman Alpha survey) team (Rhoads et al.2003) have selected Lyα emitters to $z = 5.75$. They are currently taking images for the $z = 6.6$ airglow window. Also in 2002, Hu et al. have located a cluster-lensed galaxy at the outstanding redshift of $z = 6.56$! And as these pages are completed, a Subaru group has found a faint Lyα galaxy at $z = 6.578$.

Modern research on quasars (QSOs, to be more precise), has also progressed; Osmer (1999) reviewed the situation 3 years ago, with QSOs located up to $z = 5.0$. Since then, the Sloan Digital Sky Survey has successfully pushed QSO redshifts to and beyond $z = 6$! The key here is to obtain good red and near IR photometry, in particular looking for objects with very red (I-z) colors. The Sloan results are very current; Fan et al. (2002) found SDSS J103027.1 at $z = 6.28$, and a preprint on another Sloan QSO at $z \sim 6.43$ is just available as this section is being written. So the most distant QSO to date still trails the most distant, much fainter normal galaxy by a modest margin!

The distant QSOs are likely buried in a host galaxy which itself is well-hidden in the glare of the Active Galactic Nucleus (AGN). We now assume the presence of the underlying galaxy of stars and gas, in part confirmed indirectly by the normal abundances of the elements inferred from the emission lines in the QSO spectra.

8.4 The Identification of Very Distant Galaxies

How do we go about locating the faint and distant galaxies at the heart of our exploration and this review?

We found several successful (or partly successful) methods to locate the faint targets at high redshift: none are without 'flaws'. For example, some methods are weakened by 'contaminants', be they intrinsically faint M, L, or T dwarf stars in the galactic disk, or a mis-identified (longer wavelength, smaller redshift) emission line.

Following the theme in the Stern & Spinrad (1999) review, we shall discuss several of the more successful search techniques; initially we'll review the finding of distant galaxies utilizing non-optical wavelengths. Often these techniques turn out to be 'safe' and productive.

8.4.1 Radio-Loud Galaxies

Radio galaxies at high redshift are rare but interesting guides to the location of large, mature galaxies and correlated structures - sometimes actual (rich) clusters (van Breugel et al. 1999; Lilly & Longair 1984). For some specific cases, like 4C 41.17 ($z = 3.798$), and also radio sources resembling it, we note that steep radio spectral indices and moderate flux densities correlate with high redshift and great luminosity. Such objects are visible across much of the presently observable Universe.

The stronger radio galaxies, those with fluxes $S_{408} \geq 100$ mJy, tend to follow a good Hubble relationship in the observer's near-IR bands; that is, their (K,z) magnitude–redshift correlation is linear with only a moderate scatter.

This result shows that the powerful radio galaxies, E systems in morphological appearance, have a fairly strong resemblance to a luminous 'standard candle' (van Breugel et al. 1999; Best et al. 1999). The history of the steep radio spectral index 'angle' is reviewed by de Breuck et al. (2000). Going for the steeep radio spectral counterparts also tends to minimize the 'contamination' by quasars (radio spectral indices < -1.3).

We then may inquire: are all steep radio sources luminous galaxies and quasar candidates? The answer here is mainly negative; it is the medium strength (so as not to exceed some limiting intrinsic luminosity) steep spectrum sources, identified at long wavelengths in the optical and IR that have the greatest promise in pointing out very distant spectrographic targets. These may be radio-loud stellar systems at a large redshift, say $z \geq 4$.

Somewhat tangential to our central motivation, we note that at both small and large distances, radio galaxies possess some/many of the characteristics of giant E galaxies (or luminous cluster Es). Since these E galaxies here and now have a strong correlation amplitude at small separations, we can anticipate many of the distant radio Es to also have smaller companions - perhaps in a group population. These indications of early structure are going to be valuable; the recent paper of Venemans et al. (2002) illustrates a large (2Mpc) overdense region at a redshift $z = 4.1$ located 'around' the radio galaxy TNJ1338-1942. So the radio galaxy becomes a valuable marker in such a case. We note another, less well-documented case in the HDF(N) is currently being explored by Stern, Dey, Dawson and Spinrad. Here the redshift is even greater; the first observed galaxies have $z \simeq 5.2$. No radio source takes part in that overdensity region, however. Stern et al. (2003) show a group surrounding the radio galaxy MG0442+0202 at $z = 1.11$.

The record redshift for a radio galaxy is still $z = 5.19$ (van Breugel et al. 1999), with TNJ0924-2201. Several observing groups are concentrating on the identification of deep samples showing a steep spectrum, with the expectation that some are ultra-luminous and located at $z > 5$. These are rare systems; one problem in interpretation is that it should be a fairly slow process to 'build' a large and luminous galaxy. Perhaps it requires a cosmic interval in excess of a billion years to do so, either in the model described as a 'monolithic collapse' (Eggen, Lynden-Bell & Sandage 1962), or by the accumulation of smaller structures (Searle & Zinn 1978) - a hierarchical model. With the currently popular cosmology $[H_0 = 65, \Omega_\lambda = 0.7, \Omega_m = 0.3]$, the look-back interval between $z = 4$ and (an arbitrary) $z = 20$ is only ~ 1.2 Gyr (see Fig. 8.2 again). That might be sufficient time to build a large galaxy; the implication is then a SFR of ~ 80 $M_\odot yr^{-1}$. That is a rarely observed and atypically high SFR. So it is a clue that massive radio galaxies are unlikely to be found at $z > 5$. But the near-IR Hubble Diagram of the highest-z radio galaxies plotted by van Breugel et al. (1999) continues to suggest a continuity in galaxy luminosity which we may still extrapolate to stellar (and gaseous) mass similarities.

Under standard CDM-based models of galaxy evolution, we expect the giant elliptical galaxies, which are the hosts of today's radio galaxies, to form late (at $z \sim 1$) through a process of merging of smaller sub-units. Although these models seem to be consistent with what is known so far about field galaxy evolution (e.g. Barger et al. 1999), and indeed with observations of the hosts of the radio-quiet quasar population (Ridgway 2000), it is clear that radio galaxies are an exception. They seem to only show significant evolution at $z > 2$, and still appear to be luminous galaxies at $z \sim 3$ and perhaps beyond. One possible solution is that the most massive galaxies formed first in so-called anti–hierarchical baryonic collapse. In this model (Granato et al. 2001) the high baryon densities in the centers of the most massive dark matter haloes cause them to start forming stars early. Thus, the fate of simplistic theoretical analyses suggest the need for a sharper observational analysis.

8.4.2 Galaxies With Strong X-ray Emission (Hidden AGNs)

To date many new X-ray galaxies have been located, using modern X-ray satellites such as Chandra and XMM-Newton. However, there are few X-ray-selected very distant galaxies, or AGN. To my knowledge there is one at a redshift in excess of 5; it is #174 in Barger et al. (2002) at $z = 5.186$, in the Chandra Deep-Field, North. We will return to this galaxy a bit later. There are, however, a considerable number of QSOs and other clearly noticed AGN at $z > 4$ (cf. Brandt 2002). Why are we physically interested in X-ray galaxies, anyway? As Barger et al. (2001) affirm, X-ray surveys, especially at hard (2-7keV) energies, provide a direct indication of an AGN, presumably due to an ultra-massive black hole at the galaxy nucleus. At $\gtrsim 5$ keV, absorption will play less of an obscuring role than seen for some 'hidden' AGNs at optical frequencies and soft X-ray energies. Complete samples of hard X-ray energies are now possible with the Chandra X-ray Observatory; the $1''$ X-ray positions produce robust optical identifications of the counterparts. And about half of the sources can be identified with optically bright and 'quiet' galaxies; they are at small redshifts.

With the longest integrations (say, one mega-second integrations) we begin to locate the faint X-ray population. Some of these sources are quite distant, $z > 4$ (cf. Barger et al. 2002). Their survey of the Chandra Deep-Field, North (equivalent to the HDF(N)) yielded a fair number of more-distant X-ray identifications; Table 8.1, below, puts them in $\Delta z = 0.5$ bins, and includes both narrow and broad-line (AGN) X-ray sources.

A quick inspection of Tables 8.1 and 8.2, and Fig. 8.3 suggests no dramatic physical change in the co-moving density of X-ray emitting galaxies compared to all field galaxies.

The largest redshift in the securely-identified group we discuss is B174 at $z = 5.186$. This source is associated with a moderately faint optical identifi-

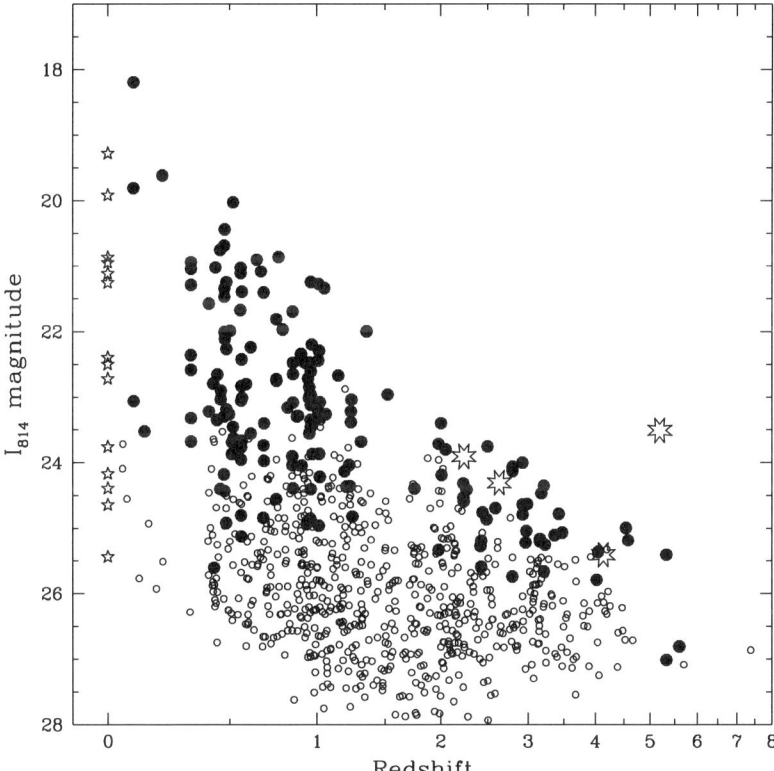

Fig. 8.3. I-band (I814) AB magnitudes versus redshift in the HDF(N), diagram by Mark Dickinson. The star-symbols at $z = 0$ are Galactic stars; filled circle symbols have spectroscopic redshifts, while the small open circles have photometric redshifts. When spectroscopic redshifts are found, the photometric point is removed. We note that three of the four distant X-ray sources (the large star-symbols) have rather normal (non-AGN?) magnitudes - like the field in general.

Table 8.1. Large Redshifts in the Chandra X-ray Sources (CDFN)

Δz	2.5 − 3.0	3.0 − 3.5	3.5 − 4.0	4.0 − 4.5	> 4.5
n	4	4	1	1	1

Table 8.2. Photometry of Narrow-Line Sources at High Redshift

Barger (2002) #	z	R	I	Note
174	5.186	24.5	23.1	Lyα, optically luminous
285	4.137	25.7	25.0	Lyα emission
287	2.638	24.4	23.9	weak Lyα
294	2.240	24.1	23.5	weak Lyα

cation - a bit too faint to classify morphologically. The near-IR I and z band photometry of this $z \sim 5.2$ source suggest its intrinsic luminosity may lie between that of luminous QSOs and an $\mathcal{L}^*$ galaxy; the AGN may be partly hidden, as the spectra of B174 does not display a broad component to its strong Lyα emission line. The other three X-ray galaxies at $z > 2$ appear to be residents in normal-luminosity host galaxies, based upon their photometry. The rough field galaxy correlation between I mag and the galaxy redshift can be seen in Fig. 8.3.

With the present generation of X-ray satellites and plausible integration times (Mega-secs), we cannot anticipate a large identification content of X-ray (AGN, or even 'starburst') galaxies beyond $z = 5$. Eventually I would speculate that some sources with fluxes in the 2-8 keV range below $10^{-16}\,\mathrm{erg\,s^{-1}\,cm^{-2}}$ may yield a few very large redshift objects.

8.4.3 Dusty Sub-mm (IR) Galaxies

In recent years it has become evident that a modest number of fairly high redshift galaxies ($z \gtrsim 2$) are most readily recognized as unusual at far-IR or sub-mm wavelengths. Our Earth's atmosphere is a substantial barrier to sub-mm research on galaxies likely to be both very dusty and also have a rapid pace of star formation. That recipe augurs for reddened high IR emission, which we can most readily discover at sub-mm wavelengths. It follows that the most-secure continuum detections are at a wavelength of 850 μm (where receivers are fairly efficient, and our atmosphere fairly transparent). The detection system of choice at the moment is called SCUBA, for Submillimeter Common User Bolometer Array. This camera, utilized on the JCMT (James Clerk Maxwell Telescope), has enabled, for the first time, deep and relatively unbiased surveys which may identify the distant dusty galaxies (and/or AGN).

It is important to clarify which galaxies (or AGN) radiate so profusely at IR and sub-mm wavelengths. They *may* be largely responsible for the Far-IR extragalactic background. With the presently available redshifts for securely identified IR/sub-mm galaxies, their integrated energy density may be quite comparable to the integrated optical (emitted UV galaxy light measured in the HDFs) energy (*e.g.*, Genzel et al. 2002).

To deal with this global question, and also to understand the limit of SFR in a huge starburst, reddened or not, the crying need is a reliable set of redshifts.

Blain et al. (1999) comment on the reason why many galaxies detected in the sub-mm spectral window are likely to be at high redshift. This is because the long wavelength side of the canonical sub-mm source spectrum has a very steep slope (cf. Blain et al. 2002). This steep long-wave side leads to a substantial negative K-correction. That is, the observer's band (850 μm) benefits from a larger redshift moving the emitted and then redshifted peak distribution into that atmospheric window. This effect compensates for the usual geometric dimming of increasing luminosity-distance at higher z. Thus comparing the sub-mm (850 μm) flux with the VLA radio flux (say, near 1.4 GHz) can yield approximate redshifts without an optical spectrum. But they are not individually robust.

Another method of deriving a more precise redshift for a sub-mm galaxy detection is to make good use of the fact that dusty systems occasionally also show strong molecular lines of CO in emission. The transitions in CO are (3-2) or (4-3) for the redshift domain of $z \sim 2.6$-2.8 (Frayer et al. 1998). But only a small minority have yielded CO molecular redshifts to date.

Very recently Chapman et al. (2003) have succeeded in obtaining good numbers of optical spectroscopic redshifts for sub-mm galaxies and AGN with precise radio positions. 16 redshifts of quality were obtained; probably one is a quasar. A few others may have some weaker AGN signal. The median redshift for the galaxies is $z = 2.4$, with a maximum redshift of $z = 3.699$. Thus one must extrapolate the 850 μm fluxes down to 1-2mJy in anticipation of future achievements in the $z \geq 5$ domain for sub-mm galaxies. That will surely require new hardware.

One sort of instrument planned for the near future is the APEX antenna (the Atacama [Chile] Pathfinder Experiment). It is a planned 12-m diameter sub-mm telescope at a high, dry site in northern Chile.

Surveys with the APEX should go deeper than the present SCUBA system. And that will be just a taste of what is to come with ALMA (the Atacama Large Millimeter Array). ALMA will be the mm/sub-mm counterpart of the VLT with 64 times the collecting area of APEX! It should make possible IR galaxy detections 100 times fainter than we now do with SCUBA and with good spatial acuity. This great array should lead to many redshifts with molecular CO lines and the [CII]158 μm line.

We end this section with an astrophysical speculation: with the Chapman et al. (2003) data we suggest a relatively high space density of very luminous and distant sub-mm ($z > 2$) galaxies. They may be 1000 times the density of similarly IR-luminous local starbursts found here and now. Hence the detailed study of a few of the powerful IR galaxies will tell us much about young galaxy SF and dust interactions.

8.4.4 Gamma-Ray Bursters

A new and exciting demonstration of extragalactic 'power' has recently emerged with the realization that Gamma-Ray-Bursters (GRBs) are apparently the most powerful cosmic explosions; observing their optical or radio afterglows can give us an indirect glimpse of a distant host galaxy. (See also Chap. 12 by G. Vedrenne and J.-L. Atteia in this volume). Not all bursters are successfully tracked for days or weeks after the outburst, but a reasonable fraction do point to distant ($z \gtrsim 1$) starburst galaxy hosts. So, for this review we note that occasional luminous afterglows may signal the locations of star-forming young galaxies at $z > 4$.

The detailed physics of the situation is unclear, but there are now believable scenarios suggesting that the GRBs originate from the collapse of a massive star or even a stellar merger. So sites of active SF may be one of the 'usual suspects', much as Type II SNe may be sited in young-star-rich locations. With the improved ability to locate GRBs we do find several opportunities annually to follow the afterglows as they decay; occasionally a redshift from an afterglow spectrum rich in UV interstellar lines (shifted to the visible) is obtained. The highest conventional spectroscopic redshift measured to date is $z = 3.42$ (Kulkarni et al. 1998).

Because many GRBs are very luminous (for a short time interval) we note that the possibility exists to derive 'photo-z's' or obtain low-resolution spectra of even more distant GRBs - perhaps with a little help from their galaxy hosts. Indeed Andersen et al. (2000) suggest a GRB at $z \sim 4.5$ from the afterglow's broad-band colors. At higher redshifts we will need photometry and/or spectroscopy in the near-IR. The J-band at $\lambda \sim 1.2$ μm will take the strong spectral discontinuity anticipated at Lyα (1216 Å, rest) to $z \sim 9$! Of course our present abilities to obtain good S/N infrared spectra would be taxed by all but the earliest bright GRB afterglows; spectroscopy in the first minutes may be needed!

8.4.5 Optical Selections of Distant Galaxies: 'Photo-z's' and Lyα Emission Lines

The case for the use of photometric redshifts – that is, redshifts based upon colors in 2 or (likely) more wave bands – has gradually strengthened since the mid-1990s. Most critically, we now expect fair precision from photometric redshifts and few catastrophic failures.

Stern & Spinrad (1999) compare spectroscopic and photometric redshifts in the HDF. The photometric redshifts are from the Stony Brook group

(Fernandez-Soto, Lanzetta & Yahil 1999), and are determined by fitting the observed galaxy colors (long wavelengths only for really distant candidates) with redshifted spectral templates. These templates may be empirical, synthetic, or a hybrid. A second approach (Connolly et al. 1995) is purely empirical - having already a relationship between previously-observed galaxy redshifts and the observed total magnitudes (m) with color information (C) to boot. Then a derived redshift can be found from the multi-dimensional (m,C) pairs, used for training. More detail on these 'template fits' can be found in the Stern and Spinrad review. Comparisons between photometric and spectroscopic determination in the HDF yield residuals typically around 0.1 for Δz at almost all redshifts.

Naturally the most important usage of such photometric redshifts is at very faint levels ($m > 26.5$). These numerous faint galaxies are well beyond the capabilities of 10-m class telescopes for spectroscopic redshifts. The danger here is that galaxies marginally detected in the red-optical I,z bands and perhaps also in J, H, K [1.2, 1.6, 2.2μm] can feign very large redshifts if their signal is just a noise incursion at I or z bands, slightly below the 1 μm observational limit of silicon-based CCDs. Since this topic is close to the kernel of this review, we note that Lanzetta et al. (1999) give some examples of faint, red photometric-z cases of difficult S/N. Their redshifts could exceed 6. Almost all of these ambiguous but potentially exciting cases have yet to be resolved. I speculate that better IR photometry (perhaps using the rejuvenated NICMOS camera on HST) would help in resolving that situation and perhaps suggest targets for future generations of near-IR spectrographs.

There is also a systematic problem at some level with color/redshift degeneracies; blue galaxies in general may show similar colors over a substantial intermediate z range. Prior information like the galaxy apparent magnitude can help decisively. This 'Bayesian' procedure is illustrated by Benitez & Broadhurst (1999) for the HDF(N).

My personal recent experience with 'I-drops' (implying a galaxy with only detectable flux at wavelengths above the I band, $\lambda \geq 8500$ Å at the red edge) is that many of the spectroscopic candidates (15 to 20 targets per slitmassk) are very difficult due to their faintness ($m \sim 25\text{-}26$ mag) at longer wavelengths. A few also turn out to be low-luminosity galactic stars; these late M, L, and T class dwarfs turn up rather frequently. Since many of the candidates come from ground-based imaging, their image structure is not a very discriminating way to separate stars from QSOs from galaxies.

Most of the I-drops show a marginally detected red-color continuum, and thus add little to our initial appraisal. It turns out that approximately a quarter of the I-drops do eventually yield a redshift; about a third of these with the continuum discontinuity at Lyα (λ_0 1216 - the Lyα 'forest'). Two-thirds of the spectroscopically detected I-drop systems (with eventual redshifts) have a noticeable to strong Lyα emission line. That usually yields an unambiguous redshift, as the reader can see with the illustrations in Weymann et al. (1998)

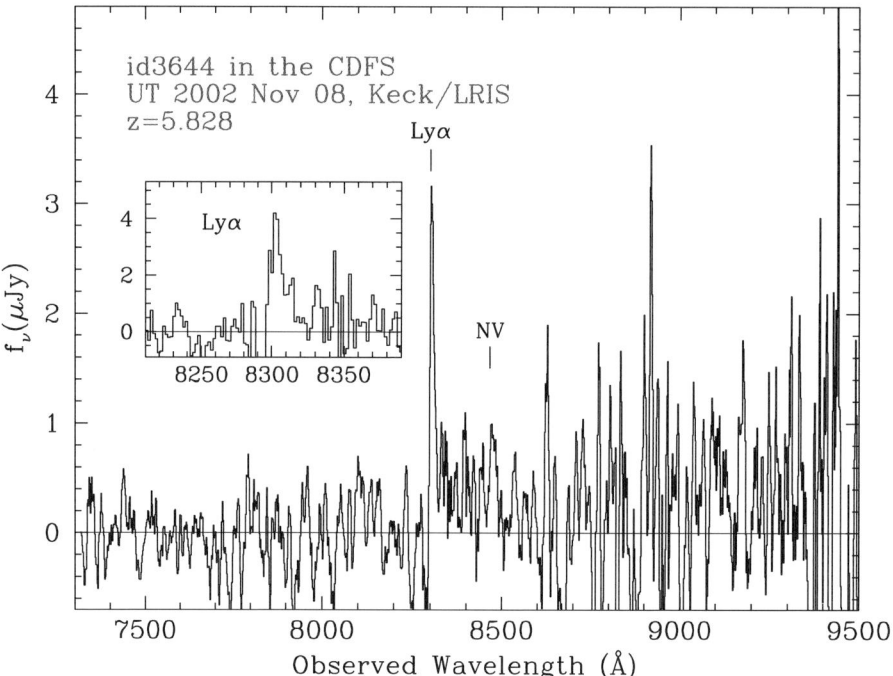

Fig. 8.4. A recent Keck spectrogram of a color-selected (I-drop) faint galaxy. The strong Lyα emission line indicates a redshift $z = 5.83$. Also note the continuum discontinuity. The 'spectral teams' were led by Spinrad and Filippenko, with reductions by Daniel Stern and Steve Dawson. This galaxy was originally selected by Mark Dickinson and the GOODS team.

and Fig. 8.4, here, by Spinrad, Stern, Dawson, Filippenko, and the GOODS team ($z = 5.83$).

The pairing of a red continuum color, a continuum discontinuity, and a fairly strong emission line usually signifies a robust Lyα redshift. The multiple-criteria spectroscopic technique has been successful to at least $z = 5.8$ and probably to $z = 6.57$. It should eventually be pushed to $z \sim 9$ with the Lyα line at (rest) 1216 Å, right in the middle of the conventional near-IR J-band. Right now that is too technically difficult.

As these pages were being written, two preprints crossed our desk. In the first, Lehnert & Bremer (2003) discovered 6 galaxies at $4.8 \leq z \leq 5.8$. These galaxies were selected as photometric 'R-drops'- that is, with little flux in the R-band and a flat spectrum at longer wavelengths. Follow-up spectroscopy with the VLT yielded accurate redshifts for these 6, with fairly strong Lyα emission. Their largest redshift was $z = 5.869$ (see Table 8.3 on page 176).

The second very timely contribution, by Kodaira et al. (2003) (a Subaru telescope team), used deep narrow-band near-IR images to locate potentially very distant Lyα galaxies. The group also obtained a few spectra which lead to two fairly certain identifications. One line, with a symmetric line shape, is assumed to be Lyα and in the other case it appears to be satisfactorily asymmetric, hence reliably Lyα (see § 8.4.6 for discussion of this point). The best spectrum is of SDFJ 132418.3 at $z = 6.578$. That would make this Lyα galaxy the largest redshift of any individual system measured to date. The redshift is only slightly greater than that of HCM 6 A ($z = 6.56$) by Hu et al. (2002), and Hu, Cowie & McMahon (2002).

These very contemporary detections of galaxies beyond the 'QSO-limit' of $z = 6.4$ show us that UV emission from galaxies is still present at the 'tail' of the 'dark ages'. A future space-desideratum will be the galaxy morphology *in the Lyα line*. We are interested in any extended neutral gas about the galaxy – via the scattered Lyα emission from the central ionizing region (Haiman 2002 and references therein).

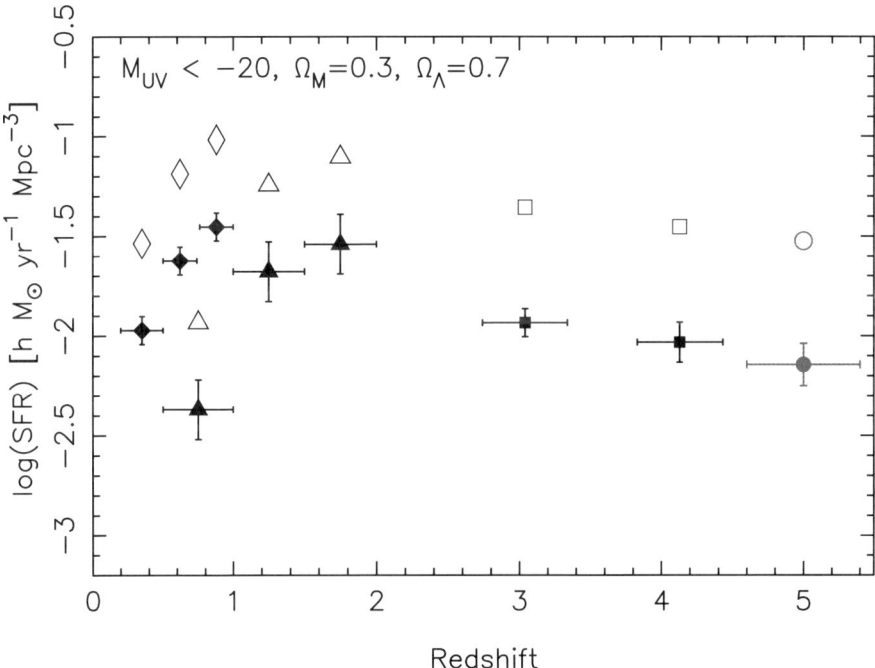

Fig. 8.5. Star-formation rate density as a function of redshift based on the UV-luminosity function with a magnitude limit $M_{UV} < -20$. Triangles and diamonds are from Connolly et al. (1997) and Lilly et al.(1996), respectively. Squares represent data from Steidel et al. (1999) at $\langle z \rangle \sim 3$ and and 4. The circle is the data of Iwata et al. (2003) at $z = 5$. Filled symbols indicate values without correction for dust extinction. Dust extinction was corrected following the prescription of Calzetti et al. (2000) with $E(B - V) = 0.15$ for all data points. Dust-corrected values are denoted by open symbols. Plot courtesy of I. Iwata.

When the luminosity function of Lyα emitters is extended to $z \sim 6.5$ (fainter galaxies have to be included) we should be able to extend the SFR density to that great distance. A sample of the near-constancy of the SFR density from $z \sim 2$ to $z \sim 5$ is illustrated in Fig. 8.5 (from Iwata et al. 2003). The galaxies going into the computation of the SFR density are photometrically selected, using the top of the UV-luminosity function ($M_{UV} - 5\log h < -20$). Interestingly, an attempt by D. Stern and the author to utilize serendipitously discovered Lyα emitters at $z \sim 5$ yields a SFR density slightly higher than that of the $z \sim 5$ Iwata point in Fig. 8.5 (with considerable uncertainty). We view this as a possible coincidence, as these two methodologies may be sampling different populations. It is somewhat surprising that the relatively slight decline of the SFR density, noted by Iwata et al. (2003) should be maintained to $z \sim 5$. At that redshift the detected objects are effectively sub-galactic in size and probably rather modest in mass. At least temporarily, their M/L ratios must be quite low. Will that be true of most small sub-galactic systems?

8.4.6 Details on the Lyα Emission Line in Very Distant Galaxies

The classic proposal by Partridge & Peebles (1967) that the Lyα emission line might carry a fair fraction of the escaping bolometric luminosity of a young-star-rich galaxy is now testable. The review by Pritchet (1994) is also strongly recommended. Of course these early predictions did not reflect the possible presence of dust. Since the 1990s various searches have been initiated for Lyα-emitting galaxies at large redshifts. Initially all of these searches led to negative results (eg. Thompson & Djorgovski 1995).

However, deeper photometric and spectroscopic searches of the last 6-7 years have yielded a modest number of 'safe' Lyα emitters - often (at the largest z's) the line being the only measurable spectral feature. The peak flux from a distant Lyα emission line galaxy can often exceed the (redward) continuum level by a factor greater than 10! Of course the line from a faint system still has to compete with the strong telluric sky emission bands of OH and O_2. Space-spectra won't deal with such a bright near-IR sky, and that will be advantageous.

Successful Lyα searches include Cowie et al. (1998); Hu et al. (1998), Pascarelle et al. (1998), Hu et al. (1999), Steidel et al. (2000), Kudritzki et al. (2000), Fynbo, Möller, and Thomsen, (2001).

There are three modes of Lyα detection used with success in the past few years. They are narrow-band photometric excesses at fixed wavelengths (redshifts), a Lyα forest (Lyman breaks in the continua) plus emission at the line, and serendipitous or fortuitous detections on multi-slit spectrograms. The issues we may face for each/all of the sub-types include the emission line strength and shape, the luminosity function of Lyα emitters (and their surface densities), the effect of widespread neutral gas and dust, and the termination of the 'dark ages' before or during the re-ionization epoch. Many of these topics have been addressed recently by Stern & Spinrad (1999); Rhoads et al. (2003); Ellis et al. (2001); Hu et al. (1999); Hu et al. (2002a), and in a predictive manner by Stiavelli (2003).

I suggest a few specific points where new observations and interpretations may be of substantial interest. For example, we'd like to confirm or deny that strong emission line Lyα galaxies ($z \geq 4$) obey the same luminosity function distribution as do photometrically selected Lyman break systems at $z = 3$ and $z = 4$ (cf. . Steidel et al. 1999; Giavalisco 2002).

The difficulty in a present-sample comparison between Lyman break galaxies and Lyα emitters is that (at high luminosities, at least) only a modest fraction of Lyman break (continuum selected) galaxies have strong Lyα emission lines ($W_0 > 20$ Å, say). Among the Lyα-emitting systems (narrow-band or serendipitous detections) many candidates have very faint continua and would be missed in normal broad-band photometry. This latter bias is stressed by Fynbo et al. (2001). Indeed, Rhoads et al. (2003) found that if they summarized the line/continuum ratio in Lyα galaxies, the equivalent widths occasionally 'rose' to $W_\lambda^0 \geq 1000$ Å, but more frequently to 190 Å. 60% of the Lyα emitters studied by Malhotra & Rhoads (2002) had observed equivalent widths (hereafter EW) > 240 Å. For Ly-break systems, Shapley et al. (2001) find their 60th-percentile line to be a marginally-detectable 20 Å EW. The Shapley galaxies are at a slightly lower redshift; that difference is not critical.

If the above trend of lower-continuum-luminosity galaxies ($z > 4$) having stronger Lyα-emission were to continue, we might diagnose this systematic as a trend toward lower metallicities for lower masses. But there are other possibilities; the Lyα-emission line may as easily depend upon physical outflows (galactic winds), which in turn could have some total mass-dependence (or merger timing).

To get some idea as to the evolution of the luminosity function of young galaxies, we can compare the surface densities of distant galaxies. Pritchet (1994) made a first approximation to this. We utilize the Steidel et al. (1999) luminosity function zero point, and the 'predictions' by Lanzetta et al. (1999) and Stern & Spinrad (1999) for a constant (with z) luminosity function. The cumulative surface density of identified $z \geq 4.5$ galaxies in the HDF(N) is about $1.5/\square'$. These galaxies constitute a sample of continuum galaxies (photo-zs) and emission line galaxies with $I_{814} \leq 26.5$. This is very close to the 'prediction' of the Lanzetta (unevolved) surface density (also see Ouchi et al. 2002).

The Lanzetta (1999) surface density curves do suggest a drop in the faint galaxy surface densities for the extreme case, $z \geq 6$; that is not surprising at about $I_{814} = 26$. Still at slightly fainter magnitude levels a measure of the $z \geq 6.0$ density by broad-band/narrow-band photometry may be a viable check on the luminosity function zero point and its shape (Lehnert & Bremer 2003).

What is the best physical interpretation of the very large EWs of Lyα often measured for galaxies at $z > 3$?

The Lyα-emitting galaxies with line EW in excess of 200 Å (rest-frame) (Malhotra & Rhoads 2002) are difficult to explain with a conventional O-B star mass function and ionizing spectra that are similar to those anticipated in extant solar-abundance models. The models rarely (and temporally) exhibit $W_\lambda^0 \geq 150$ Å (e.g., Charlot & Fall 1993). To decrease the observed Lyα EW would be easy; as the dominant resonance line it is scattered frequently, and the resulting 'random spatial walk' at the center of this line, coupled by small amounts of dust, can easily and drastically reduce the emission measure. It would, of course, also depend on the geometry.

To obtain a higher EW and/or higher flux in Lyα, one can call upon three scenarios:

(a) A 'tilted' mass function, with more O stars than found in local HII regions, as an *ad hoc* premise.

(b) We can also reduce the heavy element abundances in our models, and this allows an increase in the number of ionizing photons per O star. A recent paper by Schaerer (2003) considers the temporal evolution of the Lyα line from model stellar populations ranging down from solar metal-abundances to very low metallicities (below the abundance level of the most metal-poor stars and gas in relatively nearby star-forming systems). We amplify this discussion below.

(c) Finally, sometimes a strong Lyα emission line is the signature of an AGN. However, 'real' AGN spectra, from QSOs down to modest-luminosity accretions, usually produce a broader Lyα emission line ($\Delta v \geq 1000$ km s^{-1}) than seen in normal galaxies ($\Delta v \sim 500$ km s^{-1}). They usually, but not always, also show C IV (moderately broad) 1549 Å. So most of the narrow-line Lyα galaxies must have a line powered by the UV flux from OB stars. This is confirmed by the lack of hard X-ray flux in LALA galaxies at $z \simeq 4.5$ (Malhotra et al. 2003), indicating they are not obscured AGN.

The previously-mentioned Schaerer paper (Schaerer 2003) predicts EW of $\sim$ 240-350 Å for metallicities down to $Z = 4 \times 10^{-4}$ (down from solar by a factor of $\sim$ 50 times). Stiavelli (2003) shows even larger EW for Lyα in metal-poor OB stars. Conceivably the initial stellar mass function (IMF) could also vary and be itself slanted toward higher masses because of the lower abundances. So the pairing of low abundance and a structure favoring massive O stars might allow EW to match most of the Lyα galaxies selected by Malhotra & Rhoads (2002) and by Rhoads et al. (2003). An almost-practical spectroscopic test of this idea can be made by examining the UV HeII transition at $\lambda_0 1640$ Å. This line is much weaker than Lyα in star-forming populations - with EW ~ 5 Å anticipated at low abundances of the metals. At higher abundances (near solar) it will be even weaker. Thus higher S/N spectrograms will be required in practice to use this He II feature in Lyα 'test galaxies'.

The shape of the Lyα emission line in distant star-forming galaxies is peculiar and may turn out to be an interesting guide to the circumgalactic medium as well as to galaxian winds or sporadic outflows.

The asymmetry of the Lyα line has been noted by Kunth et al. (1998) and Pettini et al. (2001); it is also mentioned by Stern & Spinrad (1999). We have utilized the broad red wing of the Lyα line and its sharp ISM/IGM cutoff on the blue side as a secondary criterion for assuming a single strong emission line is to be identified as Lyα. This is opposed to the profile of the [O II]3727 doublet - unresolved in most lower-spectral-purity observations of faint objects. Recent work by E. Landes, S. Dawson, and the author has compared a spectal asymmetry index (a lambda-space ratio) for ten strong Lyα emission lines; this particular index is small for a symmetric line and large for a red winged emission. Out of a sample of seven medium-resolution spectra of galaxies with a 'solid' [O II] identification ($z = 1.0$) the asymmetry index averages 0.9 ± 0.1, while the 10 bonafide Lyα galaxies, with $\langle z \rangle \approx 4$ display a larger range of index, from 1.0 to 2.3, with none less than unity. Seven of the Lyα systems are clearly asymmetric with a noticeable red wing (see Fig. 8.6).

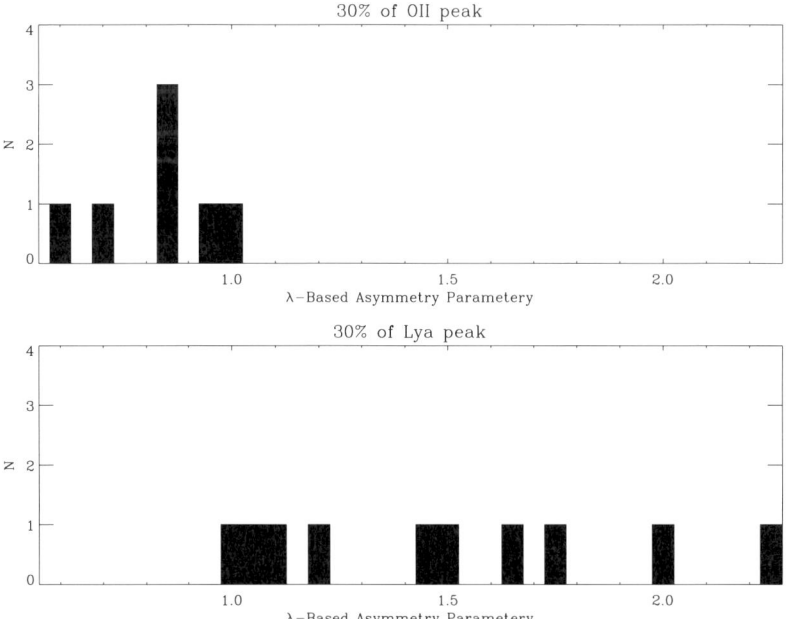

Fig. 8.6. The Lyα emission line asymmetry index, applied to [O II] emitters (upper panel) and to Lyα lines (lower panel). Ther line wavelength asymmetry is defined at 30% of the line peak; an index over unity implies a stronger red wing to the line profile. Most but not all of the strong Lyα line emitters show an asymmetric red wing, with an index ≥ 1.5. The Lyα galaxies range in redshift from $z = 3$ to $z = 5.3$. Figure and reductions by Emily Landes.

The Lyα line is usually steeply declining on its blue side; we'll soon come back to this observation. So deciding whether an emission line is [O II] at a modest z or Lyα at a large z, can often be helped by measuring the asymmetry. Of course a Lyα (bigger redshift) decision based upon a large line asymmetry index becomes a sufficient, but not necessary condition for claiming the Lyα identification.

The astrophysics behind the red wing of Lyα has been well expounded by Tenorio-Tagle et al. (1999), Ahn, Lee & Lee (2002), and Dawson et al. (2002). The scenario here is a mini-galaxy scale outflow of neutral and partly ionized matter; the blueward velocity component being absorbed by external and expanding neutral H gas between us and the outflow. The backscattered component can be sufficiently redshifted off of the receding wind, and hence avoid immediate absorption. This will impose a broadened red wing to the Lyα line.

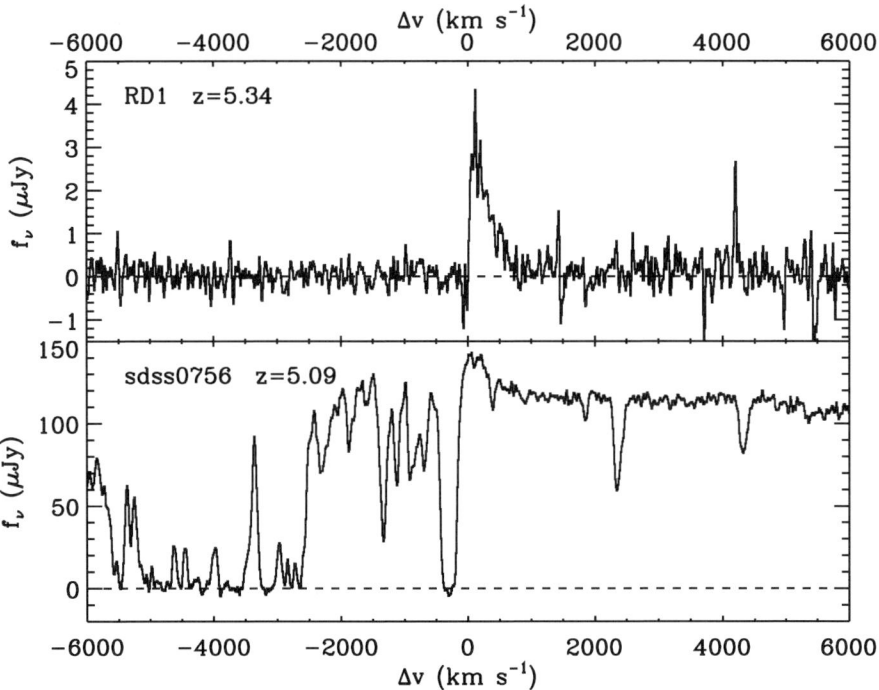

Fig. 8.7. The steepness of the ultraviolet side of the Lyα emission line in a QSO ($z = 5.09$) and a faint galaxy, RD1 ($z = 5.34$, Dey et al. 1998). The very sharp and rapid decline of the blue side in the distant galaxy may be indicative of nearby (surrounding?) neutral gas. On the other hand, the QSO presumably ionizes much of any circumgalactic H originally present (with a small Δv) Thus the QSO line and continua are detectable to $\Delta v_2 \simeq 2500$ km s^{-1}. Reductions and Figure by S. Dawson.

On the blue side of Lyα we have a rapid decrease in intensity, a very sharp cutoff to the galaxy emission line at a slightly smaller redshift. The actual galaxy systemic velocity is likely to be near but blueward of the line peak, rather than its bisector at about half of maximum intensity.

In any case the Lyα H absorption can take place in neutral circumgalactic gas, and in putative cluster gas, and also, at slightly lower redshift, neutral H clouds in the IGM - the well-studied Lyα forest.

One interesting semi-quantitative aspect of the blue side cutoff is the difference we have noticed between the blue edge of Lyα in QSO spectra and that of the normal distant galaxies, highlighted in this review (see Fig. 8.7). A new type of 'proximity effect' seems in place, in the sense that the galaxy Lyα profile on the short wavelength side is extremely steep, going from the line peak to near zero intensity in $\Delta v_1 = 100$ km s^{-1}, on our few echelle (higher spectral resolution) observations of the brightest distant systems (in their Lyα line). The profile on the blue side of the strong emission line in QSO spectra (also $z > 4$) is moderately steep, but has a typical $\Delta v_2 \approx 800$ km s^{-1}, but often > 1000 km s^{-1}.

Our interpretation of this systematic difference between UV-luminous QSOs and UV-fainter galaxies is straightforward. In proximity to the luminous ultraviolet radiation field of the QSOs H is very thoroughly ionized and thus doesn't absorb Lyα photons at small Δv. On the other hand, a galaxy's UV ionizing radiation may not escape (or fully escape - see Dawson et al. 2002). Thus the rapid decline on the blue side of Lyα may simply augur the existence of neutral gas in the circumgalactic environment near the galaxy. The effect may increase with redshift, but this is not yet well documented. This trend is potentially of interest in our present and future attempts to document the degree of IGM ionization near active objects and also on a diffuse, larger scale. Our coverage in redshift implies that we are looking back close to the re-ionization redshift, between $z = 6$ and $z = 20$, apparently.

8.4.7 Current Redshift Record Breakers With Lyα Emission or Absorption Breaks

In Table 8.3 we list published or otherwise secure 'record redshifts' for galaxies; most have prominent Lyα emission lines or at least a strong Lyα forest absorption.

We note that since 1999 astronomers have added at least 25 galaxies with $z \geq 5$. This is an impressive and useful score; however, a more physical analysis of several aspects of the pioneering effort is now an obvious and desired second approach. Also, the morphologies of the continuum and Lyα lines may provide useful information on the environs of very early galactic systems. The cut-off date for entries in Table 8.3 was 2003 February.

Table 8.3. Census of Galaxies Confirmed at $z \gtrsim 5$

z	Source	Reference	NB	LBG	ser	other	lens
6.578	SDFJ 132418.3	Kodaira et al.	x				
6.56	HCM 6A	Hu et al.	x				x
6.541	SDFJ 132415.7	Kodaira et al.	x				
5.869	BDF1:19	Lehnert and Bremer		x			
5.83	CDFS 5144	GOODS		x			
5.783	CDFS SBM03#3	Bunker et al. 2003		x			
5.746	LALA5 1-03	Rhoads et al.	x				
5.744	BDF1:10	Lehnert and Bremer		x			
5.74	SSA22-HCMI	Hu et al.	x				x
5.700	LALA5 1-06	Rhoads et al.	x				
5.69	LAE J1044-0130	Ajiki et al. 2002	x				
5.674	LALA5 1-5	Rhoads et al. 2002	x				
5.655	LAE J1044-0123	Taniguchi et al.	x				
5.649	BDF2:19	Lehnert and Bremer		x			
5.631	HDFF 36246-1511	Dawson et al. 2001			x		
5.621	Lynx R-drop	Stern et al. in prep		x			x
5.60	HDF 4-473	Weymann et al. 1998		x			
5.576	Abell 2218 lens	Ellis et al. 2001			x		x
5.46	NDFWS R-drop	Dey et al., in prep.		x			
5.34	HDF 3-951.0	Spinrad et al. 1998		x			
5.34	RD1	Dey et al. 1998			x		
5.190	HDFF ES1	Dawson et al. 2001			x		
5.19	TN J0924-2201	van Breugel et al. 1999				x	
5.19	ES1	Dawson et al. 2001			x		
5.186	HDFF Chandra source	Barger et al. 2002				x	
5.12	A1689 lens	Frye et al. 2002		x			x
5.056	BDF1:26	Lehnert et al. 2002		x			
5.018	BDF1:18	Lehnert et al. 2002		x			
4.99	Cetus R-drop	Stern et al. in prep		x			

Notes on the initial discovery techniques – NB = narrow-band selected; LBG = continuum Lyman-break/Lyman-forest break selected; ser = serendipitously identified; other = selected in other manner (e.g., radio-selected, X-ray selected); lens = known gravitational lens.

8.5 The Future

Wide-field narrow-band and broad-band imaging with large ground-based telescopes have considerable promise. Narrow-band and broad-spectral-band studies of the sky areas already earmarked for multi-wave observation is one useful approach. It is already been successful in the Hubble Deep Fields. Such imaging photometry has already turned up distant galaxies, especially at $z = 5.7$ and $z = 6.6$ (airglow windows for narrow-band studies). We know of several groups planning to search for Lyα emitters at the highest redshifts available to CCD detectors ($\lambda \approx 9200$ Å; $z_\alpha = 6.6$). A more ambitious plan would be to utilize IR detectors at the best (OH-band-free) sky windows in J band ($\lambda \approx 12,000$ Å; $z = 9$). Exploration of the interval $6.6 \leq z \leq 9$ should bring us to the edge of the re-ionization epoch where the first stars and quasars began to ionize (again) the haloes around collections of dark matter and baryons. A schematic cartoon (Pentericci et al. 2002b) is shown by Loeb & Barkana (2001). Is it realistic? We'll hope that very distant galaxy images and spectra will tell us about very early star formation at the end of the long 'dark age'. At this time it is uncertain as to whether the first luminous and ionizing objects were star-forming galaxies! But something or some process began star-formation through the darkness and led to the formation of young stars and young galaxies. We may soon barely detect these faint 'first galaxies' with our telescopes and intellects.

Acknowledgements

I would like to thank Curtis Manning, Steve Dawson, Arjun Dey, Mark Dickinson, Ikuru Iwata, Emily Landes, Scott Chapman, and Daniel Stern for their help with the science of faint galaxies and with this manuscript. I also wish to acknowledge the support from NSF Grant AST–0097163.

References

Adelberger KL et al. (2003) Galaxies and Intergalactic Matter at Redshift $z \sim 3$: Overview. Ap. J. **584** 45
Ahn S-H, Lee, H-W, and Lee, H (2002) P-Cygni Type Lyα from Starburst Galaxies. Ap. J. **567** 922
Ajiki M et al. (2002) A New High-Redshift Lyα Emitter: Possible Superwind Galaxy at $z = 5.69$. Ap. J. **576** L25
Andersen MI et al. (2000) VLT Identification of the Optical Afterglow of the Gamma-Ray Burst GRB 000131 at $z = 4.50$ Astron. and Ap. **364** 54
Baldwin J et al. (1974) An Analysis of the Spectrum of the Large-Redshift Quasi-Stellar Object OQ 172 Ap. J. **193** 513
Barger AJ et al. (2001) Supermassive Black Hole Accretion History Inferred From a Large Sample Of Chandra Hard X-Ray Sources Astron. J. **122** 2177
Barger AJ et al. (2002) X-Ray, Optical, and Infrared Imaging and Spectral Properties of the 1 Ms Chandra Deep Field North Sources Astron. J. **124** 1839
Baugh CM et al. (1998) The Epoch of Galaxy Formation Ap. J. **498** 504
Baugh CM et al. (1999) in Photometric Redshifts and High Redshift Galaxies ed R Weyman, L Storrie-Lombardi, M Sawicki and R Brunner ASP Conf. Ser. **191** p. 353
Benitez N, and Broadhurst T (1999) High-z Galaxies Seen Through Cluster Lenses, in The Hy Redshift Universe ed AJ Bunker and W van Breugel ASP Conf. Ser. **193** p. 509
Best P et al. (1999) A 98 per cent Spectroscopically Complete Sample of the Most Powerful Equatorial Radio Sources at 408MHz MNRAS **310** 223

Blain AW et al. (1999) Submillimeter-Selected Galaxies, in The Hy Redshift Universe ed AJ Bunker and W van Breugel ASP Conf. Ser. **193** p. 425

Blain AW et al. (2002) The 60μm Extragalactic Background Radiation Intensity, Dust-Enshrouded Active Galactic Nuclei and the Assembly of Groups and Clusters of Galaxies MNRAS **333** 222

Brandt WN (2001) The Chandra Deep Field North Survey. V. 1 Ms Source Catalogs Astron. J. **122** 2810

Brandt WN (2002) The Chandra Deep Field-North Survey. XI. X-Ray Emission from Luminous Infrared Starburst Galaxies Ap. J. **568** 85

Bunker AJ et al. (2003) A Star-Forming Galaxy at $z = 5.78$ in the Chandra Deep Field South MNRAS **342** L47

Calzetti D et al. (2000) The Dust Content and Opacity of Actively Star-forming Galaxies Ap. J. **533** 682

Chapman SC et al. (2003) Uncovering the Redshift Distribution of Submillimetre Galaxies Nature **422** 695

Charlot S and Fall S (1993) Lyman-α Emission from Galaxies Ap. J. **415** 580

Connolly AJ et al. (1995) Spectral Classification of Galaxies: an Orthogonal Approach Astron. J. **110** 1071

Dawson S et al. (2002) A Galactic Wind at $z = 5.190$ Ap. J. **570** 92

de Breuck C et al. (2000) A Sample of 669 Ultra Steep Spectrum Radio Sources to Find High Redshift Radio Galaxies Astron. Ap. Sup. **143** 303

Dey A et al. (1998) A Galaxy at $z = 5.34$ Ap. J. **498** L93

Ellis R et al. (2001) A Faint Star-forming System Viewed through the Lensing Cluster Abell 2218: First Light at $\simeq 5.6$? Ap. J. **560** L119

Eggen OJ, Lynden-Bell D and Sandage AR (1962) Evidence from the Motions of Old Stars that the Galaxy Collapsed Ap. J. **136** 748

Fan X et al. (2001) A Survey of $z > 5.8$ Quasars in the Sloan Digital Sky Survey. I. Discovery of Three New Quasars and the Spatial Density of Luminous Quasars at $z \sim 6$ Astron. J. **122** 2833

Fan X et al. (2002) Evolution of the Ionizing Background and the Epoch of Reionization from the Spectra of $z \sim 6$ Quasars Astron. J. **123** 1247

Fernandez-Soto A, Lanzetta K and Yahil A (1999) A New Catalog of Photometric Redshifts in the Hubble Deep Field Ap. J. **513** 34

Frayer D et al. (1998) Molecular Gas in the $z = 2.8$ Submillimeter Galaxy SMM 02399-0136 Ap. J. **506** L7

Frye B et al. (2002) Spectral Evidence for Widespread Galaxy Outflows at $z > 4$ Ap. J. **568** 558

Fynbo J Möller P, and Thomsen B (2001) Probing the Faint End of the Galaxy Luminosity Function at $z = 3$ with Lyα Emission Astron. and Ap. **374** 443

Fynbo J et al. (2002) Deep Lyα Imaging of Two $z = 2.04$ GRB Host Galaxy Fields Astron. and Ap. **388** 425

Genzel R et al (2003) patially Resolved Millimeter Interferometry of SMM J02399-0136: A Very Massive Galaxy at $z = 2.8$ Ap. J. **584** 633

Giavalisco M (2002) Lyman-Break Galaxies Ann. Revs. Astron. and Ap. **40** 579

Granato GL et al. (2001) Joint Formation of QSOs and Spheroids: QSOs as Clocks of Star Formation in Spheroids MNRAS **324** 757

Haiman Z (2002) The Detectability of High-Redshift Lyα Emission Lines prior to the Reionization of the Universe Ap. J. **576** L1

Hu E, McMahon RG and Cowie L (1999) An Extremely Luminous Galaxy at $z = 5.74$ Ap. J. **522** 9

Hu E et al. (2002) A Redshift $z = 6.56$ Galaxy behind the Cluster Abell 370 Ap. J. (L) **568** L75

Hu EM, Cowie LL and McMahon RG (2002) ERRATUM: A Redshift $z = 6.56$ Galaxy behind the Cluster Abell 370 Ap. J. (L) **576** L99

Iwata I et al. (2002) Lyman Break Galaxies at z=5 around the Hubble Deep Field in 8th Asian-Pacific Regional Meeting 259

Iwata I et al. (2003) Lyman Break Galaxies at $z \sim 5$: Luminosity Function. P.A.S.J. **55** 415

Klypin A et al. (1999) Where Are the Missing Galactic Satellites? Ap. J. **522** 82

Kodaira K, Taniguchi Y et al. (2003). The Discovery of Two Lyα Emitters Beyond Redshift 6 in the Subaru Deep Field P.A.S.J. Let. **55** L17

Kudritzki R. et al. (2000) Discovery of Nine Lyα Emitters at Redshift $z \sim 3.1$ Using Narrowband Imaging and VLT Spectroscopy Ap. J. **536** 19

Kulkarni S et al. (1998) The Afterglow, Redshift and Extreme Energetics of the Gamma-Ray Burst of 23 January 1999 Nature **393** 35

Kunth D et al. (1998) HST study of Lyα Emission in Star-Forming Galaxies: the Effect of Neutral Gas Flows Astron. and Ap. **334** 11

Lanzetta K et al. (1999) in The Hy Redshift Universe ed AJ Bunker and W van Breugel ASP Conf. Ser. **193** p. 544

Lehnert MD and Bremer M (2003) Luminous Lyman Break Galaxies at $z > 5$ and the Source of Reionization Ap. J. **593** in press, astro-ph/0212431

Lilly S and Longair M (1984) Stellar Populations in Distant Radio Galaxies MNRAS **211** 833

Lilly S et al. (1996) The Canada-France Redshift Survey: The Luminosity Density and Star Formation History of the Universe to $z \approx 1$ Ap. J. **460** L1

Loeb A and Barkana R (2001) The Reionization of the Universe by the First Stars and Quasars Ann. Revs. Astron. and Ap. **39** 19

Madau P and Pozzetti L and Dickinson M (1998) The Star Formation History of Field Galaxies Ap. J. **498** 106

Malhotra S and Rhoads J (2002) Large Equivalent Width Lyα; line Emission at $z = 4.5$: Young Galaxies in a Young Universe? Ap. J. **565** L71

Malhotra S et al. (2003) No X-Ray-bright Type II Quasars among the Lyα Emitters Ap. J. **585** 25

Osmer PS (1999) in The Hy Redshift Universe ed AJ Bunker and W van Breugel ASP Conf. Ser. **193** p. 566

Ouchi M et al. (2002) Subaru Deep Survey. II. Luminosity Functions and Clustering Properties of Lyα Emitters at $z = 4.86$ in the Subaru Deep Field Ap. J. **582** 60

Partridge RB and Peebles PJ E (1967) Are Young Galaxies Visible? Ap. J. **147** 868

Pentericci L et al. (2002a) VLT Optical and Near-Infrared Observations of the $z = 6.28$ Quasar SDSS J1030+0524 Astron. J. **123** 2151

Pentericci L et al. (2002b) The VLT and the Most Distant Quasars, ESO Messenger **108** 24

Pettini M et al. (2001) The Rest-Frame Optical Spectra of Lyman Break Galaxies: Star Formation, Extinction, Abundances, and Kinematics Ap. J. **554** 981

Press WH and Schecter P (1974) Formation of Galaxies and Clusters of Galaxies by Self-Similar Gravitational Condensation Ap. J. **187** 425

Pritchet C (1994) The Search for Primeval Galaxies PASP **106** 1994

Rhoads J et al. (2003) Spectroscopic Confirmation of Three Redshift 5.7 Lyman-alpha Emitters from the Large Area Lyman Alpha Survey Astron. J. **125** 1006

Ridgway SE et al. (2001)NICMOS Imaging of the Host Galaxies of z 2-3 Radio-quiet Quasars Astron. J. **550** 122

Schaerer D (2002) The Transition from Population III to Normal Galaxies: Lyα and He II Emission and the Ionising Properties of High Redshift Starburst Galaxies Astron. and Ap. **397** 527

Schmidt M (1965) Optical Spectra and Redshifts of 31 Radio Galaxies. Ap. J. **141** 1

Schneider DP, Schmidt M and Gunn JE (1991) PC 1247 + 3406 - an Optically Selected Quasar with a Redshift of 4.897 Astron. J. **102** 837

Searle, L and Zinn, R (1978) Compositions of Halo Clusters and the Formation of the Galactic Halo Ap. J. **225** 357

Shapley AE et al. (2001)The Rest-Frame Optical Properties of $z \simeq 3$ Galaxies. Ap. J. **562** 95

Somerville R and Primack J (1999) Semi-Analytic Modelling of Galaxy Formation: the Local Universe MNRAS **310** 1087

Spinrad H et al. (1998) A $z = 5.34$ Galaxy Pair in the Hubble Deep Field Astron. J. **116** 2617

Spinrad H, Stern D, Dawson S and the GOODS team (2003) work in progress

Steidel C et al. (1999) Lyman-Break Galaxies at $z \gtrsim 4$ and the Evolution of the Ultraviolet Luminosity Density at High Redshift Ap. J. **519** 1

Stern D and Spinrad H (1999) Search Techniques for Distant Galaxies PASP **111** 1475

Stern D et al. (2003) Spectroscopic Confirmation of a Radio-Selected Galaxy Over-Density at $z = 1.11$ Astron. J. **125** 2759

Stiavelli M (2003) Exploring the Cosmic Dark Ages with the Next Generation of Space and Ground-Based Facilities, in Future Research Directions and Visions for Astronomy ed A Dressler AM Proceedings of the SPIE 122 astro-ph/0208544

Taniguchi Y et al. (2003) The Discovery of a Very Narrow-Line Star Forming Object at a Redshift of 5.66 Ap. J. **585** L97

Thompson D and Djorgovski SG (1995) Serendipitous Long-Slit Surveys for Primeval Galaxies Astron. J. **110** 982

Tenorio-Tagle G et al. (1999) The Evolution of Superbubbles and the Detection of Lyα; in Star-Forming Galaxies MNRAS **309** 332

van Breugel W et al. (1999) A Radio Galaxy at $z = 5.19$ Ap. J. **518** 61

Venemans BP et al. (2002) The Most Distant Structure of Galaxies Known: A Protocluster at $z = 4.1$ Ap. J. **569** L11

Weinberg DH et al. (1999) in Photometric Redshifts and High Redshift Galaxies ed RJ Weymann, L Storrie-Lombardi, M Sawicki and R Brunner ASP Conf. Ser. **191** p. 341

Weymann RJ et al. (1998) Keck Spectroscopy and NICMOS Photometry of a Redshift $z = 5.60$ Galaxy. Ap. J. **505** L95

White S and Rees M (1978) Core Condensation in Heavy Haloes - A Two-Stage Theory for Galaxy Formation and Clustering MNRAS **183** 341

9 Optical Spectroscopy Today and Tomorrow

F. Watson

9.1 Introduction

In 1912, Vesto Slipher of the Lowell Observatory embarked on the first systematic survey of the radial velocities of spiral nebulæ—an undertaking we would today call a galaxy redshift survey. His results, published five years later [1], demonstrated that the 25 objects he had managed to observe showed predominantly recessional velocities, the first hint of the velocity-distance relationship confirmed in 1929 by Edwin Hubble [2]. Slipher's work, trivial though it might seem to us today, was a triumph of observational astronomy. Each of his galaxies required 20 to 40 hours of exposure time on the Lowell 0.6-m refractor, producing photographic spectra whose features were even then barely distinguishable. This was no mean feat, for only a few years earlier, the great 1.5-m reflector at Mount Wilson Observatory had needed no less than 80 hours to obtain a spectrum of the bright Virgo spiral NGC 4594 (the Sombrero Galaxy).

Today, the better part of a century later, things could hardly be more different. The contrast is perhaps best illustrated by the events of 11/12 April 2002 at the 3.9-m Anglo-Australian Telescope (AAT). On that one night, spectra of 1546 much fainter galaxies were obtained, bringing to an end a comparable five-year study of galaxy velocities. This time, however, the survey encompassed a sample of objects almost 10^4 times bigger than Slipher's. The 2dF Galaxy Redshift Survey (2dFGRS) measured 221,283 galaxies within the redshift range $0 \leq z \leq 0.3$ over 5 percent of the sky during 272 nights of allocated telescope time (see [3, 4] and references therein). Analysis of these data, coupled with supporting photometry, has provided new values for no less than eight local and large-scale cosmological constants—including constraints on the nature and distribution of dark matter in the Universe.

9.1.1 The Ingredients of Progress

It is not hard to find reasons for the striking contrast between these two epoch-making surveys. The instrumentation available to optical astronomers has improved beyond belief—and not merely in the apertures of the telescopes they use. While it is true that the AAT has some 40 times the light-collecting area of Slipher's modest instrument, it is also worth noting that a new survey of stellar radial velocities currently being planned for the period 2006–2010 (the RAVE survey [5]) will collect high-precision spectroscopic observations of some 25 million stars with only a 1.2-m telescope (the UK Schmidt).

Perhaps more important in this context than the burgeoning apertures of telescopes is the dramatic improvement in detector sensitivity. The photographic plates used by Slipher would have had a responsive quantum efficiency (RQE) of 1 percent at most—that is, for every 100 incident photons, only one would have been recorded. Today's state-of-the-art detectors are charge-coupled devices (CCDs), which have improved steadily over the last 20 years and now yield RQEs of more than 80 percent over much of the optical waveband (see, e.g., [6, 7]). Coupled with this has been a steady improvement in the optical throughput of spectrographs due to advances in optical design, materials, coatings and manufacturing technology. Taken together, these improvements mean that the scope for further significant gains in instrumental sensitivity is now quite limited.

It can be argued that the real hallmark of spectroscopic instrumentation today is its ingenuity. In Slipher's day, no-one would have dreamed of attempting to observe more than one object at a time with a spectrograph, whereas the 2dFGRS was conducted with an instrument capable of gathering data on 400 objects simultaneously over a two-degree field of view (see Sect. 9.3.1). This relatively new technique of multi-object spectroscopy has revolutionised statistical astronomy—but it is only one of several new ways of using spectrographs. All are designed to address specific scientific problems, ranging from tomography of the Universe to the detection of extra-solar planets. But all have wide areas of applicability, whether they are used with a 1.2-m telescope, a 4-m or an 8-m—or with one of the 30-m class ELTs (extremely large telescopes) now on the horizon.

9.1.2 Why Optical Spectroscopy?

While the 2dFGRS is a compelling example of modern optical spectroscopy, it is nevertheless appropriate to examine the technique briefly in a wider context. Why, in this era of access to every region of the electromagnetic spectrum, does optical spectroscopy still make such a vigorous contribution to contemporary astrophysics?

The optical spectrum of an object is defined as *the intensity of the excess of radiation from the direction of that object over the background sky, measured as a function of wavelength in some range within the approximate limits 330 nm to 1000 nm.* This range is determined by the visible atmospheric transmission window at sea level rather than the sensitivity of the human eye. Its importance to astrophysics comes from a number of basic characteristics:

- Most fundamentally, it is within this waveband that ordinary matter in the Universe at ordinary temperatures (i.e. 3000 K–10,000 K) emits most of its radiation. Thus, starlight at low to intermediate redshifts falls within the optical waveband.
- The optical waveband is rich in atomic and certain molecular features, and the optical spectrum of an object remains one of the most informative signatures of its true nature, its dynamical condition and, often, its environment.
- The optical waveband sits near the centre of the electromagnetic spectrum, frequently acting as a bridge between observations at widely-separated wavelengths.
- Finally, optical observations have a high information content in comparison with those in other wavebands, uniquely combining high sensitivity, high spatial resolution and high spectral resolution.

9.1.3 Scope of This Review

This review outlines developments in spectroscopy over the past five years or so. It concentrates on ground-based studies, since some of the most innovative work has been carried out in this regime. In particular, the chapter highlights advances in the front-end reformatting systems that make multi-object or integral-field (i.e. spatially-resolved) spectroscopy possible, with recent examples of both. It also presents advances in the design of spectrographs themselves, and describes some novel observing techniques that allow reliable data to be obtained at the limits of current instrumentation. Finally, it looks briefly to the future of optical spectroscopy in the forthcoming ELT era.

The chapter discusses only dispersive spectroscopy, since it is in this area that the most striking advances have been made in recent years. Thus, scanning Fabry-Perôt devices, in which the spatial content of the data predominates over the spectral content, are excluded (e.g. the Taurus Tunable Filter on the AAT [8]), as are Fourier-transform spectrographs of the kind favoured in the early days of infrared spectroscopy. Spectropolarimetry is also excluded.

Over the past two decades, observational techniques at near-infrared wavelengths (out to the K band at 2.6 μm) have become very similar to those in the optical waveband because of the development of two-dimensional detector arrays analogous to CCDs. Some of the techniques described in this article are therefore applicable to these longer wavelengths, although in the thermal infrared (beyond about 1.4 μm), steps must be taken to prevent unwanted background radiation reaching the detector. This usually involves cooling the spectrograph and providing a cold-stop to trap the extraneous radiation originating in the telescope and background sky.

9.2 Multi-Object Spectroscopy – Overview

The classical astronomical spectrograph comprises (1) a mask (usually a slit) in the common focus of the telescope and spectrograph collimator to isolate the area(s) of interest on the sky, (2) an optical system to recollimate the beam, (3) a dispersing element in the collimated beam (usually a plane reflectance grating, but frequently a grating-prism combination known as a grism), and (4) a camera to image the dispersed slit (and hence the dispersed sky image) onto (5) a two-dimensional detector (see Fig. 9.1).

Today, such an instrument would be called a long-slit spectrograph to emphasise its ability to image a dispersed slice through an extended object such as a bright galaxy—together with adjacent sky. Because spectra are themselves extended objects, spectrograph cameras are made with fast focal ratios, and the instrument also acts as a focal reducer on the telescope. That is to say, if the slit and dispersing element were removed, the spectrograph would form a direct image of the sky at a reduced image-scale. This attribute is used in some modern infrared imaging spectrographs such as IRIS2 on the AAT [9] and UIST on UKIRT [10].

The disadvantage of the long-slit spectrograph is that as a rule it accesses only a tiny fraction of the available field of the telescope, wasting precious imaging capability. It was this consideration that led to an upsurge of interest in both multi-object and integral-field (area) spectroscopy at the beginning of the 1980s.

9.2.1 Development of Multi-Object Spectroscopy

Technically, multi-object spectroscopy had long been practised in the form of objective-prism studies, in which elements (1) and (2) of the classical spectrograph are dispensed with, and the telescope itself becomes the camera

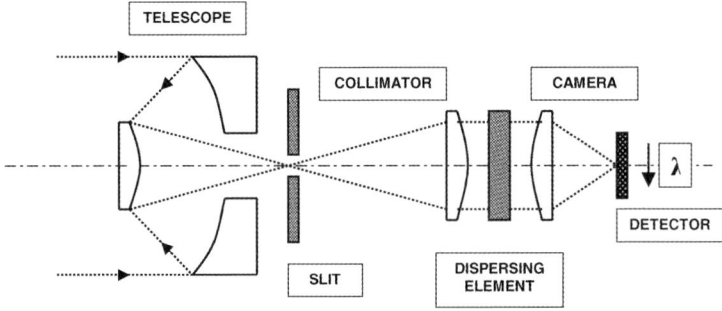

Fig. 9.1. Schematic diagram of a classical long-slit spectrograph. The dispersion direction is in the plane of the paper. In the orthogonal plane, the sky (masked by the slit) is imaged onto the detector; this is the spatial direction.

(see, e.g., [11] and the useful historical survey in [12]). During the 1980s, much useful work was carried out with the two objective prisms of the UK Schmidt Telescope (e.g. [13]). Similar multi-object work was also carried out using conventional spectrographs by simply omitting the slit (e.g. [14]). However, both these slitless techniques suffer from the severe drawback that the object spectra are superimposed on an undispersed (and hence bright) sky background, and they only survive today in a few limited applications (e.g. the highly-efficient Planetary Nebula Spectrograph [15] on the 4.2-m William Herschel Telescope).

Multi-object spectroscopy in its modern form has been developed over the past two decades or so. It uses either a multi-slit mask in place of a single slit, or multiple optical fibres to connect the telescope focal surface to the slit. A number of wide-field multi-slit spectrographs were especially designed at an early stage (e.g., for the Canada-France-Hawaii Telescope [16], the Isaac Newton Telescope [17], the Mayall Telescope [18] and the AAT [19]), attaining a reasonable degree of sophistication. In contrast, the first multi-fibre systems were essentially cobbled together by connecting existing spectrographs to telescope focal surfaces with a rudimentary fibre coupler.

Thus, we find the first example of multi-fibre spectroscopy taking place on the Steward Observatory's 2.3-m telescope in 1979, in which a standard Boller & Chivens spectrograph was spaced back from the telescope focal surface by 22 cm in order to accommodate a 20-fibre coupler called *Medusa* [20, 21]. By the early 1980s, a fully-engineered version at the AAT's 40-arcminute Ritchey-Chrétien focus was in great demand among users, highlighting the potential of the technique [22, 23].

The first truly wide-field multi-fibre system was *FLAIR* on the UK Schmidt Telescope, which covered an area of sky 6.5×6.5 deg^2. Commissioned in 1985, it was also the first multi-object system to feed a fixed (and therefore very stable) spectrograph mounted on the dome floor [24, 25]. *FLAIR* highlighted the importance of field size in multi-object spectroscopy, leading to the common use of the $A\Omega$ product as a yardstick for the suitability of particular telescopes for this work (see Table 9.1) [26].

With the implementation of the multi-fibre technique, studies were carried out on the availability of targets at given magnitudes and number-densities (e.g. [27, 28]). It was found that the technique is extremely well-matched to problems in spectroscopic survey science. Detailed attention was also paid to the properties of fibres themselves, including losses due to attenuation, focal-ratio degradation (i.e. beam-spreading in the fibre) and other effects (e.g. [29, 30]).

By the 1990s, it had become apparent that the challenges facing multi-fibre spectroscopy lay neither in its scientific applicability nor in the fibres themselves. The main concerns centred around such issues as the costly engineering needed to build facility-class equipment, the efficacy of sky-subtraction using fibres, and the data reduction techniques required to handle

Table 9.1. $A\Omega$ product for some spectroscopic survey telescopes

Telescope	Aperture (m)	Field (deg)	$A\Omega$ (m^{2}deg^2)
Gemini	8.1	0.17	1.9
VLT	8.1	0.4	10
Subaru	8.1	0.5	16
WHT	4.2	1.0	18
Sloan	2.5	3.0	56
AAT	3.9	2.0	61
UKST	1.2	6.6	63
LAMOST	4.0	5.0	400

Note: $A\Omega = $ aperture$^2 \times$ field^2

spectra hundreds at a time. Multi-slit spectroscopy was also reaching maturity during the 1990s, but here the challenges were rather different.

9.2.2 Multi-Slit Spectroscopy—VIMOS and GMOS

The appeal of a multi-slit spectrograph lies in its close relationship to the classical spectrograph. The light path from the telescope to the instrument has no additional optical elements to modify the transmission characteristics as fibres do. On the other hand, each target field observed with the instrument requires a new slit mask matching the sky positions of the target objects.

Short, multiple slits (slitlets) 8–15 arcseconds long are usually preferred to the circular apertures used by some early systems (e.g. [18]) because they allow sky samples to be taken on either side of each object. Herein lies the great advantage of the multi-slit technique over multiple fibres, because not only are these simultaneous sky samples adjacent to the object itself, but they are also imaged through an identical optical system onto adjacent pixels on the detector, providing excellent sky-subtraction.

Recent experiments with the (now decommissioned) LDSS++ multi-slit instrument on the AAT [31], however, have preferred 1–2 arcsecond square microslits, which clearly do not allow the background sky to be interpolated across a target object and require more subtle sky-subtraction methods (see Sect. 9.5.2). The advantage of the microslits is that they allow more targets to be accessed in any given field of view (e.g., ~ 900 objects in a 9-arcmin field in one LDSS++ experiment), since each spectrum is much narrower than that produced by a conventional slitlet (see Fig. 9.2).

Inherent in multi-slit spectroscopy is the problem of optimally populating the detector area with spectra for a given field of target objects, and this is routinely done with software packages designed specifically for each instrument (by modelling its imaging characteristics). The mask design (with its associated object identifications) is the output from this process; the slitlets

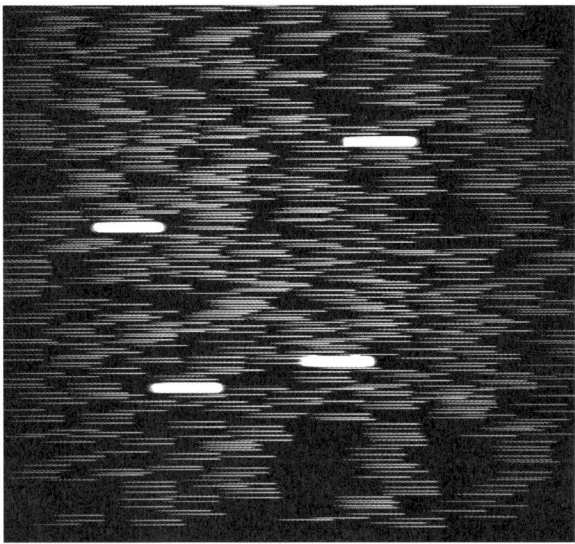

Fig. 9.2. Raw data frame from a multi-slit spectrograph (LDSS++ on the AAT) Each of the ~ 900 narrow horizontal lines is the spectrum of a galaxy imaged through a microslit. The four broad lines are the spectra of bright stars used to register the slit mask with the sky. (*Courtesy Karl Glazebrook, Johns Hopkins University.*)

are usually laser-cut at the telescope. Because of the finite detector area available, there is always a trade-off between the number of objects that can be observed and the resolution and spectral range of the observations.

Multi-slit spectrographs generally work in a fairly modest resolution regime. Defining spectral resolution in the usual way as $R = \lambda/d\lambda$, values of $R \sim 1000$ are typical. Because of their relatively low dispersion and the high-quality sky-subtraction they offer, multi-slit spectrographs are frequently designed with very faint targets in mind. These are commonly objects at relatively high redshifts (e.g. $z \sim 1$ for [OII] emission; $z \sim 3$–5 for Lyα emission). Thus, their optical systems are usually optimised for high throughput, and the production of the most efficient possible instrument is a major design consideration.

Another challenge in designing a multi-slit spectrograph comes from the fact that its field of view on the sky is likely to be limited by the field of the collimator rather than that of the host telescope. A good example is provided by the 8.1-m VLT unit telescopes of the European Southern Observatory (ESO). The 25 arcmin field of view of these telescopes at their Nasmyth foci has a linear diameter of 865 mm, and it is almost impossible to imagine a single multi-slit spectrograph on any manageable scale whose field of view would match that of the telescope.

ESO's aim has been to provide a very broad range of ancillary equipment for the VLT, including a wide field multi-slit spectrograph [32]. An instrument has therefore been designed for *Melipal* (the third VLT unit telescope) that adopts a novel approach to the problem of field coverage. VIMOS [33–35] essentially consists of four separate multi-slit spectrographs, each covering a field of 7×8 arcmin^2. The focal planes of the four collimators are butted together in the telescope's focal surface in a 2×2 mosaic, where the slit masks are located. This approach provides good coverage of the available field while maintaining tractable dimensions for the spectrographs. VIMOS was commissioned in September 2002.

No such multiplicity has been necessary for the multi-object spectrographs for the two 8.1-m telescopes of the Gemini Observatory. The Gemini telescopes have a much smaller field of view, and the 5.5×5.5 arcmin^2 field of the GMOS multi-slit spectrograph is covered with a single optical system [36, 37]. GMOS operates over the full optical waveband (~ 360–1100 nm) and offers spectral resolutions up to $R \sim 10,000$.

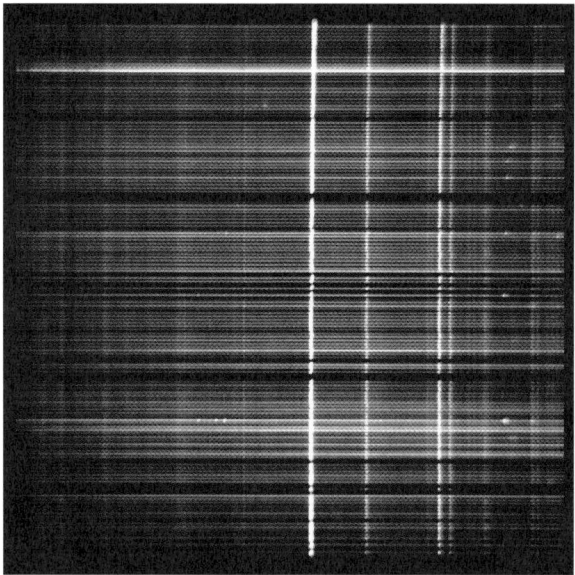

Fig. 9.3. Raw data frame from a multi-fibre system. Each horizontal band is the spectrum of a galaxy, while the vertical lines are monochromatic images of the fibre slit resulting from atmospheric emission lines. They are incomplete because of parked or disabled fibres. The 150 spectra cover the approximate wavelength range 400–700 nm, and the data frame was obtained with 6dF on the UK Schmidt Telescope (see Sect. 9.3.1).

9.2.3 Multi-Fibre Spectroscopy—SDSS

The key advantage that fibres bring to multi-object spectroscopy is the elimination of the dependence of sky coverage on collimator field diameter. Thus, the whole field of the telescope can be used, irrespective of the collimator field. That, in turn, is dictated by the number, diameter and spacing (along the slit) of the fibres themselves. Moreover, the geometry of the spectra on the detector is fixed, and usually arranged to maximise the number of objects that can be observed (see Fig. 9.3). A very broad range of spectral resolutions can be accommodated.

Like multi-slit work, fibre-optics spectroscopy requires field preparation prior to observing. In the earliest days of the technique, this consisted of the manufacture of a focal-surface mask with circular holes into which the input ends of the fibres were plugged by hand, one fibre to each object (see Fig. 9.4). Today's most notable survivor from this era is the multi-object spectrograph of the Sloan Digital Sky Survey (SDSS), used with the Sloan 2.5-m tele-

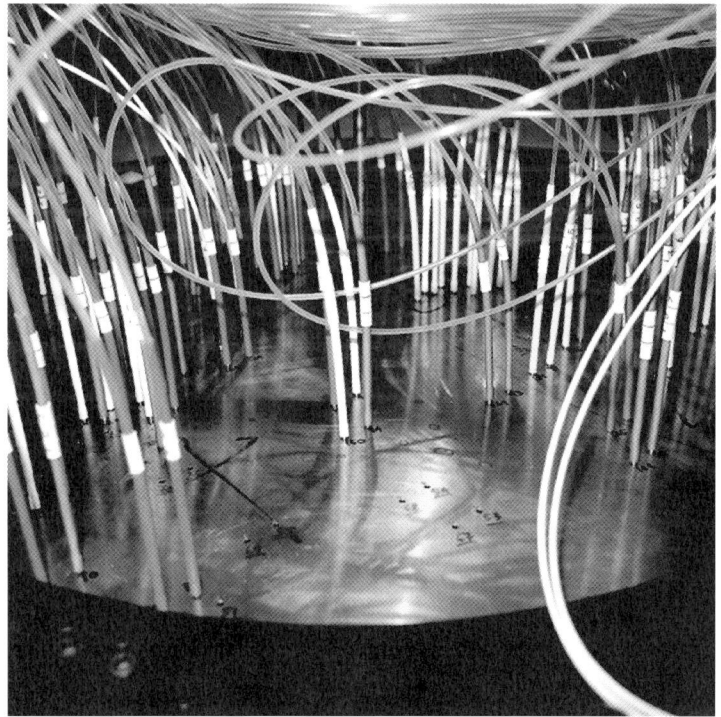

Fig. 9.4. Fibre positioning using a plug-plate, in this case with the CIRPASS near-infrared spectrometer (see Sect. 9.5.1). Because this versatile instrument can be used on several different telescopes, the simple plug-plate method is preferred for its multi-object mode. (*Photo courtesy Ian Parry and the CIRPASS team, University of Cambridge.*)

scope [38, 39]. While it was built in the late-1990s, its designers eschewed robotic technology and opted instead for the proven plug-plate method.

The SDSS system feeds 640 optical fibres 180 μm (3 arcsec) in diameter and 1.8 m long from a 795-mm (3-degree) diameter field plate to two broadband (390–910 nm) medium-resolution ($R \sim 2000$) spectrographs. The holes are drilled using a numerically-controlled milling machine, while the fibres themselves are of a type made from high OH-content silica to enhance their blue transmission. (The penalty with this commonly-used type of fibre is the presence of absorption bands from the second, third and fourth OH harmonics at 1.37 μm, 950 nm and 730 nm.) In order to maximise field turn-round times, there are nine interchangeable plug-plate cartridges (each with its own set of fibres) that can be set up during the day. The system is currently being used to carry out a redshift survey of 10^6 galaxies and 10^5 quasars in the northern galactic cap (See also Chap. 10 by A. Fairall in this volume).

9.3 Multi-Object Spectroscopy with Robots

The idea of fully-automated fibre positioners followed hot on the heels of plug-plate fibre systems. Their development was motivated by the ongoing expense of plate manufacture and the tedium of manual fibring-up. The earliest operating instrument was built at the Steward Observatory to replace *Medusa* on the 2.3-m telescope. Named *MX*, it consisted of 32 independent (r, θ) actuator arms, each with a fibre attached to its tip, protruding inwards from the periphery of the telescope's field-of-view in 'fishermen-around-the-pond' fashion [40]. Each arm had access to 20 percent of the focal surface and could be positioned simultaneously with the others (or nearly so). *MX* was successfully commissioned in 1986, but the limelight was quickly stolen by the Durham University Autofib machine, which offered a cheaper way of positioning a larger number of fibres. Nevertheless, a similar multi-actuator system (*MEFOS*) was built in the early 1990s for the 1-degree prime focus field of the ESO 3.6-m telescope at La Silla [41].

The first Autofib was built for the AAT [42]. The principle adopted was to use a circular field-plate made of ferrous material and to terminate each fibre with a 90-degree microprism and a magnetic button. Thus, the fibres themselves radiated inwards from the circumference of the field-plate, while light from the telescope mirror entered each fibre via its microprism. With a robotic gripper mounted on a carriage above the field plate, the fibre buttons could be picked up, moved to position, and placed on the field plate in turn. This 'pick-place' method obviated the need for a separate actuator for each fibre, meaning that sophisticated engineering could be lavished on a single robotic device. However, it suffered from the penalty of considerable dead time while the fibres were being reconfigured between fields.

9.3.1 Pick-Place Multi-Fibre Systems—2dF and OzPoz

The experience with 64 fibres on Autofib led to an improved version (Autofib-1.5, now at the Cananea 2.1-m telescope in Mexico [43]) and the 150-fibre Autofib-2 at the 4.2-m William Herschel Telescope [44, 45]. A similar instrument, *Hydra*, was built for the Mayall 4-m telescope [46] and later moved to the WIYN 3.5-m [47].

The Autofib principle was adopted on perhaps the grandest scale by the Anglo-Australian Observatory (AAO) for its 400-fibre 2dF (2-degree field) system on the AAT [48–51]. Conceived in the late-1980s, this system utilises massive refractive optics to allow the telescope's hyperboloidal primary mirror to form a flat 2-degree diameter image at its prime focus, the largest for any 4-m class instrument. Thus it gives the AAT a clear and unique role in the current era of much larger telescopes.

2dF avoids the problem of dead-time during the (∼1 hour) fibre set-up by having two field plates, one of which can be reconfigured while observations are taking place with the other. The plates are interchanged by a computer-controlled tumbler mechanism. The positioning robot is based on a high-precision (x, y) carriage which moves the fibre gripper. During reconfiguration, the fibres are illuminated from their output ends and a centroiding TV system within the gripper allows the control system to decide whether the fibre is in the right place or whether a further iteration is required. A typical fibre placement takes 6 seconds and is accurate to ∼20 μm. The fibres themselves are 2 arcsec (140 μm) in diameter, and are again made from high-OH fused silica.

The 2dF control system embodies integration on a scale not previously seen in a multi-fibre system. Its various tasks run on a hierarchy of interlinked computers, with a high-level control task providing a single user interface to (1) the telescope, (2) the atmospheric dispersion compensator, (3) the fibre positioner, (4) the spectrographs and (5) the spectrograph CCDs. There is also a suite of offline software utilities that allows fibres to be allocated to fields of target objects (`configure`), spectral data-frames to be reduced to individual calibrated spectra in pipeline mode (`2dfdr`), focus frames to be analysed (`drfocus`), and so on.

2dF entered service at the AAT in September 1997 after a lengthy commissioning period. As the opening paragraphs of this chapter demonstrate, it has already proved its worth as a unique facility for the AAT. It has also allowed the AAO to develop similar robotic devices under contract to other institutions—leading to the Observatory's unofficial nickname of *Fibres'R'Us*.

AAO's first major contract in this new era was for a novel type of pick-place fibre positioner for the VLT. Delivered to Cerro Paranal in February 2002, OzPoz provides an analogous function to 2dF with the FLAMES spectroscopic facility at the *Kueyen* unit telescope [52, 53]. More specifically, OzPoz feeds up to 140 objects from the 25-arcmin diameter Nasmyth field of *Kueyen* to two spectrographs, GIRAFFE and UVES, which have spectral resolutions ranging from 6000 to 47,000 in the visible wavelength range [54]. It maintains the 'double-buffered' operating mode of 2DF, in which one field-plate is being reconfigured while observations are taking place with the other (see Fig. 9.5).

OzPoz patrols a much larger physical field diameter than 2dF (865 mm compared with 440 mm) and differs from it in several respects. Most notable is the fact that the fibres are positioned on a curved field-plate (matching the 3950 mm radius of curvature of the focal surface) rather than a flat one. Thus, a simple (x,y) carriage for the gripper will not suffice. The solution adopted has been to build an (r,θ) carriage, with the r-arm curved to match the focal surface so that the gripper axis is always radial to it. Both r and θ motions are on friction-free air-bearings.

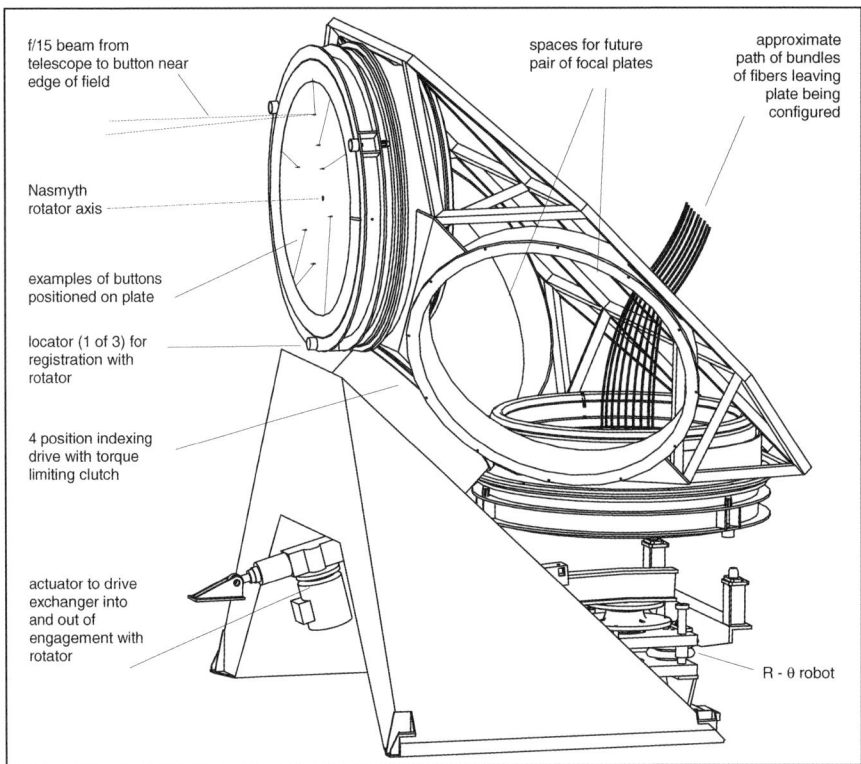

Fig. 9.5. The OzPoz multi-object fibre positioner and field-plate exchanger built by AAO for the VLT. The exchanger rotates about an axis at 45 degrees to the horizontal, and can accommodate up to four field plates, each with its own fibre feed to the spectrographs. The (r,θ) robot operates on the plate in the lowest (horizontal) position, while the plate in the observing (vertical) position coincides with the Nasmyth focus of the telescope. The assembly shown is 2.6 m high and is mounted on the telescope's Nasmyth platform. (*Courtesy Peter Gillingham and the AAO OzPoz team.*)

In order to test the air-bearing (r,θ) concept for OzPoz, a prototype fibre positioner was designed for another telescope with a curved focal surface—the AAO's UK Schmidt Telescope. This 6dF system (named for the telescope's 6-degree diameter field) was built specifically to carry out an all-southern sky galaxy redshift and peculiar-velocity survey (6dFGS: see, e.g., [26])—as well as to prove the OzPoz concept. At the time of writing (February 2003), the survey is one-third complete (see Fig. 9.6).

Because a 1.2-m telescope is considerably less spacious than an 8.1-m, the 6dF robot is housed in a separate enclosure in the dome, and two 150-fibre field-plates are manually interchanged between it and the telescope [55, 56]. Like its predecessor (*FLAIR*), 6dF feeds a stationary spectrograph in the dome, which operates at resolutions within the range $1000 \leq R \leq 9000$. It

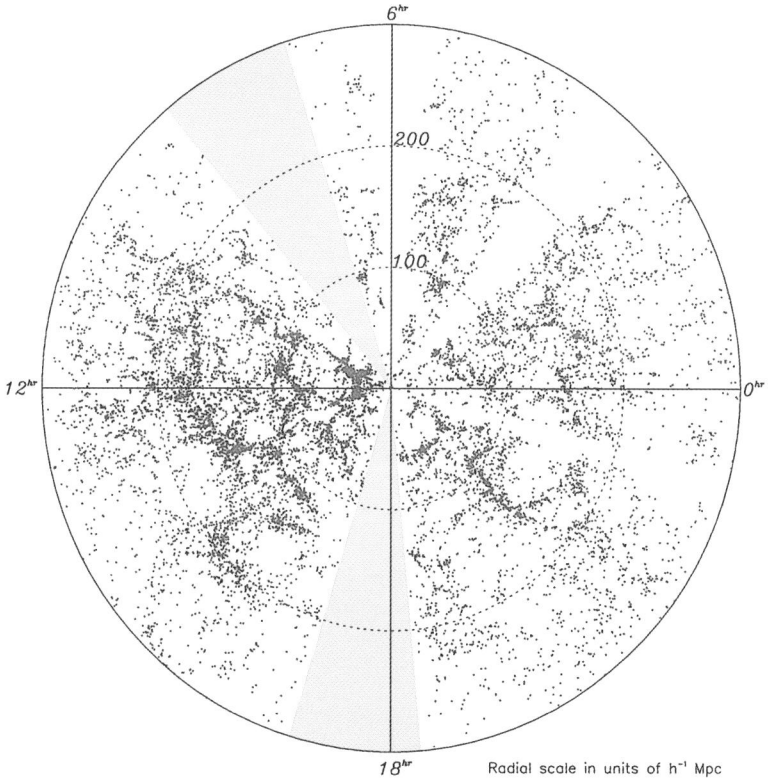

Fig. 9.6. A slice through the local Universe in a redshift map of 13,770 galaxies taken from the 6dF Galaxy Survey. The map shows the status of the survey's -30 deg declination zone in October 2002 and is still incomplete, but it clearly demonstrates the all-sky coverage possible using wide-field multi-object spectroscopy. Shaded areas represent the position of the Milky Way. The 6dFGS will contain data for $\sim 150,000$ galaxies when it is complete in 2005. (*Courtesy Lachlan Campbell and the 6dF Galaxy Survey team.*)

is also notable for the use of a recently-developed broad-band fibre material that avoids the drawbacks of high-OH silica [57].

9.3.2 Multi-Actuator Devices—Echidna and UKidna

It is only within the last few years that interest in multi-actuator systems has been rekindled, and that is partly due to the different scientific tasks that are now envisaged for multi-fibre spectrographs (see following subsection). In particular, the *Echidna* fibre positioner currently being developed for the Japanese Subaru 8.1-m telescope has pointed the way towards mass-produced fibre actuators, with considerable cost-savings over the kind of technology used by *MX* [58]. The advantages of such a system are rapid reconfiguration time for large numbers of fibres, obviating the need for double-buffered fieldplates, and an ability to operate at a much higher fibre density on the sky than pick-place systems can. Thus, it seems likely that the Chinese 4-m LAMOST telescope currently under development will also use a multi-actuator device to position its ~ 4000 fibres [59].

Echidna itself is a project of the AAO, and it will feed 400 fibres from the Subaru 30-arcmin (153 mm) diameter prime focus field to the FMOS near-infrared (0.9 to 1.8 μm) multi-object spectrograph. The name comes from an Australian hedgehog—the spiny anteater—and it conveys a fairly accurate impression of the instrument. The idea is that the first 140 mm of each fibre is enclosed in a rigid spine, and the spine can be tilted in any direction to position the fibre input end accurately (to 10 μm) within a 7-mm radius patrol area. The spines are mounted in a hexagonal array on 7-mm centres, which means that any seven fibres can just touch at their input ends. Each spine is independently driven by a quadrant-tube piezoelectric actuator (QTP), which responds to a sawtooth voltage waveform, the fibre input end moving by 10 μm with each pulse. The spines are assembled in modules of 42, and the complete system is expected to be ready for commissioning at the telescope in July 2004.

The modular design of *Echidna* lends itself to being extended to very large numbers of fibres, and such high-multiplex possibilities are being investigated at the AAO. One such study is KAOS, a collaboration between AAO and the US National Optical Astronomy Observatory (NOAO) to equip the Gemini telescopes with a wide field (~ 1.5 deg) and populate it with up to 4000 fibres. An interim step in this process is likely to be an *Echidna*-style positioner for the 1.2-m UK Schmidt Telescope to replace 6dF in 2005. The science-driver for this instrument is the RAVE survey of stellar radial velocities referred to in Sect. 9.1.1. *UKidna* will be equipped with 2250 fibres feeding a single stationary spectrograph; since the required spectral range is short (840–870 nm centred on the CaII triplet), the fibre slit will consist of three banks of 750 fibres imaging three blocks of spectra side-by-side on the detector. *UKidna* is expected to have a reconfiguration time of approximately five minutes and to be able to deliver 22,000 stellar spectra per clear night.

9.3.3 Science with Multi-Object Spectroscopy

Multi-object work comes into its own in large-scale spectroscopic surveys, several of which have already been alluded to in this chapter (2dFGRS, SDSS, 6dFGS and RAVE). In addition, the 2QZ (a spectroscopic survey of 23,424 quasars) was conducted concurrently with the 2dFGRS on the AAT. Other noteworthy surveys include CNOC 1 (galaxies in x-ray selected clusters [60]) and CNOC 2 (field galaxies [61]) of the Canadian Network for Observational Cosmology, the VIRMOS-VLT Deep Survey (redshifts out to $z \sim 5$ over 16 deg^2 [62]) and DEEP 2 (redshifts out to $z \sim 2$ with the Keck telescopes [63]). The latter two surveys will complement the SDSS and 2dFGRS by going fainter and to higher redshifts using 8-m telescopes.

Perhaps more than any other aspect of modern optical astronomy, these surveys are transforming our understanding of the structure of the local and large-scale Universe and, in the case of RAVE, will shed unprecedented light on the kinematics of our own Galaxy. On a more mundane level, they also highlight the importance of large astrometric catalogues of target objects, without which multi-object spectroscopy on this scale would be impossible.

Large-scale surveys are only part of the picture, however. An impressive range of science has come from multi-object systems, embracing virtually all aspects of astronomy except Solar System studies. It includes the dynamics of the galactic disk [64], tidal streams in the local galactic neighbourhood [65], intra-cluster planetary nebulæ [66] and stars [67] in galaxy clusters, the discovery of a new class of dwarf galaxies [68] and the evolution of the galaxy luminosity function [69].

Multi-slit systems are particularly well-suited to studies of distant galaxy clusters (e.g. for star-formation rates [70]) because of their modest fields and high sensitivities. Multi-fibre systems are also frequently used to observe clustered objects (e.g. globular cluster stars and clusters of galaxies with 2dF; the Magellanic Clouds with 6dF). The geometry of a pick-place system lends itself well to such work.

An *Echidna*-like system, on the other hand, is better suited to a more uniform distribution of objects. Simulations have demonstrated that *Echidna* itself will have a 90-percent success rate in allocating fibres to randomly scattered targets [71]. Since it is destined for the prime focus of an 8-m class telescope, it is likely to be used for faint objects in the field, both galactic and extragalactic. Similarly, the stars that will be observed with *UKidna* in the RAVE survey will, in general, be relatively uniformly distributed.

9.4 Integral-Field Spectroscopy

It was with more than a touch of whimsy that a recent reviewer highlighted the two considerations that make integral-field or area spectroscopy a necessity [72]:

- the bad behaviour of astronomical objects, whose shapes do not match the slits of classical spectrographs;
- the bad behaviour of detectors, which offer only two dimensions to record the three variables $(\alpha, \delta, \lambda)$.

The traditional approach to this double dose of delinquency has been to introduce time as a third dimension, either by scanning a long-slit spectrograph progressively across the extended object or by using a wavelength-scanning device of the type mentioned in Sect. 9.1.3. However, in recent years, techniques have been developed that reformat a two-dimensional telescopic image into the linear array required by a spectrograph, and that is what is now usually meant by integral-field spectroscopy. It has been adopted very widely, and there are numerous examples of the technique. There are three basic methods.

9.4.1 Image-Slicing

The first is a development of the early image-slicers used to improve the efficiency of large coudé spectrographs when the seeing disk was larger than the slit-width. An image-slicer infrared spectrometer built on this principle by the Max-Planck Institut für Extraterrestrische Physik (MPE) and named simply 3D [73] has demonstrated the potential of the technique. It uses two sets of Zerodur mirrors to decompose a 16×16 pixel array in the telescope image plane into a stair-like succession of 256 slitlets. The spectrometer works in the infrared H and K bands with spectral resolutions up to $R \sim 2000$. Its success on several different telescopes has led to the development of SPIFFI, a 32×32 pixel version for the VLT to capitalise on the high spatial resolution offered by adaptive optics [74].

The second method is essentially a refractive equivalent of the image-slicer. It is exemplified by the TIGER instrument, formerly at the Canada-France-Hawaii Telescope [75]. Here, the telescopic image is decomposed into individual pixels by means of a two-dimensional array of microlenses, each of which creates an isolated image of the telescope pupil (called a micropupil) on the detector. It is the micropupils that are dispersed into the unfilled portions of the detector, creating a two-dimensional array of spectra.

9.4.2 Fibre-Coupled Integral-Field Systems

Microlenses also play a major role in the third and largest family of integral-field units (IFUs), which use optical fibres to reformat the telescope image. The first generation of such devices were, like their multi-object equivalents, very crude, consisting simply of bundles of fibres hexagonally-packed at their input ends and aligned on the slit at the output end (e.g. [22, 76]). No microlenses were used. The disadvantages of this method are (1) the packing fraction is low (i.e., the telescopic image is relatively poorly sampled because of gaps between the fibre cores) and (2) light propagates through the fibres at the focal ratio delivered by the telescope, which may not be optimum for either the fibres or the spectrograph[1]. A recent (1997) fully-engineered example of such a system is INTEGRAL on the William Herschel Telescope (WHT), which offers a range of fibre bundles covering up to 30×30 arscec^2. The packing fraction is only 65 percent, and the system suffers from light losses due to overfilling of the f/8 collimator of the WYFFOS spectrograph [77].

Both these problems can be alleviated by feeding the fibres via a microlens array, each element of which is aligned with a fibre. Usually, the array images the pupil of the telescope onto each fibre. It can be made with a very high packing fraction, and is commonly of monolithic construction (often replicated in epoxy), although at least one early system (SPIRAL) used a fabricated array of individual lenslets [78]. Sometimes, a linear microlens array is also used on the output slit to match the fibre output to the spectrograph focal ratio; for example, in TEIFU, the Thousand-Element IFU [79].

TEIFU was designed specifically for use with adaptive optics on the WHT, and thus fulfils a different role from INTEGRAL. It covers an area 7.8×7.0 arscec^2 at a spatial sampling scale of 0.25 arscec^2. Another instrument designed for high spatial resolution, though not necessarily with adaptive optics, is the 256-element IFU of the PMAS spectrograph built by the Astrophysikalisches Institut Potsdam and currently in use at the Calar Alto 3.5-m telescope [80, 81]. Many other wide-field spectrographs now incorporate lenslet-array IFUs, including both VIMOS and GMOS (see Sect. 9.2.2).

9.4.3 Science with Integral-Field Spectroscopy

The scientific applicability of IFU spectroscopy is very wide and, unlike multi-object spectroscopy, extends to the Solar System (e.g. in observations of the night-side of Venus with 3D on the AAT [82]). More typical applications include studies of nearby planetary nebulæ [83], galaxy kinematics [84, 85], galaxy mergers [86] and gravitational lens sytems [87]. IFU work has also been carried out on shocked gas around HII regions, and the ejecta around SN1987a [82].

The use of multiple fibre-coupled IFUs provides a way of carrying out spatially-resolved multi-object spectroscopy with wide-field telescopes. In implementing this so-called 'deployable IFU' mode, there is usually a trade-off between the number of objects that can be observed and the number of resolution elements in each object.

Finally, in a melancholy footnote to this section, we remark that one of the casualties of the savage bushfire that destroyed Mt Stromlo Observatory (MSO) in Canberra on 18 January 2003 was an integral field spectrometer. NIFS (Near-IR Integral-Field Spectrograph) was an advanced instrument

---

[1] Unlike most wide-field multi-fibre systems, IFUs are usually fed at slow focal ratios, which are not ideal for fibre transmission.

being built by MSO for the Gemini North telescope. It had been nearing completion when the fire devastated the Observatory's detector laboratory.

9.5 Efficiency by Design

While multi-object and integral-field techniques are among the more spectacular developments in optical spectroscopy, a number of other factors have contributed to recent gains in observing efficiency and sensitivity. They relate both to the design and fabrication of spectrographs, and to the way in which they are used at the telescope. This section describes some improvements in both categories.

9.5.1 Instrumental Gains

Spectrograph design. Wide-field spectrographs of the kind used in multi-object spectroscopy are frequently reflective systems. For example, 2dF uses two spectrographs to accommodate its 400 fibres, each having an off-axis Maksutov collimator and a Schmidt camera, and imaging 200 fibres onto a 1k×1k detector. Field-flattening lenses are required at the camera focal surfaces. Systems such as these have relatively high efficiency and image quality, and a forthcoming upgrade to 2dF (known as $AA\Omega$) will also use a reflective system—this time stationary in the dome (for high stability) and able to accept IFU as well as multi-object input [88].

High-efficiency coatings make all-refracting systems attractive in terms of optical throughput, however. The absence of a central obstruction can also be an advantage in fibre-coupled applications. An all-refracting instrument of this kind, corrected for 350 nm $\leq \lambda \leq$ 900 nm, has been developed for PMAS, for example [80].

Double-beam spectrographs, in which the beam is separated by a dichroic and fed to red- and blue-optimised arms, are favoured for high-sensitivity systems. Such an instrument has been developed for the 2×8-m Large Binocular Telescope [89]. $AA\Omega$ will also use a double-beam layout. More complicated spectrographs have been developed when a high level of versatility is required. For example, the Baranne white-pupil design introduces a relay mirror to re-image the spectrograph pupil (which is normally located at the dispersing element), allowing useful control of its size and position. WYFFOS at the WHT is an example of this versatile but complex system [90, 91].

More complex still is the OH-suppression design, originated in Japan [92] and exemplified today in the travelling Cambridge fibre-fed instrument CIRPASS [93]. Here, an intermediate medium-dispersion ($R \sim 3100$) spectrum is formed on a mask which blocks the intense atmospheric OH emission-lines in the wavelength range 1.0 to 1.9 μm. These lines are numerous but very narrow, and removing them with the mask leaves an uncontaminated spectrum that can be re-imaged at lower resolution. In CIRPASS, this facility can be used in both multi-fibre and IFU modes (see Figs. 9.4 and 9.7).

At the other extreme is the 'spectrograph on a chip', suggested as a means of obtaining the highest possible optical efficiency in a low-resolution ($R \sim 500$) fibre-coupled system [94]. The design uses multiple planar (slab) waveguides, one for each fibre, with a CCD or infrared array butted against the output face of the stack. Because of the difficulty in fabricating suitable waveguides, it has not yet appeared in an astronomical application.

Gratings. Until very recently, spectrographs typically employed blazed plane reflection gratings as the dispersing element. They are replicated from ruled masters, and their optical efficiency depends on the effectiveness of the blaze at the working wavelength and the reflectivity of the coating (which is usually Al). Peak efficiencies in the range 40 to 70 percent are common.

In 1998, it was pointed out that volume-phase holographic (VPH) gratings (in which spectra are formed by Bragg diffraction in a transmissive layer with a periodic three-dimensional refractive-index modulation) offer several advantages [95]. In particular, they can be tuned to peak at various wavelengths, and they can exhibit very high efficiencies (> 90 percent). They also operate in a Littrow (symmetrical) mode, which can enhance spectro-

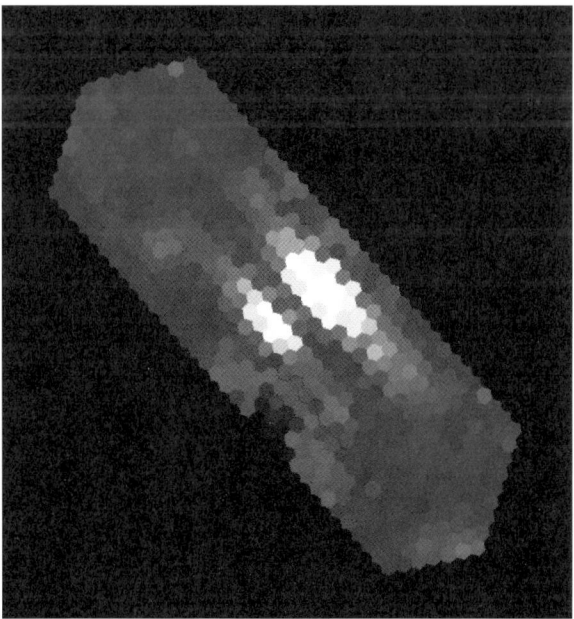

Fig. 9.7. Reconstructed image of a proto-planetary nebula made with the 490-element fibre-coupled IFU of CIRPASS on the Gemini South Telescope. Each hexagonal element of the image is 0.36 arcseconds across and contains a spectrum from 1069 to 1302 nm at 0.45 nm resolution in the original data-cube. (*Courtesy Ian Parry and the CIRPASS team, University of Cambridge.*)

graph imaging performance. VPH gratings have been adopted for the 6dF spectrograph at the UK Schmidt [96], and will be used for the AAT's $AA\Omega$ spectrograph [88].

Optical Coatings. Mirror coatings consisting of Al or Ag with a multi-layer overcoating to increase reflectivity are now fairly commonplace in spectrograph optics. For transmissive elements, however, single-layer MgF anti-reflective coatings are still preferred to multi-layer stacks as they work over a wider range of wavelengths. These vacuum-deposited coatings tend to be expensive to apply, particularly with large optical elements, and alternatives have been sought. Experiments with chemical deposition go back at least 60 years [97], but recent work has produced a porous silica coating that is both effective and durable [98]. This sol-gel coating is applied in liquid form with the optical element rotating on a turntable. It can then be hardened and made waterproof by further chemical processing. The coating is even more effective when applied to a surface that is already MgF-coated.

9.5.2 Observing Techniques

Sky subtraction. Both long-slit and multi-slit spectrographs are capable of high-precision sky subtraction ($\ll 1$ percent accuracy in the sky-subtracted sky) for the reasons outlined in Sect. 9.2.2. With microslits and multi-fibres, however, there is usually no possibility of interpolating between simultaneous sky samples on either side of the object, and those sky samples that are obtained are typically imaged through different fibres or slits onto a different part of the detector with accompanying uncertainties in sensitivity. 'Conventional' sky-subtraction with fibres, for example, yields accuracies in the region of 2 percent, compromising faint-object spectroscopy (see, e.g. [99] and references therein).

The situation has recently been greatly improved with the development of an observing technique dubbed 'va et vient' by its French originators [100] but now known as 'nod & shuffle' at the AAO, where it has been further developed and exploited. It involves repeated short offsets of the telescope to introduce adjacent sky samples into the fibres or microslits (nodding), accompanied by a synchronised shuffling of the charge on the CCD chip. Thus, adjacent sky and object samples are obtained through the same fibres or slits and imaged onto the same CCD pixels. If the process is repeated at reasonably short intervals (~ 1 minute), systematic errors due to temporal sky variations and other sources are greatly reduced (see Fig. 9.8).

The main drawback of the technique is that half the exposure time is spent observing empty sky, but it has proved highly effective with both multi-slit [101] and multi-fibre [102] observing. A recent analyisis of nod & shuffle data obtained with 2dF demonstrates that it yields Poisson-limited sky subtraction, and the technique will be adopted *ab initio* for faint-object work with the $AA\Omega$ spectrograph [103]. The technique has also been commissioned on the Gemini GMOS spectrograph.

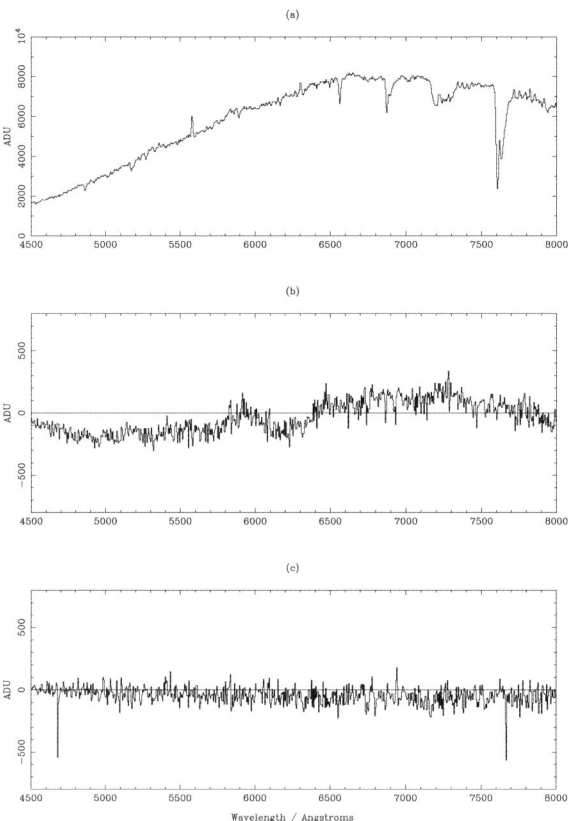

Fig. 9.8. Experimental sky subtraction with 2dF on the AAT, in which the residuals in the sky-subtracted sky spectrum are investigated (see [102]). (a) Raw sky spectrum, before sky subtraction. (b) Spectrum after standard sky subtraction (by a mean-sky method) showing strongly systematic residuals. (c) Spectrum after nod & shuffle sky subtraction, showing residuals limited mainly to Poisson noise. (The two downward spikes in spectrum (c) come from cosmic ray events, which are usually eliminated by combining multiple exposures.)

High-resolution spectroscopy. The instrumentation used to carry out high-dispersion spectroscopy ($R \sim 100,000$) is relatively well-established. The function is usually provided by a large, stationary echelle spectrograph fed by coudé mirrors or fibres. (A recent proposal to mount a high-dispersion spectrograph at the Cassegrain focus of the Gemini telescopes produced investigations into actively-stabilised spectrograph elements [104, 105], but was eventually abandoned.) Unlike conventional grating spectrographs, which use only the first or second orders of diffraction, echelles work at very high orders. The spectra are prevented from overlapping by means of a cross-dispersing prism.

New demands have been placed on such equipment by the search for extrasolar planets. In particular, very stable, high-precision wavelength calibration is required to provide absolute ~ 3 ms^{-1} velocities over periods of many months [106]. One of the innovations that have allowed extrasolar planet discoveries to reach their current routine level is an iodine cell in the beam. This produces many fine absorption lines that provide the necessary stability, so long as the cell temperature is sufficiently well-controlled (at 50.0 ± 0.1 deg C). A further step that has been taken with the HARPS spectrograph for the ESO 3.6-m telescope is to enclose the spectrograph in a vacuum vessel [107].

9.6 Future Challenges in Optical Spectroscopy

In the face of the enormous technical and observational advances oulined in this chapter, astronomers are having to learn how best to use the facilities now available to them. This is particularly true of the large-scale multi-object surveys, which demand entirely new strategies in order to allow the best science to be extracted. One way of expediting this is by on-line databases. For example, public data-releases of the complete 2dFGRS and 2QZ surveys are imminent, while an early data release of the 6dFGS [108] took place in November 2002. A new release of SDSS data is also scheduled for early 2003.

In parallel with such developments is the evolution of the global virtual observatory (VO), now in its infancy but set to expand rapidly over the next few years. Providing direct internet access to the sky in unprecedented detail by federating data-sets across all wavebands, VO will revolutionise the way optical spectroscopists formulate their observing campaigns.

New technologies are also emerging with implications for instrumental development. For example, in recent years, there has been much interest in superconducting tunnel junction (STJ) array detectors, which offer the prospect of detecting photon energy as well as position. At present, these devices are limited to formats of $\sim 10 \times 10$ pixels and spectral resolutions of $R \sim 20$ at best, but they have already demonstrated their potential for optical spectrophotometry [109, 110].

Without question, however, the biggest challenge facing optical spectroscopy is the coming generation of extremely large telescopes (ELTs) with apertures in the range 30 to 100 m. While the extrapolation of segmented-mirror technology to such apertures is, in itself, a major challenge, there seems little doubt that within little more than a decade, the first ELTs will be under construction.

9.6.1 Spectroscopy with Extremely Large Telescopes

One of the science drivers for ELTs is follow-up spectroscopy of objects discovered with the James Webb Space Telescope (JWST) for $\lambda < 3\mu$m. (Beyond this wavelength, the low infrared background of the JWST makes it more sensitive than a ground-based spectroscopic 30-m telescope.) The kind of spectroscopy envisaged for such large-aperture telescopes ranges from absorption features in the atmospheres of extra-solar planets (including life-diagnostics) to studies of the first luminous sources at redshifts greater than $z \sim 8$. Even on a 30-m class telescope, these are very ambitious projects.

Spectroscopy with ELTs also brings fundamental problems in instrument design. For example, if D is the aperture of the telescope, d is the aperture of the spectrograph collimator (i.e., the pupil diameter) and ϕ is the angular spatial resolution element on the sky, then the spectrograph resolution varies as

$$R \propto \frac{d}{D\phi}.$$

Moreover, if $\mathcal{F}$ is the focal ratio of the spectrograph camera and p is the linear pixel size of the detector, then

$$\mathcal{F} \propto \frac{p}{D\phi}.$$

(See, e.g. [111].) Thus, for a given spectral resolution, the collimator diameter must be made as large as possible and the image diameter as small as possible. Likewise, the image diameter must be made small in order to secure an attainable focal ratio for the spectrograph camera. Instruments such as 2dF and 6dF already have cameras working at $\sim f/1$, and significantly faster designs present real problems in the control of aberrations.

To some extent, these challenges have already been faced by the designers of instruments for 8–10-m class telescopes, but there seems little doubt that the spectrographs of tomorrow will have optics the size of today's medium-class telescope mirrors. They will also rely heavily on adaptive optics (AO) to keep ϕ as small as possible.

There is little prospect of AO (even multi-conjugate AO) providing significant compensation for atmospheric turbulence over fields larger than a few arcminutes, so there will also need to be a new generation of wide-field instruments designed for use in natural seeing. Thus, one such proposal for the Giant Segmented Mirror Telescope (GSMT) is MOMFOS, a wide-field multi-fibre spectrograph using an *Echidna*-type positioner [112].

9.6.2 Epilogue

A survey of such a major branch of astronomy as optical spectroscopy must necessarily be incomplete, and is likely to out of date by the time it appears in print. Despite those inadequacies, it can at least indicate the overall health of the field by the vigour with which development is taking place and the range of endeavours that are in progress. If this chapter presented no other yardsticks than those, there would still be little doubt that optical spectroscopy not only has an illustrious past, but also an energetic and very bright future.

Acknowledgments

I am indebted to Terry Bridges (AAO) for comments on this chapter. It is also a pleasure to thank Chris Ramage (AAO) for invaluable help in preparing the figures.

References

[1] Slipher, V.M., 1917. Proc. Am. Phil. Soc., **56**, 403.
[2] Hubble, E.P., 1929. Proc. Nat. Acad. Sci., **15**, 168.
[3] Colless, M., and the 2dFGRS team, 2002. AAO Newsletter, No.100, 16.
[4] Schilling, G., 2003. Sky & Telescope, **105**, 32.
[5] Steinmetz, M., for the RAVE Science Working Group, 2003. In GAIA Spectroscopy, Science and Technology, ASP Conference Series, p.381.
[6] Groom, D.E., 2000. In Optical and IR Telescope Instrumentation and Detectors, Proc. SPIE, **4008**, 634.
[7] Beletic, J.W. and Amico, P., 2002. In Instrument Design and Performance for Optical/Infrared Ground-Based Telescopes, Proc. SPIE, **4841**.
[8] Bland-Hawthorn, J. and Jones, D.H., 1998. In Optical Astronomical Instrumentation, Proc. SPIE, **3355**, 855.
[9] Tinney, C., for the IRIS2 team, 2002 AAO Newsletter, No. 99, 16.
[10] Ramsay Howat, S., Ellis, M., Gostick, D., Hastings, P., Strachan, M. and Wells, M., 2000. In Optical and IR Telescope Instrumentation and Detectors, Proc. SPIE, **4008**, 1067.
[11] Fehrenbach, Ch., 1966. Advances in Astron. and Astrophys., **4**, 1.
[12] Hearnshaw, J.B., 1986. The analysis of starlight: one hundred and fifty years of astronomical spectroscopy, Cambridge, Chapter 6.
[13] Clowes, R.G., Cooke, J.A. and Beard, S.M., 1984, MNRAS, **207**, 99.
[14] Lelièvre, G., 1983. In Instrumentation in Astronomy V, Proc. SPIE, **445**, 151.
[15] Douglas, N.G., Arnaboldi, M., Freeman, K.C., Kuijken, K., Merrifield M.R., Romanowsky, A.J., Taylor, K., Capaccioli, M., Axelrod, T., Gilmozzi, R., Hart, J., Bloxham, G. and Jones, D.. 2002, PASP, **114**, 1234.
[16] Fort, B., Mellier, Y., Picat, J.P., Rio, Y. and Lelièvre, G., 1986. In Instrumentation in Astronomy VI, Proc. SPIE, **627**, 321.
[17] Breare, J.M., Ellis, R.S., Purvis, A., Miller, W., Webb, D.A., 1986. In Instrumentation in Astronomy VI, Proc. SPIE, **627**, 278.
[18] Butcher, H., 1982. In Instrumentation in Astronomy IV, Proc. SPIE, **331**, 296.
[19] Taylor, K.T. and Ellis, R.S., 1986. AAO Newsletter, No.38.
[20] Hill, J., Angel, J.R.P., Lindley, D., Scott, J., Hintzen, P., 1980. In Optical and IR Telescopes for the 1990s, Tucson, 370.
[21] Hill, J., Angel, J.R.P., Lindley, D., Scott, J., Hintzen, P., 1980. Ap.J., **242**, L69.
[22] Gray, P.M., 1983. In Instrumentation in Astronomy V, Proc. SPIE, **445**, 57.
[23] Ellis, R.S., Gray, P.M., Carter, D., Godwin, J., 1983. MNRAS, **206**, 285.

[24] Watson, F.G., 1986. In Instrumentation in Astronomy VI, Proc. SPIE, **627**, 787.
[25] Watson, F.G., 1988. In Fiber Optics in Astronomy, ASP Conference Series, **3**, 125.
[26] Watson, F.G., Bogatu, G., Saunders, W., Farrell, T.J., Russell, K.S., Hingley, B.E., Miziarski, S., Gillingham, P.R., Parker, Q.A. and Colless, M.M., 2001. In The New Era of Wide Field Astronomy, ASP Conference Series, **232**, 421.
[27] Watson, F.G., 1987. Multi-object Astronomical Spectroscopy with Optical Fibres, PhD Thesis, University of Edinburgh, Chapter 1.
[28] Watson, F.G., 1995. In The Future Utilisation of Schmidt Telescopes, ASP Conference Series, **84**, 71.
[29] Barden, S.C., 1998. In Fiber Optics in Astronomy III, ASP Conference Series, **152**, 14.
[30] Watson, F.G. and Terry, P., 1995. In Fiber Optics in Astronomical Applications, Proc. SPIE, **2476**, 10.
[31] Glazebrook, K., 1998. AAO Newlsetter, No. 87, 11.
[32] Monnet, G.J., 2000. In Optical and IR Telescope Instrumentation and Detectors, Proc. SPIE, **4008**, 8.
[33] Le Fèvre, O., Vettolani, P., Maccagni, D., Mancini, D., Picat, J.P., Mellier, Y., Mazure, A., Saisse, M., Cuby, J.G., Delabre, B., Garilli, B., Hill, L., Prieto, E., Arnold, L., Conconi, P., Cascone, E., Mattaini, E. and Voet, C., 1998. In Optical Astronomical Instrumentation, Proc. SPIE, **3355**, 8.
[34] Mancini, D., Le Fèvre, O., Caputi, O., Ferragina, L., Fiume Garelli, V., Mancini, G., Tommasi Mavar, P., Parrella, C., Perrotta, F., Russo, M., Russo, R., Schibeci, M. and Schipani, P., 2000. In Optical and IR Telescope Instrumentation and Detectors, Proc. SPIE, **4008**, 256.
[35] Le Fèvre, O., et al. (32 authors), 2000. In Optical and IR Telescope Instrumentation and Detectors, Proc. SPIE, **4008**, 546.
[36] Simons, D.A., Gillett, F., Oschmann, J., Mountain, M. and Nolan, R., 2000. In Optical and IR Telescope Instrumentation and Detectors, Proc. SPIE, **4008**, 28.
[37] Crampton, D., Fletcher, J.M., Jean, I., Murowinski, R., Szeto, K., Dickson, C., Hook, I., Laidlaw, K., Purkins, T., Allington-Smith, J.R. and Davies, R.L., 2000. In Optical and IR Telescope Instrumentation and Detectors, Proc. SPIE, **4008**, 114.
[38] Siegmund, W.A., Owen, R.E., Granderson, J., French Leger, R., Mannery, E.J., Waddell, P. and Hull, C.L., 1998. In Fiber Optics in Astronomy III, ASP Conference Series, **152**, 92.
[39] Owen, R.E., Buffaloe, M.J., French Leger, R., Mannery, E.J., Siegmund, W.A., Waddell, P. and Hull, C.L., 1998. In Fiber Optics in Astronomy III, ASP Conference Series, **152**, 98.
[40] Hill, J.M., Angel, J.R.P., Scott, J.S., Lindley, D. and Hintzen, P., 1982. In Instrumentation in Astronomy IV, Proc. SPIE, **331**, 279.
[41] Avila, G., Guerin, J., Fernandez, A., Belleger, R., Dreux, M., Schmidt, R., Rousset, G., Felenbok, P and Collin, A., 1995. In Fiber Optics in Astronomical Applications, Proc. SPIE, **2476**, 77.
[42] Parry, I.R. and Gray, P.M., 1986. In Instrumentation in Astronomy VI, Proc. SPIE, **627**, 118.

[43] Carrasco, B.E., Vázquez, S., Ren, D., Sharples, R.M., Langarica, R., Lewis, I.J. and Parry, I.R., 1998. In Fiber Optics in Astronomy III, ASP Conference Series, **152**, 117.

[44] Parry, I.R., Lewis, I.J., Sharples, R.M., Dodsworth, G.N., Webster, J., Gellatly, D.W., Jones, L.R. and Watson F.G., 1994. In Instrumentation in Astronomy VIII, Proc. SPIE, **2198**, 125.

[45] Lewis, I.J., Sharples, R.M., Parry, I.R., Jones, L.R. and Watson F.G., Barker, S.A. and Rees, P.C.T., 1996. In Optical Telescopes of Today and Tomorrow, Proc. SPIE, **2871**, 1318.

[46] Barden, S.C., Armandroff, T., Massey, P., Groves, L., Rudeen, A.C., Vaughnn, D. and Muller, G., 1993. In Fiber Optics in Astronomy II, ASP Conference Series, **37**, 185.

[47] Barden, S.C. and Armandroff, T., 1995. In Fiber Optics in Astronomical Applications, Proc. SPIE, **2476**, 56.

[48] Gray, P.M., Taylor, K., Parry, I.R., Lewis, I.J., and Sharples, R.M., 1993. In Fiber Optics in Astronomy II, ASP Conference Series, **37**, 145.

[49] Taylor, K., Cannon, R.D. and Watson, F.G., 1996. In Optical Telescopes of Today and Tomorrow, Proc. SPIE, **2871**, 145.

[50] Lewis, I.J., Glazebrook, K. and Taylor, K., 1998. In Fiber Optics in Astronomy III, ASP Conference Series, **152**, 71.

[51] Lewis, I.J. et al. (23 authors), 2002. MNRAS, **333**, 279.

[52] Gillingham, P., Miziarski, S. and Klauser, U., 2000. In Optical and IR Telescope Instrumentation and Detectors, Proc. SPIE, **4008**, 914.

[53] Gillingham, P., Popovic, D., Waller, L. and Farrell, T., 2002. In Instrument Design and Performance for Optical/Infrared Ground-Based Telescopes, Proc. SPIE, **4841**.

[54] Pasquini, L. et al. (29 authors), 2002. ESO Messenger, No. 110, 1.

[55] Parker, Q.A., Watson, F.G. and Miziarski, S., 1998. In Fiber Optics in Astronomy III, ASP Conference Series, **152**, 80.

[56] Watson, F.G., Parker, Q.A., Bogatu, G., Farrell, T.J., Hingley, B.E. and Miziarski, S., 2000. In Optical and IR Telescope Instrumentation and Detectors, Proc. SPIE, **4008**, 123.

[57] Schötz, G.F., Vydra, J., Lu, G. and Fabricant, D., 1998. In Fiber Optics in Astronomy III, ASP Conference Series, **152**, 20.

[58] Moore, A.M., Gillingham, P.R, Griesbach, J.S., Arridge, C., Noakes, K. and Akiyama, M., 2002. In Instrument Design and Performance for Optical/Infrared Ground-Based Telescopes, Proc. SPIE, **4841**.

[59] Wang, G., 2000. In Optical and IR Telescope Instrumentation and Detectors, Proc. SPIE, **4008**, 922.

[60] Carlberg, R.G., Yee, H.K.C. and Ellingson, E., 1997. Ap.J., **478**, 462.

[61] Yee, H.K.C, Morris, S.L., Lin, H., Carlberg, R.G., Hall, P.B., Sawicki, Marcin, Patton, D.R., Wirth, G.D., Ellingson, E. and Shepherd, C.W., 2000. Ap.J. Supp., **129**, 475.

[62] Le Fèvre, O. et al. (33 authors), 2002. In Where's the Matter?, Ed. Frontières, 83.

[63] Davis, M., 2002. In Discoveries and Research Prospects from 6- to 10-Meter Class Telescopes II, Proc. SPIE, **4834**.

[64] Gilmore, G., Wyse, R.F.G. and Norris, J.E., 2002. Ap.J. Letts., **574**, 39.

[65] Odenkirchen, M., Grebel, E.K., Rockosi, C.M., Dehnen, W., Ibata, R., Rix, H.-W., Stolte, A., Wolf, C., Anderson, J.E., Jr., Bahcall, N.A., Brinkmann, J., Csabai, I., Hennessy, G., Hindsley, R.B., Ivezi, E., Lupton, R.H., Munn, J.A., Pier, J.R., Stoughton, C. and York, D.G., 2001. Ap.J. Letts., **548**, 165.
[66] Freeman, K.C., Arnaboldi, M., Capaccioli, M., Ciardullo, R., Feldmeier, J., Ford, H., Gerhard, O., Kudritzki, R., Jacoby, G., Méndez, R.H. and Sharples, R., 2000. In Dynamics of Galaxies: from the Early Universe to the Present, ASP Conference Series **197**, 389.
[67] Odenkirchen, M., Grebel, E.K., Dehnen, W., Rix, H.-W. and Cudworth, K.M., 2002. A.J., **124**, 1497.
[68] Drinkwater, M.J., Jones, J.B., Gregg, M.D. and Phillipps, S., 2000. Publ. Astron. Soc. Aust., **17**, 227.
[69] Ellis, R.S., Colless, M., Broadhurst, T., Heyl, J. and Glazebrook, K., 1996. MNRAS, **280**, 235.
[70] Couch, W.J., Balogh, M.L., Bower, R.G., Smail, I., Glazebrook, K., Taylor, M., 2001. Ap. J., **549**, 820.
[71] Gillingham, P., Miziarski, S., Akiyama, M. and Klocker, V., 2000. In Optical and IR Telescope Instrumentation and Detectors, Proc. SPIE, **4008**, 1395.
[72] Vanderreist, C., 1998. In Fiber Optics in Astronomy III, ASP Conference Series, **152**, 123.
[73] Krabbe, A., Thatte, N., Kroker, H, Tacconi-Garman, L.E. and Tecza, M., 1996. In Optical Telescopes of Today and Tomorrow, Proc. SPIE, **2871**, 1179.
[74] Tecza, M., Thatte, N., Eisenhauer, F., Mengel, S., Röhrle, C. and Bickert, K., 2000. In Optical and IR Telescope Instrumentation and Detectors, Proc. SPIE, **4008**, 1344.
[75] Bacon, R., Adam, G., Baranne, A., Courtès, G., Dubet, D., Dubois, J.P., Emsellem, E., Ferruit, P., Georgelin, Y., Monnet, G., Pecontal, E., Rousset, A. and Say, F., 1995. A & AS, **113**, 347.
[76] Barden, S.C. and Wade, R.A., 1988. In Fiber Optics in Astronomy, ASP Conference Series, **3**, 113.
[77] Arribas, S., Cavaller, L., García-Lorenzo, B., García-Marín, A., Herreros, J.M., Mediavilla, E., Pi, M., del Burgo, C., Fuentes, J., Rasilla, J.L., Sosa, N., Carter, D., Jones, L., Edwards, R., Gentles, B., Pollacco, D. and Rees, P., 1998. In Fiber Optics in Astronomy III, ASP Conference Series, **152**, 149.
[78] Parry, I., Kenworthy, M. and Taylor, K., 1996. In Optical Telescopes of Today and Tomorrow, Proc. SPIE, **2871**, 1325.
[79] Murray, G.J. Allington-Smith, J.R., Content, R., Dodsworth, G.N., Dunlop, C.N, Haynes, R., Sharples, R.M. and Webster, J., 2000. In Optical and IR Telescope Instrumentation and Detectors, Proc. SPIE, **4008**, 611.
[80] Roth, M.M., Bauer, S.-M., Dionies, F., Fechner, T., Hahn, T., Kelz, A., Paschke, J., Popow, E., Schmoll, J., Wolter, D., Laux, U. Altmann, W., 2000. In Optical and IR Telescope Instrumentation and Detectors, Proc. SPIE, **4008**, 277.
[81] Roth, M.M., Laux, U. and Heilemann, W., 2000. In Optical and IR Telescope Instrumentation and Detectors, Proc. SPIE, **4008**, 485.
[82] Tacconi-Garman, L., 1999. AAO Newsletter, No. 90, 7.
[83] Roth, M.M. and Laux, U.., 1998. In Fiber Optics in Astronomy III, ASP Conference Series, **152**, 168.

[84] Bershady, M.A., Anderson, D., Ramsey, L. and Horner, S., 1998. In Fiber Optics in Astronomy III, ASP Conference Series, **152**, 253.
[85] García-Lorenzo, B., Mediavilla, E., Arribas, S. and del Burgo, C., 1998. In Fiber Optics in Astronomy III, ASP Conference Series, **152**, 185.
[86] Chatzichristou., E.T., 1998. In Fiber Optics in Astronomy III, ASP Conference Series, **152**, 141.
[87] Serra-Ricart, M., Mediavilla, E., Arribas, S., del Burgo, C., Oscoz, A., Alcalde, D., García-Lorenzo, B., Buitrago, J. and Goicoechea, L.J., 1998. In Fiber Optics in Astronomy III, ASP Conference Series, **152**, 155.
[88] Bridges, T., on behalf of the $AA\Omega$ team, 2002. AAO Newsletter, No. 100, 20.
[89] Osmer, P.S., Atwood, B., Byard, P.L., DePoy, D.L., O'Brien, T.P., Pogge, R.W. and Weinberg, D., 2000. In Optical and IR Telescope Instrumentation and Detectors, Proc. SPIE, **4008**, 40.
[90] Bingham, R.G., Gellatly, D.W., Jenkins, C.R. and Worswick, S.P., 1994. In Instrumentation in Astronomy VIII, Proc. SPIE, **2198**, 56.
[91] Worswick, S.P., Gellatly, D.W., Ferneyhough, N.K., King, D.L., Weise, A.J., Bingham, R.G. and Oates, A.P., 1995. In Fiber Optics in Astronomical Applications, Proc. SPIE, **2476**, 46.
[92] Maihara, T., Iwamuro, F., Hall, D.N.B., Cowie, L.L., Tokunaga, A.T. and Pickles, A.J., 1993. In Infrared Detectors and Instrumentation, Proc. SPIE, **1946**, 581.
[93] Parry, I.R., Dean, A., Johnson, R.A., King, D.L., Mackay, C.D., Horton, A., Pritchard, J.M., McMahon, R.G., Medien, S.R., Sharp, R. and Ramaprakash, A.N., 2002. In Instrument Design and Performance for Optical/Infrared Ground-Based Telescopes, Proc. SPIE, **4841**.
[94] Watson, F., 1996. In Optical Telescopes of Today and Tomorrow, Proc. SPIE, **2871**, 1373.
[95] Barden, S.C., Arns, J.A. and Colburn, Willis, S., 1998. In Optical Astronomical Instrumentation, Proc. SPIE, **3355**, 866.
[96] Saunders, W., for the 6dF upgrade crew, 2002. AAO Newsletter, No. 101, 14.
[97] Dimitroff, G.Z. and Baker, J.G., 1945. Telescopes and Accessories, Blackiston, Philadelphia, 273-4.
[98] Stilburn, J.R., 2000. In Optical and IR Telescope Instrumentation and Detectors, Proc. SPIE, **4008**, 1361.
[99] Watson, F.G., Offer, A.R., Lewis, I.J., Bailey, J.A. and Glazebrook, K., 1998. In Fiber Optics in Astronomy III, ASP Conference Series, **152**, 50.
[100] Soucail, G., Cuillandre, J.C., Picat, J.P. and Fort, B., 1995. In New Developments in Array Technology and Applications, IAU Symposium No. 167, 263.
[101] Glazebrook, K. and Bland-Hawthorn, J., 2001. PASP, **113**, 197.
[102] Glazebrook, K., 1999. AAO Newsletter, No. 90, 11.
[103] Bridges, T., 2002. 2dF Nod & Shuffle Tests: Analysis and Conclusions, AAO Report.
[104] Diego, F., Brooks, D., Charalambous, A., Crawford, I., D'Arrigo, P., Dryburgh, M., Jamshidi, H., Radley, A., Savidge, T. and Walker, D., 1996. In Optical Telescopes of Today and Tomorrow, Proc. SPIE, **2871**, 1126.
[105] D'Arrigo, P., Diego, F. and Walker, D.D., 1996. In Optical Telescopes of Today and Tomorrow, Proc. SPIE, **2871**, 1306.

[106] Butler, R.P., Marcy, G.W., Williams, E., McCarthy, C., Dosanjh, P. and Vogt, S.S., 1996. PASP, **108**, 500.
[107] Pepe, F., Mayor, M., Rupprecht, G. and the HARPS team, 2002. ESO Messenger, No. 110, 9.
[108] Read, M., Colless, M., Saunders, W. and Watson, F., for the 6dFGS team, 2002. AAO Newsletter, No.101, 16.
[109] Rando, N., Verveer, J., Verhoeve, P., Peacock, A., Andersson, S., Reynolds, A., Favata, F., Perryman, M.A.C. and Goldie, D.J., 2000. In Optical and IR Telescope Instrumentation and Detectors, Proc. SPIE, **4008**, 646.
[110] Martin, D.D., Verhoeve, P., Peacock, A.J. and van Dordrecht, A., 2002. In Instrument Design and Performance for Optical/Infrared Ground-Based Telescopes, Proc. SPIE, **4841**.
[111] McLean, I.S. and Chaffee, F.H., 2000. In Optical and IR Telescope Instrumentation and Detectors, Proc. SPIE, **4008**, 2.
[112] AURA New Initiatives Office, 2002. Enabling a Giant Segmented Mirror Telescope for the Astronomical Community, CD-ROM.

10 Large-Scale Structures in the Distribution of Galaxies: The 2dF and Sloan Surveys

A.P. Fairall

10.1 Introduction

The spatial distribution of galaxies reveals an underlying cosmic texture that, aside from its own intrinsic interest, carries information on the nature and evolution of the Universe. Such a distribution is mapped in redshift space, whereby cosmic distances are assessed according to the well-known Hubble relation. In Hubble's day it took many nights of telescope time to obtain a single redshift, but the improvement in the sensitivity of photographic emulsions accelerated their acquisition. By the 1960s, over two thousand were known. The process was further accelerated by the introduction of image intensifiers and eventually CCDs, and optical fibres that enabled multiple objects to be observed simultaneously. Currently it is possible to acquire more than two thousand redshifts in a single night of observation.

Two major surveys - each a collaborative effort by a large number of researchers - have been organised to take advantage of this situation. They are the '2dF Galaxy Redshift Survey' and the 'Sloan Digital Sky Survey' (referred to here as the '2dF' and 'Sloan' surveys). Both seek to map the spatial distribution of galaxies on scales of an order of magnitude beyond anything previously available. In short the 2dF survey has measured towards a quarter million galaxy redshifts, while the Sloan survey is aiming to catalogue some 10^7-10^8 galaxies and obtain close to a million redshifts. Both surveys also have associated quasar surveys that penetrate to deep space. The two surveys so dominate the field that it is considered appropriate to devote this article to a review of their progress

Observations for the 2dF survey were completed in April 2002, while the Sloan survey still has far to go. Nevertheless the data already available to the 2dF and Sloan teams - including that put out as the 2dF '100K' release and the Sloan Early Release (14 million objects with 54000 spectra) in 2001 - have already given rise, *inter alia*, to a great number of papers conveying significant findings concerning galaxies and cosmology. This review (written late 2002/early 2003) attempts to synthesise highlights of over 60 papers (conference papers excluded), and the author apologises in advance for any misjudgement in coverage. This review does not include the quasar surveys and it says nothing of the Sloan stellar survey.

The 2dF survey is a British-Australian collaboration based around the existing 3.9 metre Anglo-Australian Telescope; its team of collaborators numbers 32. By contrast the Sloan survey is a collaboration mainly based in the United States but with significant Japanese involvement. It has a purpose-built 2.5-metre telescope in New Mexico and a team that appears to number no less than 143 people. Both surveys sample the Universe to a median depth of approximately $z = 0.1$ (400 Mpc), though clearly the American effort will produce a greater quantity and quality of data. Since both seek to extract the same cosmological parameters from their samples (though not necessarily over the same patch of sky), they are somewhat competitive to one another. The 2dF team have clearly been keen to get their results out first, while the Sloan researchers, and many groups, have been eager to get their hands on their early data.

2dF stands for 'two-degree field', produced by a special corrector lens system (designed by the late Charles Wynne) that so enlarges the usable field of view of the Anglo-Australian Telescope that it may survey a significant fraction of the entire sky. Similarly, the Sloan telescope is a modified version of a Ritchey-Chretian design, with an oversized secondary and two corrector lenses, that allows for a usable three-degree field. Such fields might include up to several hundred target galaxies, the light from which is intercepted by individual optical fibres.

The two surveys differ in the way they position the fibres. Each of the fibres in the 2dF system has a tiny prism on its front end, which sits on a magnetic button base. A computer robot sets it into position by grasping the button, drawing out its fibre and then placing the button with precision on a metal plate (to coincide with the position of the target galaxy in the focal plane of the telescope). Initially this went rather slowly, but its speed was improved until the robot could position all 400 fibres in about an hour. While this is going on, the telescope would be observing the previous field. Once ready to switch to the new field, a tumbler mechanism at the top of the telescope turns the new field, with its positioned fibres, into place, while the old field is turned ready for the robot to reset. The Sloan survey uses the older less-sophisticated system whereby a metal plate is prepared for each field with drilled holes coinciding with the positions of the target galaxies. The fibres are simply inserted by hand. In the past this was a painstaking process as particular fibres had to go into particular holes and human error sometimes crept in. With Sloan, however, the 640 fibres per field are set up anyhow, but then a moving strip of light scans across the plate in perpendicular directions. A photometric device monitors the sequence in which the output end of the fibres light up and thereby automatically takes care of the identifications. Both 2dF and Sloan feed their fibres to double spectrographs.

10.2 Target Galaxies

The essential foundation of any redshift survey is a catalogue of target galaxies, but the number of faint galaxies involved in 2dF and Sloan far exceeds that in any existing catalogue, so much so that creating a target list becomes a major project in itself. In the case of the Sloan survey (the Sloan Digital Sky Survey) the creation of the photometric catalogue by the same telescope that subsequently does the follow up spectroscopy is a major effort and product of the project.

The target galaxies for the 2dF survey are derived from a greatly extended 'APM' (Automatic Plate Measuring) survey of fields of the photographic SERC IIIa-J Sky Survey carried out (in the 1980s) with the UK Schmidt Telescope, also at the Anglo-Australian Observatory. The spectral response of the J emulsion peaks in the green, with a long tail to the blue, and the magnitudes of the targets derived from the area and density of the emulsion are expressed as b_J. Objects identified automatically as galaxies and with magnitudes brighter than $b_J = 19.4$ are considered as targets for the redshift survey. Magnitudes extracted from photographic plates are not ideal. They typically have errors of tenths of a magnitude and can even be wildly off if the core of the galaxy has saturated emulsion.

A further complication, which applies to both surveys, is the separation of galaxy images from those of stars. Unresolved stars should show a symmetrical point spread function, while galaxies will deviate from this. The APM survey has long developed its own separation technique (Maddox et al. 1990), but for instance close double stars, with images overlapping, are a significant contaminant. The greatest form of incompleteness in 2dF is the classification criteria (Colless et al. 2001). The Sloan survey, however, does have the advantage of multi-colour photometry which provide a further effective discriminator, and claims to be better than 99% complete (Strauss et al. 2002).

Foreground galactic stars can also be incorrectly identified as galaxies and included in the target list, but redshift observations easily sort them out. The 2dF survey reports that 6% of targets turned out to be foreground stars, while Sloan indicates less than 1%.

10.3 The Sloan Photometric Camera

A major aspect of the Sloan survey is its imagery, using CCDs, which are far preferable to photographic plates. While CCDs normally only image a very small part of the sky, but the Sloan survey uses a mosaic of no less than thirty large 2048×2048 CCDs (Gunn et al. 1998). The CCDs are mounted on an Invar base and arranged in a 6 by 5 array. Each of the six columns observes in drift scan mode, whereby the starfield drifts across the array, and the data is read out continuously from the CCDs at an appropriate matching rate. The standard set is the sidereal rate resulting in each star or galaxy taking 54 seconds to cross one of the CCDs (hence a 54 second exposure time). Rather than following bands of constant declination, the telescope is made to follow great circles in the sky, so that the exposure time is kept uniform over the survey area (York et al. 2000).

Each star or galaxy image passes over a sequence of five CCDs, each equipped with a different filter. The filters are the 'Petrosian' u', g', r', i', z' that span from the violet to the near infrared. Combined with the CCD, peak response is in the r-band. The magnitudes of all objects are extracted in those five colours. Since the CCDs are of necessity spaced apart from one another, the intervening strips of sky that miss being recorded are picked up on a subsequent scan when the telescope is offset accordingly. There is a small overlap between the strips from the original scan and those of the secondary scan. The array of photometric CCDs is complimented by 24 smaller CCDs, that record the same sky strips for astrometic purposes. The survey therefore extracts precise positions and 5-colour magnitudes for close on 10^8 objects (thereby exceeding all existing stellar photometric measurements), most of which are foreground stars, but from which galaxies (and quasar candidates) are extracted by the selection algorithms.

Given that the survey is expected to identify approximately 5×10^7 galaxies but only 10^6 of these can be followed up spectroscopically, two different sorts of targets are selected. The larger sample (perhaps 9×10^5) will consist of all identified galaxies brighter than r' = 17.77 and surface brightnesses above 24.5 magnitudes per square arcsecond. The additional sample (10^5 galaxies) will comprise red galaxies down to g' = 21.4. The latter are most likely giant elliptical, or cD galaxies, that tend to be the dominant members of clusters; their selection should approximate a deeper volume-limited sample out to z = 0.38, significantly deeper than 2dF.

For comparison purposes Gaztanaga (2002) has shown that combinations of Sloan data can be used to approximate APM (2dF) data.

10.4 Sky Coverage

The 2dF survey covers approximately 5% of the entire sky, the Sloan approximately 25%. Figure 10.1 shows the regions of sky covered. Both surveys steer clear of the foreground obscuration of the Milky Way. Even so, there are still some patches with slight obscuration; these have to be taken into account as they affect the magnitudes of the galaxies and the completeness of the surveys.

2dF mainly concentrates on the Southern Galactic Hemisphere. Its fields there cover 10 strips of declination, extended in right ascension. In addition to the main area, a hundred scattered fields extend the coverage in that hemisphere. In the Northern Galactic Hemisphere, the area covered is 6 strips and there are no scattered fields.

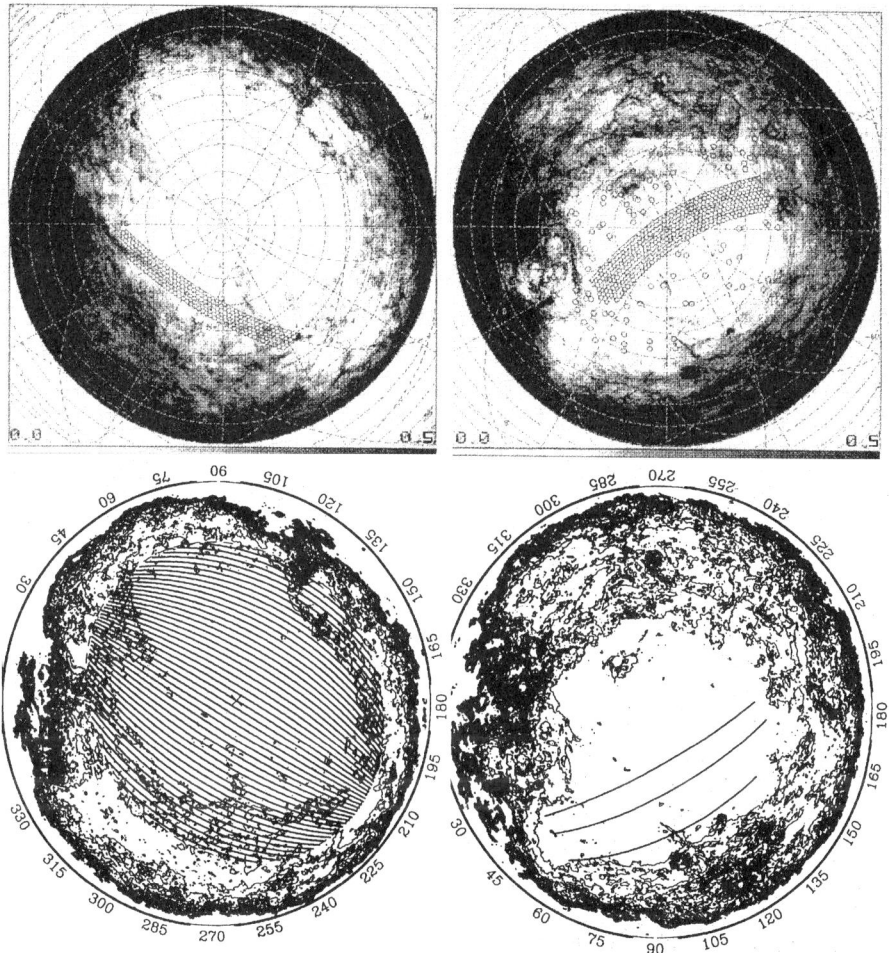

Fig. 10.1. The regions of sky covered by 2df (top) and Sloan (bottom) in the northern (left) and southern (right) galactic caps. Galactic obscuration from Schlegel et al. (1998) is shown.

The Sloan survey concentrates on almost complete coverage of the unobscured portion of the Northern Galactic Hemisphere and only token coverage - three strips - in the south. Note that the strips of the survey are along great circles in the sky.

10.5 Spectroscopy

Both surveys have to address the same problem: Given that the sky distribution of galaxies is irregular (due to the presence of large-scale structures and clusters), how are the fields for spectroscopic observations chosen so to maximise the number of target galaxies observed for the minimum number of exposures. The 2dF fields allow for almost 400 galaxies to be observed simultaneously, and the Sloan fields 640. Furthermore, 2dF does not allow fibres to be closer than 2.2 arcmin and Sloan 0.9 arcmin, so crowded fields require multiple exposures. The aim of 2dF was to obtain towards a quarter million redshifts, so for this reason the target selection criterion (b_J brighter than 19.4) was set so that ample targets were available and fibres would not go unused; In all some 390 000 targets were identified. Both surveys developed tiling algorithms to maximise coverage by adjusting the spacing of the fields (unlike the imagery, Sloan spectroscopy cannot obviously be done in drift scan mode). Sloan claims to reach better than 92% of its targets.

Sample spectra from both 2dF and Sloan surveys are shown in Fig. 10.2. They include both galaxies with the common absorption-line spectra (integrated starlight when star formation is quiescent) and those with emission-line spectra (interstellar gas excited by the ultra-violet radiation of hot young stars). From the point of view of extracting redshifts, emission lines make the job very easy while, with absorption spectra, it is a matter of pushing the signal to noise ratio such that absorption features can be recognised beyond reasonable doubt. Since the exposure times cannot be varied for individual

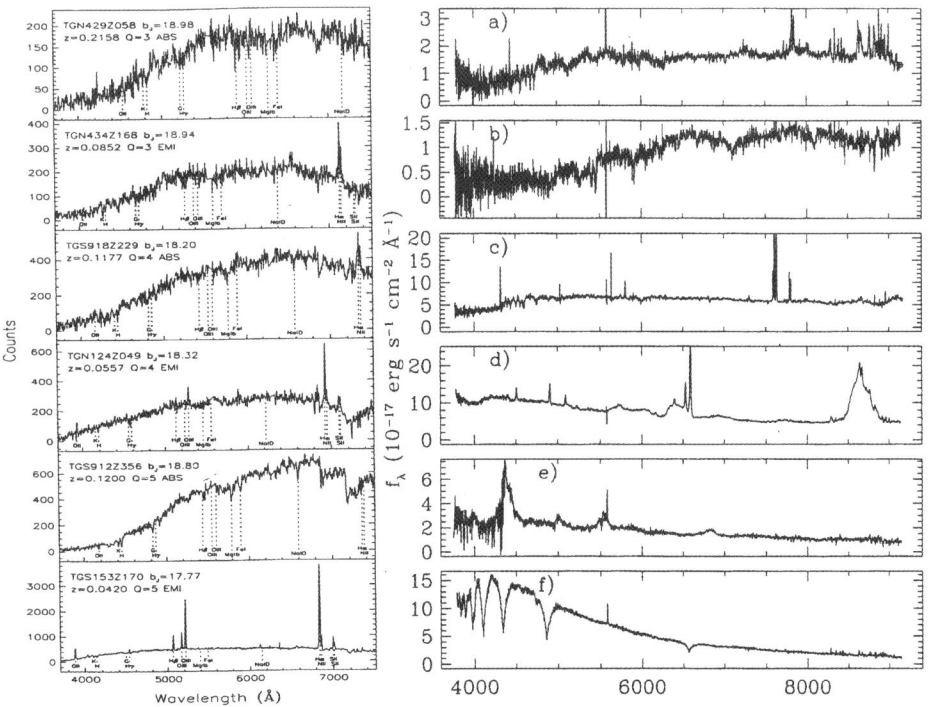

Fig. 10.2. Sample spectra from 2dF (left) and Sloan (right). (Diagrams have been slightly distorted to match scales)

galaxies, the general exposure times chosen (45-60 minutes in total) is such that signal to noise will be adequate for most targets. However, there are obviously marginal cases.

The 2dF survey assigns a quality parameter from 1 to 5 for each of the spectra. The small number of low quality spectra (Q1 and Q2) are rejected and only those of higher quality (Q3 and above), where there is a greater than 90% chance that the extracted redshift is correct, are accepted. In almost all cases the redshift is extracted automatically by cross-correlation software.

Both 2dF and Sloan (Colless et al. 2001, Castander et al. 2001) have been very careful to assess how their redshifts compare to existing data. 2dF also observed some 8000 galaxies twice over to assess the reliability of the redshifts extracted (and from which they claim an rms error of single observations at 85 km s^{-1}).

10.6 Redshift Maps and Large-Scale Structures

Figure 10.3 shows a histogram of the 120 000 2dF redshifts, from which it can be seen that the median redshift of the survey is z = 0.11. The departures of the data from a smooth distribution are an indication of the existence of large-scale structures. Perhaps it is a little curious that there are three conspicuous peaks uniformly spaced. The spacing however, approximately $\Delta z = 0.025$, does not match that of the 'BEKS' peaks claimed in years gone by.

The most visible product of any redshift survey are plots showing the distribution of the galaxies in redshift space. That for the 2dF survey is shown in Fig. 10.4. The texture it reveals is already familiar, but it has never before been seen on so large a scale. Regions of high and low density alternate to produce a complex frothy or filamentary structure. As with the its predecessor - the Las Campanas Redshift Survey - it could be said to be 'seeing beyond bigness'; while there are significant very large-scale inhomogeneities present, the sample volume is far larger and shows a repetition of such structures. This is significant in terms of standard cosmological models which assume homogeneity on a large scale. However the left-hand side of the diagram shows one particular strong clump (at z = 0.08), while there is a deficiency of clusters (at z = 0.05) on the right side (DePropris et al. 2002).

The geometry of the sample regions lends itself to being plotted in this style, but the slices in declination are nevertheless not that thin, and the view the reader sees is looking through quite a thick slice, such that features are superposed upon one another. Plots that separate the volume into thinner slices, or allow the data to be visible in three dimensions, are still needed to assess the character of the texture. In particular, it is not clear whether the large-scale structures are predominantly filamentary in nature, or whether the texture is more 'frothy' where surfaces of galaxies enclose voids somewhat like soapsuds. Past investigations of the 'genus' of local large-scale structures

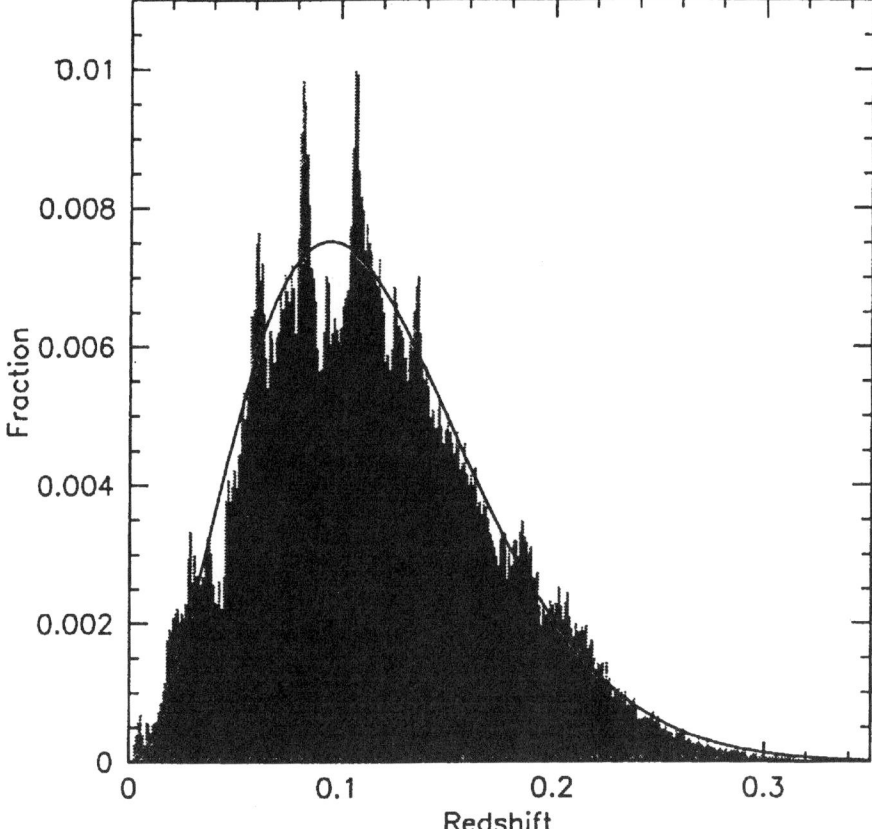

Fig. 10.3. The distribution of 2df redshifts.

have indicated a tendency towards zero, representing a 'bath sponge' topology with interconnected networks of both high density regions and voids. Hoyle, Vogeley and Gott (2002) report than an analysis of the two dimension distribution shows a slight tendency for the Northern Galactic Hemisphere (left-hand side of diagram) towards 'meatball' topology (positive genus) and 'bubble-like' in the Southern Galactic Hemisphere (right-hand). This is not obvious to the eye, and might easily be swayed by the presence of the dense clump (meatball) already mentioned.

The Sloan survey has not (at the time of writing) produced equivalent plots, but an early plot based on only 29 000 data points is shown in Fig. 10.5. Being much thinner than the 2dF plots, it is effectively a cross section through the large-scale structures and shows something of a 'soapsud' texture. Early studies of Sloan structures and genus (Doroshkevich et al. 2002, Hikage et al. 2002) report a compatibility with those for standard flat, Λ-CDM cosmological models, but note some systematic differences (probably marginal) between the northern and southern galactic caps.

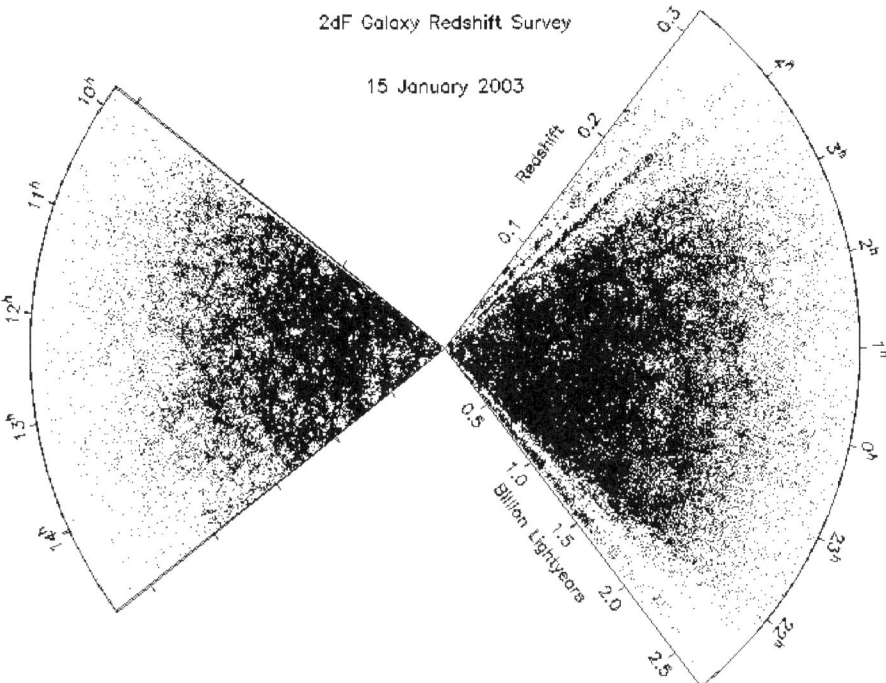

Fig. 10.4. The large-scale distribution of galaxies in the 2dF survey.

10.7 Galaxy Properties and Spatial Distribution

The photometric colours of the Sloan survey galaxies (Blanton 2002, Hogg et al. 2002, Strateva et al. 2001, Shimasaku et al. 2001) clearly separate two general populations of galaxies: The redder population is comprised chiefly of elliptical and S0 galaxies, with concentrated (de Vaucouleurs $r^{0.25}$) profiles, generally of higher luminosity. The bluer population has spiral and irregular galaxies, with exponential profiles and generally lower luminosity. While this trend is well known, never before has such a distinct separation been seen. An alternative approach to galaxy morphology is offered by Abraham et al. (2003).

An examination of the luminosity density of the red galaxies (Hogg et al. 2002) finds that they account for one fifth of the number density but produce two fifths of the luminosity. Bernardi et al. (2003a, 2003b, 2003c, 2003d) have performed detailed analyses of 9000 of these early-type galaxies. Amongst other things, they find that at any given redshift, the intrinsic distribution of luminosities, sizes and velocity dispersions are all approximately Gaussian, and that evolutionary effects are weak, with colours and chemical abundances consistent with formation 9 Gyrs ago. They are also able to determine a 'fundamental plane' relationship. Measurements of their axial ratios (Khairul Alam & Ryden 2002) show that they cannot be a population of oblate ellip-

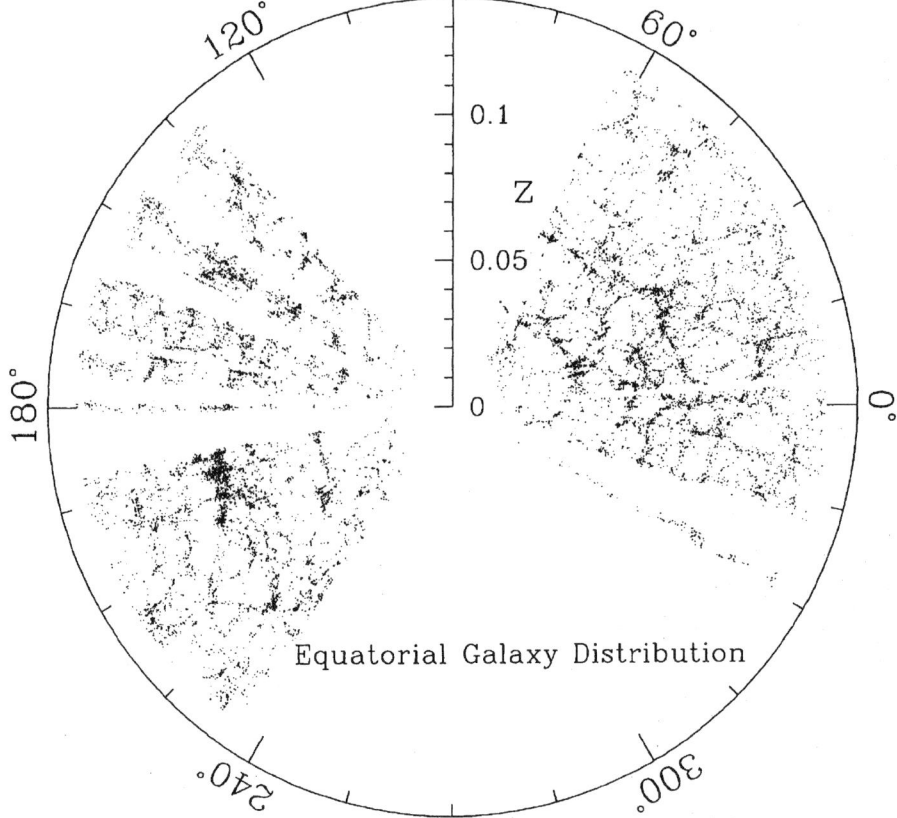

Fig. 10.5. A plot of the distribution of some Sloan data.

soids (randomly orientated) - further evidence that ellipticals are more likely tri-axial ellipsoids. While the lack of colour in the 2dF survey does not allow a photometric separation between the red and blue populations, spectral type serves as a substitution.

Not surprisingly, the well known tendency for the E/S0 galaxies to be strongly clustered in regions of high density similarly separates their spatial distributions: An analysis of 29 000 Sloan galaxies (Zehavi et al 2002) finds the red galaxies have a stronger steeper correlation function and that luminosity bias is independent of scales for $r \leq 10\ h^{-1}$ Mpc.

The 2dF team (Norberg et al. 2001) have looked at the luminosity dependence of galaxy clustering in their survey, by means of measuring 2-point correlation functions for different intervals of absolute magnitudes. In the process the investigators also determine a general correlation function for the survey for which they found a correlation length of $4.9 \pm 0.3\ h^{-1}$ Mpc and a slope $\gamma = 1.71 \pm 0.06$. They find that the most luminous galaxies are 3.0 times as clustered as the norm. In a further paper (Norberg et al. 2002a) they explore the dependence of clustering on both luminosity and spectral type. The team (Madgwick et al. 2002, Folkes et al. 2002) also use 75 000 2dF galaxies to establish luminosity functions according to spectral class. For those galaxies with mainly emission spectra (the blue galaxies), they find a steepening of the function towards the faint end. For galaxies with mainly absorption spectra (red) they find a poor fit at the faint end to the standard 'Schechter' fit, which might suggest the presence of a significant dwarf population, though this is denied in a follow up paper (Cross et al. 2001).

A luminosity function derived from very early Sloan data (Blanton et al. 2001) reported luminosity densities 2.0 times that found (in the same wavelength range) by the Las Campanas Redshift Survey and 1.4 of that found by 2dF (different wavelength range). A subsequent 2dF paper (Norberg et al. 2002b) however claimed excellent agreement with Sloan. The 2dF team (Cole et al. 2001) also used magnitudes from the 2MASS catalogue with the 2dF data to obtain a near-infrared luminosity function which has the usual Schechter form.

The 2dF team (Baldry et al. 2002) compare their (red galaxy) absorption-line data to star formation models of population synthesis, and conclude that there was a peak of star formation in the past of at least three times the current value. Their emission-line data (blue galaxies) shows a correlation between star formation rate and local projected density (Lewis et al. 2002), a result confirmed by Sloan data (Gomez et al. 2003). Similarly, star-formation histories, derived separately from 2dF and Sloan data, are found to be broad agreement (Glazebrook et al. 2003). A sample of galaxies described as 'post-starburst' have also been extracted from the Sloan data (Goto et al. 2003b).

Kauffmann et al. (2002) have also studied star formation in the Sloan data. They find a family of lower mass galaxies (the blue galaxies) having young stellar populations and a family of high mass galaxies (the red galaxies) where star formation has long since ceased. They have also looked at size distribution (Shen et al. 2003).

10.8 Clusters

The 2dF team (De Propis et al. 2002) have investigated several hundred clusters of galaxies, previously identified (as Abell, APM and EDCC clusters), in the 2dF data, and are able to establish levels of contamination by foreground and background galaxies in each of the classes.

The Sloan team uses a 'cut and enhance' method (Goto et al. 2002a) to identify clusters from their imagery. Using photometric redshifts for distant clusters (Goto et al. 2002b), they determine a composite luminosity function. They also determine a mass function (Bahcall et al. 2002), and confirm the 'Butcher-Oemler' effect whereby more distant clusters have an increasing fraction of 'late' type galaxies (Goto et al. 2003a). Basilakos (2003) has detected superclusters by percolations from Sloan clusters and finds that filamentary structures are dominant shapes.

10.9 Redshift Space Distortion, Biasing and the Mass Density of the Universe

The generally accepted scenario for the formation of large-scale structures is that they grew primarily by gravity from very mild over- and under-densities that were present in the early Universe. Inflationary theory - that supposes the volume of the Universe expanded exponentially for a brief instant of time - has it that the fluctuations that seeded the structures are quantum in origin and Gaussian in distribution. The mild, but highly significant, fluctuations detected in the Cosmic Microwave Background (early Universe) are presumed to be the embryonic structures.

If so, the present day structures have formed from the migration and motion of galaxies. Such motions should still be apparent today, and be apparent in the 2dF and Sloan data, since they measure the distribution of galaxies in redshift space, rather than normal space. Since the peculiar motions of galaxies are superposed on the cosmological expansion, they cause the galaxies to deviate from the linear Hubble relation. Such distortion only occurs in a radial direction. Thus plots of the correlation function in both radial and non-radial directions reveal the magnitude of the redshift space distortion, both that on a small scale caused by clusters (the well known 'finger of God') and that on a larger scale caused by the condensing large-scale structures.

Figure 10.6 shows the redshift space distortion found by the 2dF team (Peacock et al. 2001) and that from very limited Sloan data (Zehavi et al. 2002). Kaiser (1987) has shown that this distortion allows one to measure the parameter $\beta = \Omega^{0.6}/b$ where Ω is the mass density of the Universe (in units of the cosmological critical mass) and b is the bias of galaxies to mass. The 2dF team, in a paper to *Nature*, report that they have measured it at 0.43 ± 0.07, a more precise measurement than any previous made and, when combined with Cosmic Microwave Background data, a significant constraint on the mass density of the Universe, which they find to be $\Omega_m \approx 0.3$.

Verde et al. (2002) have similarly analysed 80 million triangular configurations and found the bias factor 'consistent with unity' suggesting that optically selected galaxies do trace the underlying mass distribution, and $\Omega_m = 0.27 \pm 0.06$. Biasing in 2dF is also discussed further by Lahav et al. (2001).

The Sloan cluster mass function (Bahcall et al. 2002) is also compared against cosmological simulations which finds a best fit of $\Omega_m = 0.19 \pm 0.08$.

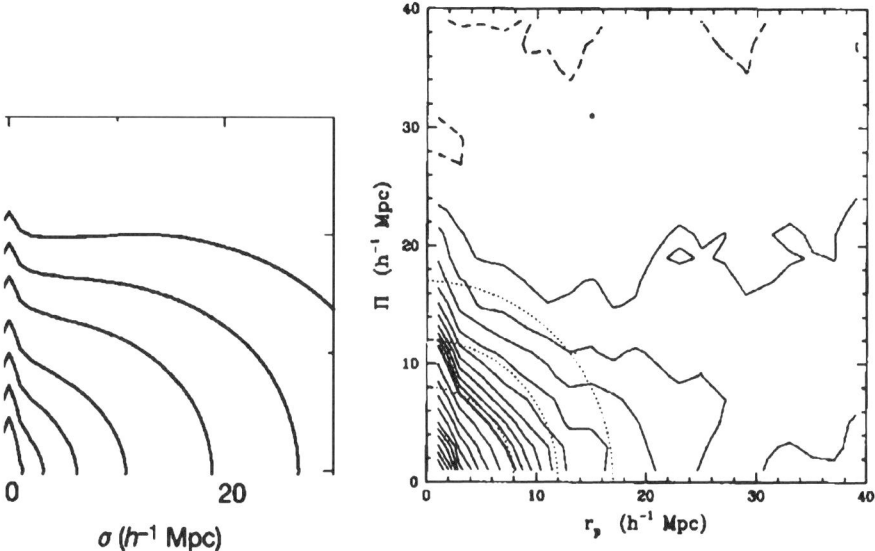

Fig. 10.6. Distortions in redshift space from 2dF (left) and Sloan (right).

10.10 Power Spectra

The most commonly used analytical tool for spatial distributions on extragalactic scales is the correlation function, or its Fourier transform, the power spectrum. In the same fashion that a power spectrum of a time sequence (such as a pulsating variable star) reveals characteristic frequencies and overtones, so a power spectrum of a spatial sequence will reveal characteristic spatial separations, and how their relative power varies with wave number (the equivalent parameter to frequency).

Recent years have seen intense and competitive determinations of that corner stone of cosmology, the power spectrum of the Cosmic Microwave Background (See also Chap. 5 by A. Jaffe in this volume). Major progress has been made with the prediction of acoustic peaks in the spectra at higher wave numbers, and their subsequent discovery by the BOOMERANG and WMAP probes. If that was the power spectrum then, measurements of the power spectrum now should reveal how the Universe has subsequently evolved and constrain various cosmological parameters. Numerous computer (n-body) simulations have attempted this and produced theoretical power spectra that vary according to the cosmological parameters selected.

There is of course a difference in the style of the data. The Cosmic Microwave Background is a two-dimension continuous image that reveals a weak pattern of irregularities (anisotropies) from which the power spectrum is derived. In contrast, the computer simulations have data sets of points distributed in three-dimensional space.

Redshift surveys, such as 2dF and Sloan, map the distribution of galaxies in three-dimensional redshift space (which closely approximates their distribution in normal space). Power spectra can therefore be extracted in similar fashion to the simulations. This has been done in the past for previous major redshift surveys (for example those from Center for Astrophysics, Southern Sky Redshift Survey, Las Campanas Redshift Survey, etc.). The observed power spectra show a general accordance with the theoretical ones. However they are most poorly determined at the shorter wave numbers - which hold the greatest cosmological interest - because of the limited spatial extent of the surveys and the limited availability of data. Consequently one of the main quests of both the 2dF and Sloan surveys was to push to greater depths, and to acquire a far greater amount of data than had been previously available, so as to map the current power spectrum more accurately to lower wave numbers. Many investigators would see this as the most important purpose of the two surveys. Accordingly there has been great enthusiasm in extracting power spectra, as we shall now describe.

The 2dF survey (Percival et al. 2001) has been first across the line with a power spectrum based on the first 160 000 redshifts available. The procedure has been to assign each of the galaxies to a cell in a 512 by 512 by 256 grid. Power spectra (like correlation functions) compare the observed distri-

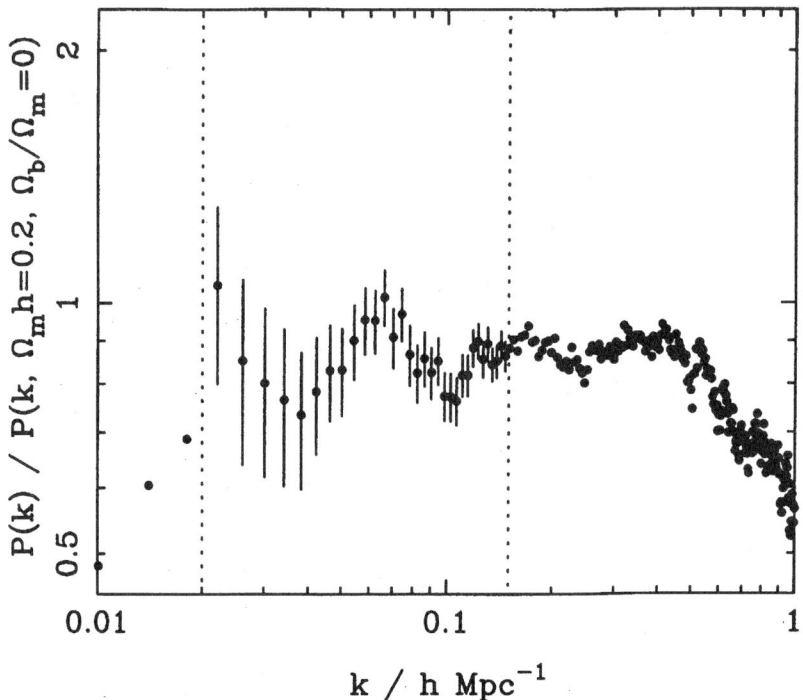

Fig. 10.7. The 2dF power spectrum (see text for explanation).

bution to a random distribution. (A random distribution shows no tendency for points to associate with one another on any scale and would therefore produce a flat power spectrum.) The 2dF teams constructed some random distributions, with 800 000 data points, against which to measure the power spectrum. The simulations, although random, have to nevertheless incorporate any observational biases, such as the thinning of data with increasing redshift, that may affect the observational data, but which have nothing to do with the character of the Universe. Other considerations come in. The limited volume surveyed and its varying levels of incompleteness create a window function.

Figure 10.7 shows the 2dF power spectrum still convolved with the window function. Instead of presenting a traditional 'humpback' power spectrum (that normally peaks around 0.07 h Mpc-1), the authors have chosen to show the data relative to the power spectrum of a linear-theory cold dark matter model (with $\Omega_m h = 0.2$ and Ω_b - the baryonic contribution - at zero). At high wavenumbers, the departure of the data from a horizontal line is due to the non-linear development of large-scale structures on small scales. However, the region of greatest cosmological interest is left of centre between the two broken lines (the small portion further left would be too affected by the window function to be relevant). As pointed out by the 2dF team, although the error bars almost accommodate a level function, it is curious and probably

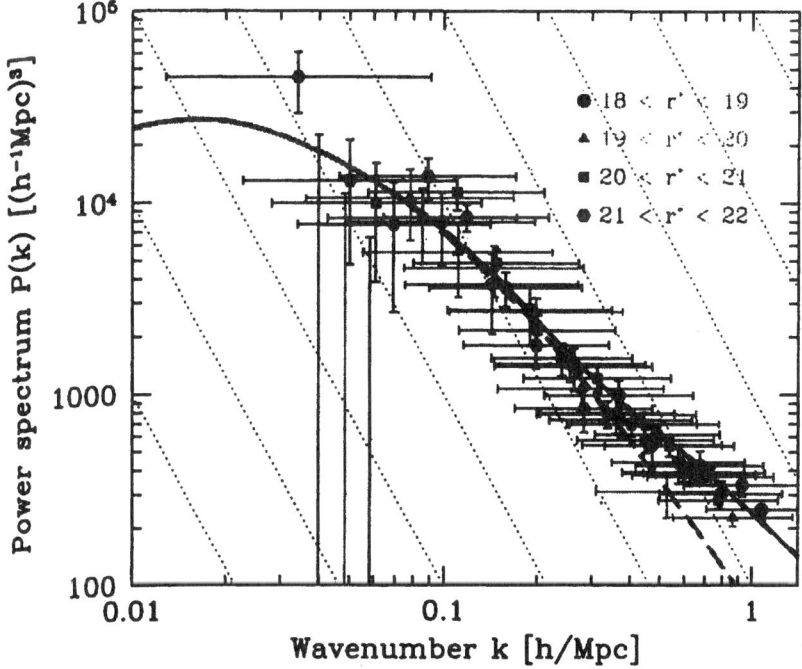

Fig. 10.8. The angular power spectrum of the Sloan survey.

significant that the spectrum shows such oscillations. Such oscillations appear to carry information regarding the baryonic content of the Universe, and are found to best fit a model with $\Omega_b/\Omega_m = 0.15$. It will be very interesting to see if the Sloan survey is able to confirm this finding. Further comparison to cosmological models has allowed Elgaroy et al. (2002) to put a limit on the neutrino mass of 1.8 eV.

Whilst it is still too early for the Sloan team to have produced a comparable three-dimensional power spectrum, a two-dimensional angular power spectrum can be produced from the imagery (Scranton et al. 2002, Tegmark et al. 2002, Connolly et al. 2002). Such a power spectrum would normally be blurred by the varying distances of the galaxies, but by dividing the data into four magnitude intervals, four separate functions can be derived, as shown in Fig. 10.8. By using estimates of how the number of galaxies would vary with redshift, a preliminary 3-D power spectrum can be extracted (Dodelson et al. 2002) and certain parameters estimated (Szalay et al. 2001). Extended investigations include the angular clustering of galaxy pairs (Infante et al. 2002) and the determination of higher order moments (Szapudi et al. 2002).

10.11 Λ Cosmology

If the large-scale structures measured in the power spectrum today have grown from the fluctuations of the same sort as represented in the power spectrum of the Cosmic Microwave Background, then the reconciliation between the two is dependent on nine cosmological parameters. The 2dF team (Efstatiou et al. 2002) apply maximum likelihood analyses to this problem. Under the assumption that the fluctuations are Gaussian and adiabatic, and

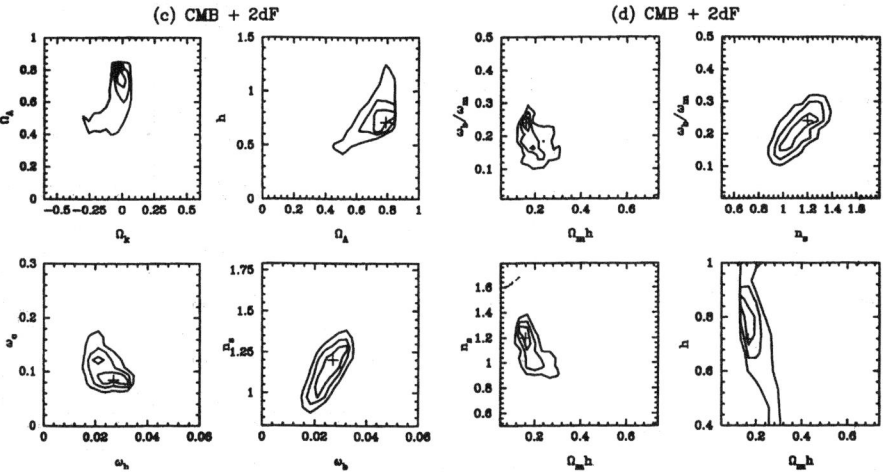

Fig. 10.9. Results of the likelihood analysis applied to 2dF and CMB data.

that the power spectrum on large scales is proportional to the linear matter spectrum, their work leads to a set of diagrams - Fig. 10.9 - that show likelihood contours for pairs of parameters, with crosses indicating the position of maximum likelihood.

The most significant finding is that in the first (top-left) plot: Ω_Λ is seen to be in the range of 0.65 to 0.85. Here is confirmation of a positive Cosmological Constant that is currently causing the expansion of the Universe to accelerate. The evidence comes independent of the well-publicised supernova data which has revolutionised cosmology in recent years. The plots also find $\Omega_m h$ to be in the 0.16 to 0.21 range, which combined with the previous finding for Ω_m puts h in the range 0.60 to 0.86 (hence the Hubble parameter H_0 60 to 86 km s^{-1} per Mpc). These figures are revised somewhat in a more recent paper (Percival et al. 2002) which narrows h to 0.665 ± 0.047 and therefore Ω_m to 0.313 ± 0.055. They also receive some support from the quasar survey (Outram et al. 2001).

10.12 Concluding Remarks

This limited review has not touched on some of the miscellaneous other products of the Sloan survey such as weak lensing (Sheldon et al. 2001, McKay et al. 2002), number counts (Yasuda et al. 2001) or the discovery of Seyfert galaxies (Williams et al. 2002). Also the 2dF redshifts have allowed analyses of radio galaxies samples (Sadler et al. 2002, Magliocchetti et al. 2002).

It is, however, obvious that both 2dF and Sloan have opened a new era in extragalactic astronomy and have already earned a place in history. We can still expect further analyses of the complete set of 2dF data, while the Sloan survey is still going to produce several hundred papers at least, and we shall see whether its cosmological parameters agree with the 2dF ones above. For imagery alone, there will be great pressure for Sloan to be extended to cover the southern galactic cap in the decade ahead.

References

Full lists of publications are maintained on respective websites:
   www.mso.anu.edu.au/2dFGRS   and   www.sdss.org/publications/index.html
Most references below have very large numbers of co-authors; for these only the first author is shown.

Abraham, R., van den Bergh, S. & Nair, P. (2003) A new approach to galaxy morphology: I. Analysis of the Sloan Digital Sky Survey Early Data Release astro-ph/0301239.

Bahcall, N. et al. (2002) The cluster mass function and cosmological implications. Ap.J. **580** 663.

Baldry, I.K. et al. (2002) The 2dF Galaxy Redshift Survey: Constraints on cosmic star formation history from the cosmic spectrum. Ap.J **569** 582.

Basilokos, S. (2003) Shape statistics of the Sloan Digital Sky Survey superclusters astro-ph/0302596.

Bernardi, M. et al. (2003a) Early-type galaxies in the SDSS - I: The sample astro-ph/0301631.

Bernardi, M. et al. (2003b) Early-type galaxies in the SDSS - II: Correlations between observables astro-ph/0301624.

Bernardi, M. et al. (2003c) Early-type galaxies in the SDSS - III: The fundamental plane astro-ph/0301626.

Bernardi, M. et al. (2003d) Early-type galaxies in the SDSS - IV: Colors and chemical evolution astro-ph/0301629.

Blanton, M.R. et al. (2001) The luminosity function of galaxies in SDSS commissioning data. A.J. **121** 2358.

Blanton, M.R. et al.(2002) The broad-band optical properties of galaxies with redshifts $0.0 < z < 0.2$.astro-ph/0209479.

Castander, F.J. et al. (2001) the first hour of extragalactic data of the Sloan Digital Sky Survey spectroscopic commissioning: The Coma Cluster. A.J. **121** 2331.

Cole, S. et al. (2001) the 2dF Redshift Survey: Near infrared galaxy luminosity functions. MNRAS **326** 255.

Colless, M. et al. (2001) The 2dF Redshift Survey: spectra and redshifts. MNRAS **328** 1039.

Connolly, A.J. et al. (2002) The angular correlation function of galaxies from early Sloan Digital Sky Survey data. Ap.J. **579** 42.

Cross, N. et al. (2001) The 2dF Galaxy Redshift Survey: The number and luminosity density of galaxies. MNRAS **324** 825.

De Propris, R. et al. (2002) The 2dF Galaxy Redshift Survey: a targeted study of catalogued clusters of galaxies. MNRAS **329** 87.

Dodelson, S. et al. (2002)The three-dimensional power spectrum from angular clustering of galaxies in early Sloan Digital Sky Survey data. Ap.J. **572** 140.

Doroshkevich, A., Tucker, D.L. & Allam, S.(2002) Large scale structure in the SDSS Galaxy survey. astro-ph/0206301.

Efstatiou, G. et al. (2002) Evidence for a non-zero Λ and a low matter density from a combined analysis of the 2dF Galaxy Redshift Survey and the cosmic microwave background anisotrophies. MNRAS **330** L29.

Elgaroy, O. et al. (2002) A new upper limit on the total neutrino mass from the 2dF Galaxy Redshift Survey. Ph.Rv.L. **89f** 1301.

Folkes, S.R. et al. (1999) the 2dF Galaxy Redshift Survey: Spectral types and luminosity functions. MNRAS **308** 459.

Gaztanaga, E. (2002) Large scale structures in the early SDSS: Comparison of the north and south galactic strips. Ap.J. **580** 144.

Glazebrook, K. et al. (2003) The Sloan Digital Sky Survey: The cosmic spectrum and star-formation history astro-ph/0301005.

Gomez, P. et al. (2003) Galaxy star-formation as a function of environment in the early release data of the Sloan Digital Sky Survey. Ap.J. **584** 210.

Goto, T. et al. (2002a) The cut & enhancemethod: selecting clusters of galaxies from the SDSS commissioning data. A.J. **123** 1807.

Goto, T. et al. (2002b) Composite luminosity functions of the Sloan Digital Sky Survey cut & enhance galaxy cluster catalog. PASJ **54** 515.

Goto, T. et al. (2003a) The morphological Butcher-Oemler effect in the SDSS Cut&Enhance Galaxy Cluster Catalog astro-ph/0301302.

Goto, T. et al. (2003b) Hdelta-selected galaxies in the Sloan Digital Sky Survey I: The catalog astro-ph/0301305.

Gunn, J. et al. (1998) The Sloan Digital Sky Survey Photometric Camera. A.J. **116** 3040.

Hoyle, F., Vogeley, M.S. & Gott, J.R. (2002) Two-dimensional topology of the Two-Degree Field Galaxy Redshift Survey. Ap.J. **570** 44.

Hikage, C. et al. (2002)Three-dimensional genus statistics of galaxies in the SDSS early release data. PASJ **54** 707.

Hogg, D.W., Blanton, M. & Strateva, I.(2002) The luminosity density of red galaxies. A.J. **124** 646.

Infante, L. et al. (2002) The angular clustering of galaxy pairs. Ap.J. **567** 155.

Kaiser, N. (1987) Clustering in real space and in redshift space. MNRAS **227** 1.

Kauffmann, G. et al. (2002) The dependence of star formation history and internal structure on stellar mass for 10^5 low-redshift galaxies. astro-ph/0205070.

Khairul Alam, S.M. & Ryden, B.S. (2002) The shapes of galaxies in the Sloan Digital Sky Survey. Ap.J. **570** 610-617.

Lahav, O. et al. (2001) The 2dF Galaxy Redshift Survey: The amplitudes of fluctuations in the 2dFGRS and the CMB, and implications for galaxy biasing. MNRAS **333** 961.

Lewis, I. et al. (2002) The 2dF Galaxy Redshift Survey: The environmental dependence of galaxy star formation rates near clusters. MNRAS **334** 673.

Maddox, S.J., Efstathiou, G., Sutherland, W.J. & Loveday, J. (1990) The APM Galaxy Survey I - APM measurements and star-galaxy separation. MNRAS **243** 692.

Madgwick, D.S. et al. (2002) The 2dF Galaxy Redshift Survey: galaxy luminosity functions per spectral type. MNRAS **333** 133-144.

Magliocchetti, M. et al. (2002) The 2dF Galaxy Redshift Survey: the population of nearby radio galaxies at the 1-mJy level. MNRAS **333** 100-120.

McKay, T.A. et al. (2002) Dynamical confirmation of SDSS weak lensing scaling laws. Ap.J. **571** L85.

Norberg, P. et al. (2001) The 2dF Galaxy Redshift Survey: luminosity dependence of galaxy clustering. MNRAS **328** 64.

Norberg, P. et al. (2002a) The 2dF Galaxy Redshift Survey: the dependence of galaxy clustering on luminosity and spectral type. MNRAS **332** 827.

Norberg, P. et al. (2002b) the 2dF Galaxy redshift Survey: The b_J-band galaxy luminosity function and survey selection function. MNRAS **336** 907.

Outram, P.J. et al. (2001) The 2dF QSO Redshift Survey - VI. Measuring Λ and Beta from redshift-space distortions in the power spectrum. MNRAS **328** 174.

Peacock, J.A. et al. (2001) A measurement of the cosmological mass density from clustering in the 2dF Galaxy Redshift Survey. Nature **410** 169.

Percival, W.J. et al. (2001) The 2dF Galaxy Redshift Survey: the power spectrum and the matter content of the Universe. MNRAS **327** 1297.

Percival, W.J. et al. (2002) Parameter constraints for flat cosmologies from cosmic microwave background and 2dFGRS power spectra. MNRAS **337** 1068.

Sadler, E.M. et al. (2002) Radio sources in the 2dF Galaxy Redshift Survey - II. Local radio luminosity functions for AGN and star-forming galaxies at 1.4 Ghz. MNRAS **329** 227.

Schlegel, D.J., Finkbeiner, D.P. & Davis, M.(1998) Maps of dust infrared emission for use in estimation of reddening and and Cosmic Microwave Background radiation foregrounds. Ap.J. **500** 525.

Scranton, R. et al. (2002) Analysis of systematic effects and statistical uncertainties in the angular clustering of galaxies from early SDSS data. Ap.J. **579** 48.

Sheldon, E.S. et al. (2001) weak lensing measurements of 42 SDSS/RASS galaxy clusters. Ap.J. **554** 881.

Shen, S., Mo, H.J., White, S.D.M., Blanton, M.R., Kauffmann, G., Voges, W., Brinkman, J. & Csabai, I. (2003) The size distribution of galaxies in the Sloan Digital Sky Survey astro-ph/0301527.

Shimasaku, K. et al. (2001) Statistical properties of bright galaxies in the SDSS photometric sysytem. A.J. **122** 1238.

Strateva, I. et al. (2001) Color separation of galaxy types in the Sloan Digital Sky Survey imaging data. A.J. **122** 1861.

Strauss, M. et al. (2002) Spectroscopic target selection in the Sloan Digital Sky Survey: The main galaxy sample. A.J. **124** 1810.

Szalay A.S. et al. (2001) KL estimation of the power spectrum parameters from the angular distribution of galaxies in early SDSS data.astro-ph/0107419.

Szapudi, I. et al. (2002) Higher order moments of the angular distribution of galaxies from early Sloan Digital Sky Survey data. Ap.J. **570** 75.

Tegmark, M. et al. (2002) the angular power spectrum of galaxies from early Sloan Digital Sky Survey Data. Ap.J. **571** 191.

Verde, L. et al. (2002) The 2dF Galaxy Redshift Survey: The bias of galaxies and the density of the Universe. MNRAS **335** 432.

Williams, R.J., Pogge, R.W. & Smita, M. (2002) Narrow-line Seyfert 1 galaxies from the Sloan Digital Sky Survey early data release. A.J. **124** 3042.

Yasuda, N. et al. (2001) Galaxy number counts from the Sloan Digital Sky Survey commissioning data. A.J. **122** 1104.

York, D.G. et al. (2000) The Sloan Digital Sky Survey: Technical Summary. A.J. **120** 1579.

Zehavi, I. et al. (2002) Galaxy clustering in early Sloan Digital Sky Survey redshift data. Ap.J. **571** 172.

11 Active Galactic Nuclei and Supermassive Black Holes

I. Robson

11.1 Introduction and the Big Picture

11.1.1 The Discovery of AGN

The study of Active Galactic Nuclei, or AGN for short, began with Carl Seyfert in 1943 with his discovery of a handful of galaxies with bright nuclei and highly ionized emission lines – both an unknown phenomenon for galaxies at the time. These were the first AGN and were later classified as Seyfert galaxies. Following the initial discovery and early study of AGN in the optical, the newly emerging field of radio astronomy made a major breakthrough with the discovery of radio galaxies and subsequently quasars in the early 1960s. From then on the subject exploded into a golden age of observational and theoretical discovery, finally resulting in an agreement of the general physical picture – the 'standard model' – in which we believe we understand the global workings of AGN.

This was not a smooth process however, and for many years there was a major dispute, sometimes bitter, in which one side proclaimed quasars as being very distant and super-luminous galaxies, while the other side preferred to believe their recession velocities were not due to the expansion of the Universe and that they were much closer objects with much reduced luminosities. The first camp rapidly won out with quasars being recognised as the most luminous of the AGN and all lying at cosmological distances. However, a further word of caution; while the big picture is almost certainly correct, many of the details, even some of the most fundamental ones, still remain hazy. In this review I will give a broad overview of the workings of AGN and the major discoveries in the past decade or so, before reviewing exciting discoveries of the past year. Future years will provide regular updates on progress.

11.1.2 The Classification 'Zoo'

During the 1960s and 70s much effort went into observationally classifying 'types' of AGN, as the differing signatures of the central engine were probed in exquisite spectroscopic detail. From out of this wealth of study the classes of Seyferts, quasars, QSOs, radio galaxies, BL Lac objects, blazers, LINERS and N-galaxies became common terms. Yet another twist came with the observation that Seyferts themselves could be spectroscopically split into at least two distinct sub-categories: type 1 and type 2 and the picture seemed to be taking on an ever increasing confusion of detail. Yet out of this detail would emerge clarity and a unification of all the categories with a single explanation. In a nutshell, the differences between the type 1 and type 2 spectra provided the key and because of this importance it is worth spending a little time describing the differences. In type 1 objects, the spectra are dominated by strong permitted lines, mostly of hydrogen, with very broad, full-width half-maxima (FWHM) corresponding to 1,000 to 10,000 km s^{-1}. Also present in the spectra are forbidden lines; these are much weaker and lack the broad wings, with FWHM of a few hundred km s^{-1}. In the type 2 spectra on the other hand, the permitted and forbidden lines are of much more equal strength, with a distinct absence of broad wings. Clearly the ionization states and velocity conditions of the gas clouds producing the emission are very different; the lines in the type 1 objects originate in two distinctly different regions in the galaxy, whereas the lines in the type 2 objects come from the same region. These are now universally referred to as the broad-line region (BLR) and the narrow-line region (NLR) respectively, with type 1 objects revealing both regions, but type 2 objects lacking a BLR. The burning question was why?

11.1.3 Polarimetry to the Rescue: The Obscuring Torus

While it had been clear for some time that the favoured power-source for the central engine for all AGN was a supermassive black hole, why this gave rise to so many different types of object remained a puzzle. The solution came with a breakthrough observation by Antonucci and Miller (1985) investigating polarized light from the Seyfert 2 galaxy NGC 1068. This dramatic observation showed that while in ordinary light the galaxy presented a typical type 2 spectrum (narrow lines, etc.), in contrast, the polarised light spectrum revealed broad-lines indicative of a type 1 object. Now it had been known for some time that the narrow-lines came from a region much farther away from the nucleus than the broad-lines and so the simple solution to the type 1 and type 2 separation came down to the presence of an obscuring 'torus' of molecular gas and dust in the galaxy and the orientation of the observer with respect to the central engine. A cartoon depicting how the various types of AGN are unified by orientation with a torus is shown in Fig. 11.1, although on most of these cartoons the scale-lengths are usually incorrect (see Sect. 11.1.4).

Where the observer lies in the plane of the torus, the broad-lines are buried inside the torus and are well and truly hidden from view while the narrow-lines remain readily visible. As the viewing angle to the plane increases, then the broad-lines eventually come into sight, finally followed by the radiating accretion disk – if we can recognise its spectral signature (see Sect. 11.5). For

11 Active Galactic Nuclei and Supermassive Black Holes

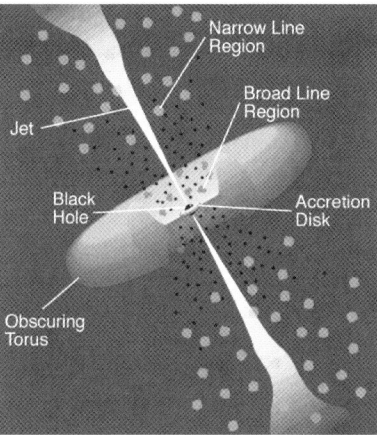

Fig. 11.1. One of the most famous cartoons in astrophysics taken from Urry and Padovani (1995) showing the scheme whereby AGN of differing types are unified by orientation due to the presence of an obscuring torus. This is very much a schematic and size scales are not representative. (Courtesy Meg Urry)

those rare sources that show strong radio-jets, the side-on view is of a double-lobed radio-galaxy and as the face-on view approaches the object becomes a radio-loud quasar and ultimately a blazer, when we are looking close to the line-of-sight of the jet. However, while this gives the general hand-waving picture, questions such as the range of luminosity and radio-jet generation need much further study.

The nuclear-enshrouding torus became a bandwagon of study in its own right and the race was on to find supporting evidence for its existence. Numerous studies of the narrow-line region of relatively nearby Seyfert galaxies showed that a distinct ionization cone can be traced out, expanding outwards from the central zone (e.g. Falke, Wilson and Simpson 1998). This fitted well with the idea that the torus defined this radiation cone. However, what was really needed was a picture of a torus, and this is where we take a short break in the search for the torus, and try to get a handle on the sizes of the zones in question.

11.1.4 The 'Standard Model'

In the so-called 'standard model' for AGN the central power source is a supermassive black hole, located at, or close to the nucleus of a galaxy. The black hole generates power by conversion of gravitational energy into radiation or particle energies; we shall only consider the former. The black hole is fed by an accretion disk, the inner parts of which are extremely hot. Surrounding this inner zone is the broad-line region, which itself is almost certainly highly stratified and believed to be composed of tens of thousands of clumps of hot gas distributed in a sort of coronal halo.

For a 10^8 $M_\odot$ black hole the Schwarzschild radius is of order 2 AU, and the inner edge of the accretion disk no closer than $\sim$6 AU. The broad-line clouds surrounding the black hole occupy a volume of radius much less than 1 pc. (Indeed, the BLR region could be very much smaller and still provide the enormous luminosity needed to account for the observed line emission.) The lateral extent of the accretion disk is not well known, but distances of up to parsecs have been postulated. Beyond some point the disk presumably becomes much more fragmented and less organised as mass infall flows into the very central region. Surrounding this ensemble is the putative unifying molecular torus, the obscuring shield that effectively defines the type 1 and type 2 categories for a distant observer. Depending on the model (see below) this has a diameter ranging from tens to a few hundred pc. Finally, well beyond the accretion disk – in a fan-shaped pattern defined by the ionizing radiation field constrained by the torus – lies the narrow-line region, again extending from a few hundred pc to many kpc in the most extreme cases. For the heavily obscured and edge-on systems, the presence of a NLR may be the only optical clue that an AGN is present (see Sects. 11.4.3 and 11.4.4).

Given that the very largest galaxies extend up to 100kpc in diameter, it is clear that the central engine of the black hole, the BLR and the accretion disk are extremely compact compared to the surrounding galaxy, and are going to be incredibly difficult to probe by direct imaging even for the nearest galaxies (but see Sect. 11.2.2). The NLR and the molecular torus on the other hand are more readily open to such study and one of the most famous images of what might be a torus is shown in Fig. 11.2. But remember, at a distance of a Virgo cluster galaxy ($\sim$16Mpc), a linear size of 100pc subtends an angle of

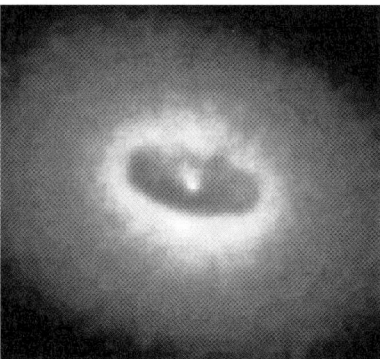

Fig. 11.2. An HST image of the central regions of the nearby galaxy NGC 4261 showing a disk of gas and stars lying perpendicular to the radio jet axis. Although not trivial to figure out what is going on (especially in black and white) the diameter of the dark dusty disk is just over 100pc and it is tempting to believe that this may be linked to the putative torus, although it is on the large size for the compact tori anticipated from those of Fig. 11.1 (Courtesy STScI)

just over an arcsecond, so even here, the situation is challenging. Indeed, it is fair to say that solid evidence for tori is still very thin on the ground.

Ionization cones favour a rather compact torus, but this hits snags from other observations. Large warped or flared disks were the vogue for some time although these also have problems (see, e.g. Fadda et al. 1998 and references therein). An entire article could easily be devoted to the unifying torus but suffice to say that although the evidence for unification through obscuration is sound, the details of what this obscuration is; whether a compact torus as in the cartoons, a flared disk or a wind-driven extended toroidal structure remains to be seen. Maybe different objects have different structures just to confuse the picture for the observational astronomer. In the near future next-generation, mid-infrared instruments, like MICHELLE on Gemini, should make major inroads into investigations of the dusty torus in nearby galaxies.

In spite of the difficulty of imaging, many other observational tools are available to probe the inner secrets of AGN, spanning all the wavelengths from radio through γ-rays and in this realm spectroscopy comes to the fore. This review will touch on some of the results from the most recent techniques in elucidating our understanding of the central engine.

11.1.5 A Note of Caution

Before getting into the review itself I should reinforce the earlier comment that although we believe in the big picture as described above, many of the details remain unexplained. We have already noted the question mark about the torus, and another example is the accretion disk. Not only is the shape of this (flat or puffed-up) very uncertain, even clear observational evidence for the existence of an accretion disk remains worryingly sketchy (although this may sound like heresy). Likewise, there is still concern that the energy budget in terms of identified radiating components does not completely stack up against the observations. As an excellent example for students to ponder over, the reader is referred to an article by Andy Lawrence (2003) intriguingly called 'The Ghost in the Machine', in which he tackles from first principles the question of the evidence for accretion disks (see Sect. 11.5) and the subsequent problem of the energy budget.

11.2 The Evidence for Supermassive Black Holes

11.2.1 Bulge Galaxies House Supermassive Black Holes

One of the key discoveries of the last decade has been that all the local galaxies with a bulge harbour supermassive black holes (e.g. Kormendy and Gebhardt 2002). That the mass of the black hole is directly linked with the mass of the central bulge is now an accepted fact (see Fig. 11.3 and Magorrian et al. 1998). This exciting link has had a major influence on observational

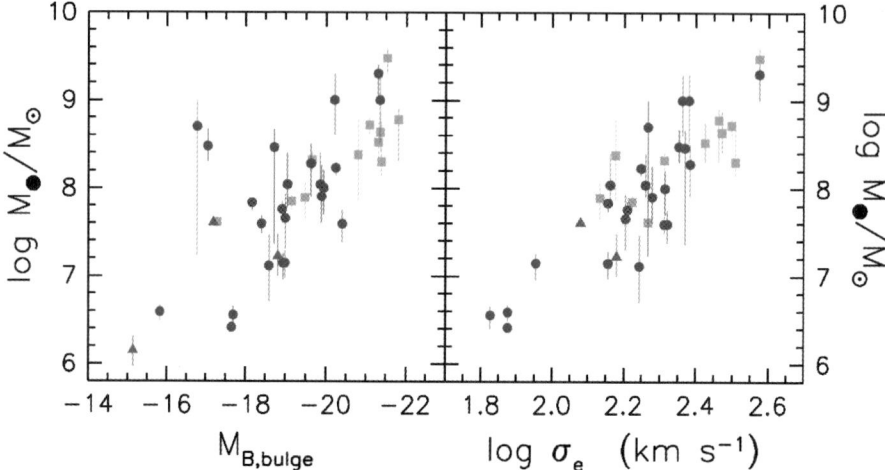

Fig. 11.3. The correlation of black hole mass with the bulge luminosity of the host galaxy (*Left*). Because the bulge mass is directly related to the bulge luminosity, this means that the black hole mass is correlated to the bulge mass, which is indicative of the mass of the galaxy – so more massive galaxies house more massive black holes. The data for the black hole mass are obtained from a variety of techniques. (*Right*) This shows the correlation between the black hole mass and the average line-of sight velocity of the *main body* of stars in the host galaxy – indicating that the mass of the black hole is related to the velocity dispersion of the stars in that galaxy, which in turn is related to the formation process of the galaxy. (courtesy of John Kormendy: http://chandra.as.utexas.edu/ kormendy/bhsearch.html).

studies over the past five years and has driven an entirely new understanding of the connection between galaxy and AGN evolutions. The observational evidence on which the mass relationship is based stems from very careful measurements of the velocity dispersion of stars in the very inner parts of the galaxy and then comparing the deduced mass concentration with the observed luminosity to obtain a mass of unseen (or non-radiating) matter. This gives a very tight correlation and although the black hole mass to bulge mass correlation is not as good, nevertheless, for the purposes of this article it is adequate to portray the overall picture we wish to paint. Perhaps an equally important observation is that galaxies *without* a central bulge do *not* seem to house a black hole.

11.2.2 Key Observations: Masers and X-rays to the Fore

Two other major observational techniques have provided conclusive evidence for the presence of supermassive black holes in galaxies. The first uses the kinematics of maser emission observed with sub-milli-arcsecond resolution by radio Very Long Baseline Interferometry (VLBI) (see Fig 11.4). These observations trace out the motions of masers in the nuclear regions of a galaxy

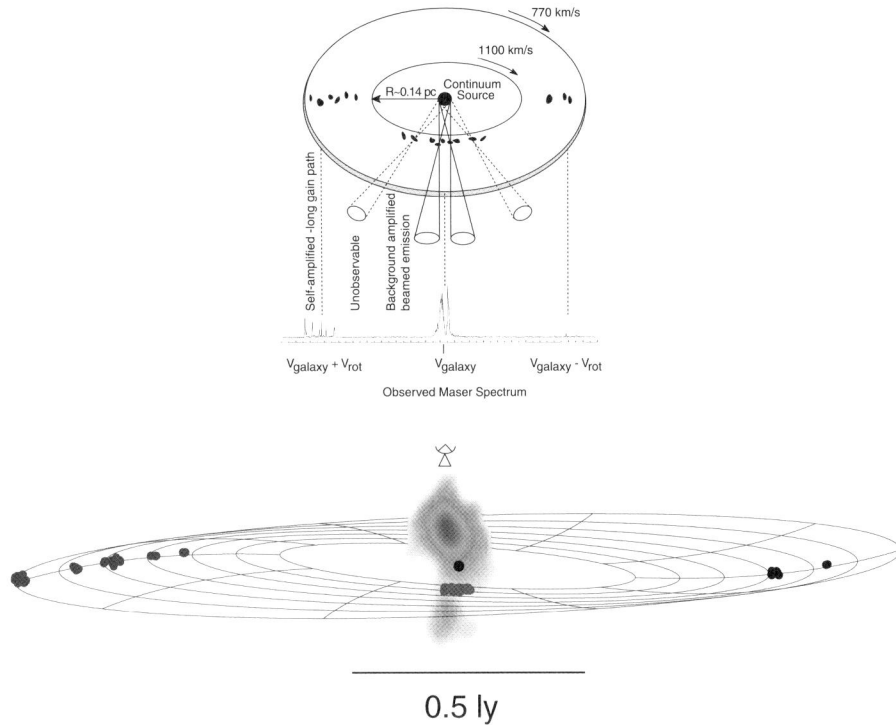

Fig. 11.4. Water vapour masers in the galaxy NGC 4258: (**a**) a cartoon of the emission from the masers, which are deduced to lie at the three locations. Their corresponding spectral signatures are also shown, (**b**) diagram of the suspected warped disk that satisfies all the observations of the maser emission over a period of many months of observations. The central dot marks the dynamical centre of the disk but not emission whilst the shaded 1.3 cm continuum 'jet' lies orthogonal to the plane of the disk (Courtesy of Lincoln Greenhill).

enabling a kinematic picture to be determined and hence the gravitational field and thereby the mass of the black hole. The pioneering work was reported by Myoshi et al. (1995), where a 'line' of masers in the galaxy NGC 4258 was used to trace out a rotating (and warped) disk of material in Keplerian motion about 0.1 pc from a massive, albeit dark, central mass (see also Greenhill et al. 1995).

The second method is X-ray based and relies on the 6 keV iron fluorescent line (see Fig. 11.5). The line is highly broadened and is asymmetric to the low-energy side, which is very consistent with gravitational redshift produced in the immediate surroundings of a supermassive black hole. The breakthrough observation was reported by Tanaka et al. (1995) for the galaxy MCG-6-30-15. Over the past year this technique has been extended to a few other galaxies with data from the two new X-ray satellites of XMM-Newton (e.g. Reeves 2003) and Chandra (e.g. Yaqoob and Padmanabhan 2003). We should

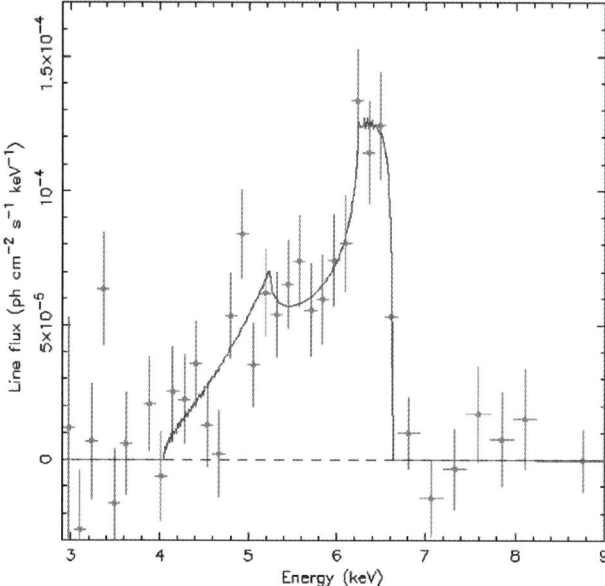

Fig. 11.5. The highly broadened X-ray iron K-alpha line from the ASCA source MCG-6-30-15. This is the residual spectrum after removal of a power-law and reflected component. The broadening of the emission line of around 100,000 km s^{-1} along with the asymmetry towards energies lower than the rest energy is in excellent agreement with a gravitational redshift of gas orbiting a supermassive black hole. The solid line shows a model fit to material in an accretion disk orbiting between 3 and 10 Schwarzschild radii. The original data are from Tanaka et al., 1995, Nature, 375, 659.

note that this evidence is from a very much closer distance to the black hole than the maser data (see Sect. 11.5.4)

11.2.3 The Galactic Centre

If all local galaxies with a central bulge possess a black hole, what about our own Galaxy? We know our Galaxy possesses a bulge, so why not a black hole? A very compact radio source, called Sgr A*, lies very close to the centre of the Galaxy and while it has an extremely low radio luminosity compared to radio-loud AGN, it is known to be variable on timescales of less than a day and has been determined to be less than 16 AU in extent (from precise VLBI measurements). Unfortunately, in spite of the Galactic Centre being a mere 8 kpc distant, it is hidden behind some 30 magnitudes of visible light extinction due to dust in the Galactic Plane. Therefore, observations probing the region around Sgr A* need to use wavelengths that are not as susceptible to extinction. Radio and infrared have been the main tools, focusing on a group of infrared sources and the highly ionized gas in the region. From a study of the orbits of the stars, mass-loss and luminosity arguments, suggestive evidence for a $\sim 2 \times 10^6$ M$_\odot$ supermassive black hole has been assembled over the past decade.

However, the last piece of the jigsaw that finally cemented a black hole as the culprit for the central mass resulted from a painstaking ten-year, near infrared study of the orbital motions of the stars closest to Sgr A*. This was probably the major AGN observation of 2002 and a highlight for adaptive optics (Schödel et al. 2002). The team made careful astrometric observations of stars in the immediate vicinity of Sgr A*. At the distance of the Galactic Centre, one arcsecond equals 0.039 pc and one of the major tasks was determining the relative astrometry between Sgr A* and the infrared field (in which the radio-bright Sgr A* is invisible). This was achieved using seven SiO masers in the same field, resulting in a positional uncertainty of only 10 milli-arcseconds for the position of Sgr A* with respect to the infrared stars.

From the ten year's worth of data and the highly refined position of Sgr A*, one of the stars being monitored was found to have a closed and bound Keplerian orbit about Sgr A* with a period of 15.2 years (Figs. 11.6

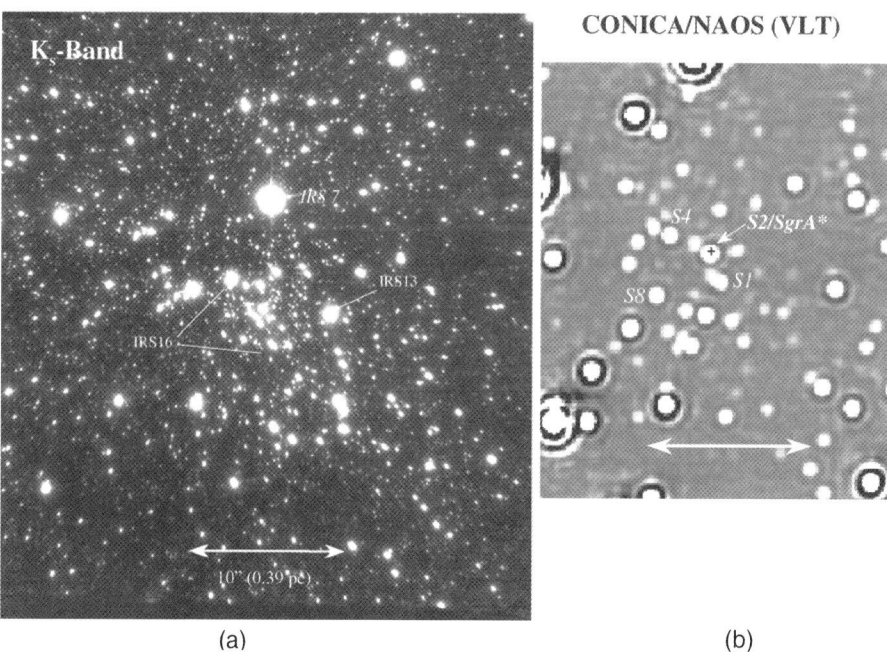

(a) (b)

Fig. 11.6. (a) A diffraction limited (56 mas), 2.18 micron image of the Galactic Centre region taken with the NAOS/CONICA adaptive imaging camera on the VLT-4. The open circles identify the seven SiO maser sources used for astrometry, while the prominent infrared sources are also labelled. (b) The central 2 arcseconds in which the position of the radio source Sgr A* is indicated. The star whose orbit is shown in Fig. 11.7 is labelled as S2. The rings around the brighter objects are artefacts. Further details are found in Schödel, R. et al., 2002, Nature, 419, 694.

240 I. Robson

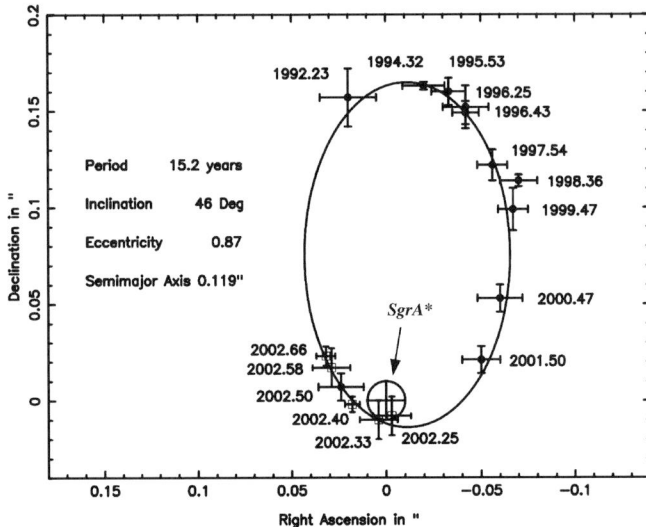

Fig. 11.7. Spectacular reconstruction of the ten-year orbit of the star, S2, shown in Fig. 11.6b relative to the position of Sgr A*. The orbital elements show that the best fit is given by a central black hole mass for Sgr A* of 3.7×10^6 M$_\odot$.

and 11.7). The orbit is highly elliptical and at its closest approach comes within only 17 light hours of Sgr A*, allowing a very accurate determination of the latter's mass of $3.7 \pm 1.5 \times 10^6$ M$_\odot$. Crucially, the observations rule out two of the three remaining contenders for an explanation; a cluster of dense and non-radiating objects (low-mass stars, neutron stars or black holes) or, even more exotically, a sphere of 'dark particles' – in this case, massive fermions. The only remaining contender in the theoreticians' bag is a sphere of alternate 'dark particles' – bosons. However, it is fair to say that most astronomers by far are quite happy to plump for a supermassive black hole as the most likely explanation. We might also note in passing that even at the close distance of 17 light hours, the star is still over 2,000 Schwarzschild radii from the black hole, where gravitational (tidal) disruption forces are expected to be small, even for a black hole mass of 10^6 M$_\odot$.

11.3 Black Holes and the Relation to the AGN

11.3.1 Mass, Luminosity and Accretion Rates

The work on galaxy bulge mass and black hole mass has been extended to include AGN as shown in Fig. 11.8 (e.g. McLure and Dunlop 2002 and references therein), and we see that a good correlation between the bulge luminosity and the black hole is maintained. This paper also provides an excellent discussion of many of the parameters of black hole mass determinations and implications for AGN that space does not permit to be accommodated in this review. Considering the form of Fig. 11.8, one might suspect that AGN activity scales with the mass of the black hole; the highest luminosity quasars housing the most massive black holes. Indeed, some of the earliest determinations of black hole mass seemed to support this proposition and has theoretical support when we consider what drives the AGN luminosity.

Remembering that gravity is the underlying mechanism, working through accretion of material towards the black hole, there are two critical parameters to consider: the accretion rate, and the efficiency of conversion of accreted mass to output photons. The second is very difficult to determine; however we know that it cannot exceed 40%, and is additionally expected to be very much greater than the nuclear energy generation rate of 0.7%. For computational purposes the efficiency factor is usually attributed to 5-10%, giving acknowledged factors of a few uncertainty. With observed luminosities of 10^{39} W and above, the accretion rates then become of order a few solar masses per year and to sustain this luminosity for up to $\sim 10^8$ y requires the consumption of a significant fraction of the gas mass of normal galaxies. While this in itself is not an insurmountable problem, how to get the mass to the very centre of the galaxy and onto the black hole presents enormous difficulty. In fact the easiest way out of this conundrum is to suppose that the duty cycle times are much less and indeed there is other suggestive evidence that this is indeed the case (e.g. Combes 2003).

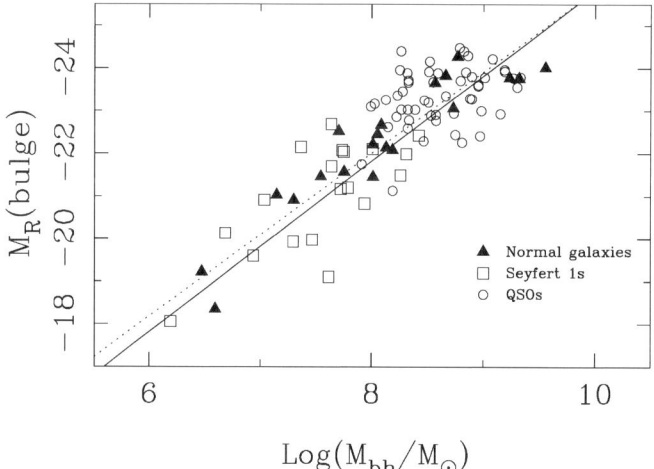

Fig. 11.8. Black hole mass v galactic bulge luminosity for a sample of normal galaxies, Seyfert 1s and quasars. The bulge luminosity is given as the absolute R-band magnitude and the AGN black hole masses are deduced from their Hβ line-widths. The dotted line is the best-fit linear relation whilst the solid line is the best formal fit. The figure is from McLure, R.J. & Dunlop, J.S. (2002) Courtesy of MNRAS.

Regarding the accretion rate, we know that apart from very special and perhaps contrived circumstances it cannot exceed the Eddington limit – the limiting case whereby the maximum accretion rate is reached when accretion is balanced by the radiation pressure of the outward photon flux. For a simple geometry the relation between the mass and luminosity for objects radiating at the Eddington limit is given by

$$L/L_\odot \sim 3.4 \times 10^4 M/M_\odot$$

which clearly predicts that the luminosity scales with the mass. Indeed, a very luminous quasar of $L \sim 10^{12}$ $L_\odot$ needs a black hole of mass $\sim 3 \times 10^7$ $M_\odot$ assuming it is radiating at the Eddington limit, and very much more if it is radiating at a significantly lower rate. However, very recent work by Woo and Urry (2002) using a large (over 200) sample of black hole masses in AGN, derive the surprising conclusion that there is little correlation between black hole mass and AGN luminosity. This work also reveals that for a given black hole mass a wide range of luminosity is observed, indicating that there is a large range of accretion rates. Indeed, the upper bound to the accretion rate from their work is at (or even above but within observational uncertainties) the Eddington limit, showing that some objects do radiate at, or close to this level, but that for many others, the rate is much less. So, we are left with the overall conclusion that based on these data the mass of the black hole seems to have little impact on the luminosity of the AGN we observe. Unfortunately, selection effects beset the samples (especially for low luminosity sources at all redshifts) and we may not be seeing the overall picture clearly yet. Nevertheless, this shows great potential for the study of the formation and evolution of black holes and perhaps the difference between radio-loud and radio-quiet AGN.

11.3.2 Accretion Rate the Key?

If, as we have just seen, the mass of the black hole is not the driving factor for the luminosity, then we are left with the mass accretion rate. Indeed, this has recently been cited as a reason for explaining the split in the spectral energy distribution of BL Lac objects into two categories: those that have emission luminosities that peak at long wavelengths (the original BL Lac objects peaking in the short radio) from those objects discovered much later that peak at short wavelengths (UV-X-ray). Now we should remember that when we look at BL Lac objects we are actually seeing emission from the jet (and not too far from the line-of-sight) rather than the underlying galaxy of stars or the AGN core. (Note that the jets in BL Lac objects are not generally highly relativistic (compared to those in quasars) and considerations of the energetics of the jet and statistical arguments suggested that orientation might be the answer to the BL Lac split (see, e.g. Urry and Padovani 1995) and this subsequently became the popular viewpoint.) However, with larger samples and more observations it is now becoming clearer that the real case is probably much more of a continuous distribution of spectral energy types rather than a bimodal split (although the radio-loud BL Lacs still dominate numerically).

Recent, and technically difficult observations of the host galaxies of a sample of radio and X-ray BL Lac objects (Scarpa et al. 2000) revealed that both were normal giant elliptical galaxies, again with no clear separation into two categories or clear correlations. This is not surprising in terms of our current suspicions of galaxy evolution and morphology, but if these galaxies also obey the black hole-bulge mass relationship, then it must be something else that drives the difference in output spectral energy distribution. Rather than the long accepted idea of orientation, the mass accretion rate has recently emerged as the favourite parameter. Wang, Staubert & Ho (2002) conclude that based on host galaxies and emission-line luminosities (as a marker of the AGN luminosity) the radio BL Lacs have significantly higher accretion rates than their X-ray counterparts, but both are notably sub-Eddington, albeit with assumptions about accretion disk modeling. The observed peak luminosities, which remember are luminosities of the jet, and peak wavelength are also correlated with the accretion rate (higher accretion rate peaking at shorter wavelengths) which gives support to the notion that the accretion disk and the jet are closely coupled. Given equal other parameters, the radiative efficiency of the jet then becomes a driver for explaining the differences; whereby the radio objects have lower efficiencies than those objects radiating preferably at shorter wavelengths. The X-ray jets are somehow able to convert the kinetic energy of the electrons in the jet into radiation more efficiently even though they have lower accretion rates onto the black hole.

These are early days for these exciting studies, which should be treated with some caution as the number of sources in the samples is small, the assumptions about derivation of luminosity are non-trivial and the extraction of independent parameters remains tricky. Nevertheless, this shows a very promising way forward in terms of explaining observational consequences from basic physical parameters of the central engine. Indeed, the above is just a taster to one of the fundamental questions for AGN research; the difference between radio-loud and radio-quiet AGN. This has long been one of the puzzling differences that have defied simple explanation, and while accretion rate has recently become one of the targets for investigation, we shall leave this topic for a future review.

11.4 AGN Evolution and Black Hole Growth

11.4.1 Black Holes Are Ubiquitous

As we noted in the previous section, a number of far reaching implications follow from the discovery that all local galaxies with a central bulge house a supermassive black hole. Assuming that the relationship between the central bulge mass and black hole mass is a universal factor rather than a local anomaly, then the number density of supermassive black holes in the present Universe is, to a first approximation, enough to match the AGN population at various epochs. This directly leads to speculation that the formation of the central bulge and the black hole are intimately linked; the million dollar question right now is which came first, the central bulge or the black hole? Either way, this is clearly telling us about galaxy formation in the early Universe. Also, the fact that so many local galaxies have supermassive black holes, yet the galaxies are as un-AGN-like as you can find, with zero evidence of any central engine activity even when probed to exceedingly faint levels, dramatically strengthens the view that that the fuelling of the black hole is a time dependent phenomenon and that activity is a transient stage in the life of a galaxy. Many large galaxies have been quasars in the past.

Indeed, the last few years has seen an explosion in the study of galaxy evolution through extensive data from major surveys being available from new instruments and techniques. We know that while quasars are an extremely rare phenomenon in the local Universe, they were much more common at earlier epochs, peaking at a redshift of ~ 2.5, where, for the most luminous quasars, their number density was $> 10^3$ times higher. We also know that they declined very steeply at earlier epochs. On the other hand, data from the Sloan Digital Sky Survey have shown that luminous quasars still exist at redshifts of greater than 6 (Fan et al., 2001, 2003), albeit being exceedingly rare.

This is a very important finding because for a reasonable cosmology this corresponds to an age of around 1 Gy. We note from Sect. 11.3.1 regarding quasar black hole masses that if quasars radiate at the Eddington limit then the required black hole mass is large, and even larger if the radiation is sub-Eddington. For the $z \sim 6$ quasars arguments suggest that a black hole mass of around 10^9 M$_\odot$ is needed. Therefore, the formation of supermassive black holes so very early in the Universe poses special theoretical problems and the linking of the bulge evolution and AGN evolution becomes very tempting, if not necessary. A detailed discussion of the complexities surrounding black hole growth and galaxy evolution can be found in Yu & Tremaine (2002).

With the recent appearance of hard X-ray data we can now add intermediate luminosity AGN (Seyfert-like) to the cosmic evolution of the AGN number density (Barger et al. 2003). And we find they are different to that of quasars. It is known that lower luminosity quasars studied from optical surveys do not show such a drastic fall-off to the present epoch (e.g. Boyle et al. 2000) and indeed, the lower luminosity X-ray selected AGN fail to show much of a drop-off at all. Instead they reveal a much flatter distribution, indicating that the lower-mass accretion rates continue to the present time with little, if any decline. This implies that unless something else is changing, the growth of supermassive black holes is healthy and strong at the present day for this luminosity category of objects.

To conclude this topic we should note that while not linked to the huge mega-surveys in the optical and IR, recent submillimetre observations seem to be hinting that dust is extensive at medium-to-high redshifts, both in non-AGN as measured by normal criteria (e.g. Ivison et al. 2002) and for quasars (Robson et al. 2003, Priddey et al. 2003) who find large dust masses even in the highest redshift ($z>6$) quasars. This means that heavy element formation also proceeded rapidly in the very early Universe and furthermore, dust obscuration may play a significant role in biasing samples obtained at optical/UV wavelengths.

11.4.2 Fuelling of the Central Engine

Returning to quasars at the peak of their evolution, it follows that the fuelling was much more efficient than at later epochs. This turns on two factors: the available fuel supply and the ability of the fuel to penetrate to the accretion disk. Gravitational perturbations are an obvious method of channelling gas to the central engine, so galaxy encounters and collisions are candidates for triggering AGN activity. Indeed, very deep imaging of quasars and other AGN has revealed that many of the hosts are perturbed objects and this is now becoming part of the standard picture. Another possibility of providing a conduit to the centre was thought to be bars in galaxies as these distort the otherwise uniform potential. However, in the relatively local Universe the latest studies seem to show that barred galaxies do not favour AGN cores, but that AGN appear in about equal numbers for barred and unbarred systems (see e.g. Combes 2003).

11.4.3 Starbursts and AGN: Missing Link?

Recent studies of AGN and extensive star formation activity in galaxies suggest that rather than these being very separate phenomena they might be related and the triggering mechanism for both could be interactions or merging of two gas-rich galaxies (e.g. Combes 2002, Fadda et al. 2002). In further support of this link Franceschini et al. (1999) point out that the decline of luminous AGN (quasar) activity from the peak epoch of redshift around 2.5 to the present epoch closely mimics the decline in the star formation rate in galaxies. (However, we note from Sect. 11.4.1 the comments regarding the much reduced fall-off for the lower luminosity AGN.) To prove this link for AGN in general is non-trivial because both AGN activity and star formation are expected to occur in high density regions, with hydrogen gas column densities exceeding 10^{23-24} cm^{-2}. However, such regions are also associated with extensive dust, and as we noted above, this poses a major obstacle to ready detection of a buried AGN. The very high resulting dust opacity completely prevents observations of the nucleus from optical through low-energy X-ray emission, and in very dust-enshrouded cases, even infrared emission lines become invisible.

As hinted in Sect. 11.4.1, the salvation to the problem lies in observing high energy X-ray emission (E> 6 keV) as these very energetic photons can penetrate most regimes, at least up to a column density of gas of 10^{25} cm^{-2}, when even hard X-ray emission can no longer be detected. Using this technique on local samples has shown that at least for some galaxies previously classified as LINER or starburst galaxies (both being originally not in the AGN category – or at the very low end of the AGN luminosity scale) they not only contain a hidden AGN core, but surprisingly, one which has a luminosity typical of the more luminous (quasar) objects (eg Vignati et al. 1999). While the number of objects so far observed is small, the picture is steadily growing of starburst galaxies housing buried AGN. This is supported by very recent observations by the BeppoSAX X-ray telescope, which showed that Arp 299 – an interacting starburst system – possesses a deeply buried and powerful AGN core (Della Ceca et al. 2002). Even so, it remains the case that the luminosity of the AGN is a factor of 50 lower than that of the far infrared (FIR) luminosity, and even taking into account the assumption that the bulk of the AGN luminosity might be radiated in the UV region, it is still inadequate to provide the bulk of the FIR luminosity, which must be coming from the star formation process. Thereby, we have at least one good example of powerful star formation and AGN activity in the same galaxy, which has until recently been classed as a non-AGN.

The general case for starburst and buried AGN activity is difficult to separate but with near-to-far infrared wavelength coverage (such as from the ISO satellite) modelling of the spectral energy distribution can give strong insights into the contributions of each component (e.g. Rowan-Robinson 2000). A very good example of the benefit and deductions of using infrared and X-ray studies is given by Wilman et al. (2002). In this paper, the authors describe deep XMM-Newton observations of two starburst galaxies. From the infrared observations alone both were suspected to have AGN cores, with about equal starburst and AGN luminosities. In fact one was known to have a Seyfert 2 optical spectrum. However, while the starburst luminosities of the galaxies are very similar, the X-ray luminosity is very different. Does this indicate a difference in the AGN power, or is it variable dust obscuration? This deserves some amplification as it is an excellent example of the difficulties of piecing all the parts of the jigsaw together.

Taking the Seyfert 2 galaxy first, the anticipated soft X-ray emission from the starburst is indeed seen, but the higher energy X-ray emission – showing a flat continuum and an iron K absorption edge – suggests that this is seen as a reflected component rather than by direct emission. Hence, the line-of-sight to the primary X-ray source is fully obscured and Compton absorbed. The authors conclude that in order to account for the total luminosity, the material reflecting the X-rays must subtend a rather large angle to the primary X-ray source, but not inconsistent with that expected from a Seyfert 2 geometry with the torus lying in the plane of the line-of-sight. On the other hand, in the second galaxy both the soft and hard X-ray emission is weaker than expected, while the hard X-ray shows complete obscuration but in this case with no reflected component. So, is the AGN weaker, or is there a difference in geometry of the reflected component, such as a smaller reflector or more obscuration? The authors opt for the middle course; the reflection component presenting a much smaller solid angle to the primary X-ray source. This paper clearly elucidates the gains to be had from the new X-ray facilities but demonstrates the challenges still to be faced. Overall we are left with the strong suspicion that obscuration and geometry hold the key to many of the observational differences we see across a wide spectral energy range.

11.4.4 Small Samples and Selection Effects

Given the discussion of the previous section, it is obvious that selection effects are a clear problem for AGN studies, particularly at high redshifts, where inevitably the more extreme objects are preferentially selected. Given the problems associated with dust in general, one expects that the obscuration problems become even more severe for type 2 objects. Indeed, only very recently has it been possible to investigate the type 2 AGN or radio galaxy AGN at redshifts around 3. New numbers suggest that these narrow-lined AGN are some 50-100 times more numerous than the readily-observed radio galaxy population. To give some feel for what this means, the overall number density of AGN at redshift 3 is then of order 400 per square degree per unit redshift (Steidel et al. 2002).

A particularly noteworthy paper by Moran, Filippenko and Chornock (2002) vividly demonstrates the difficulty of observational selection effects. This paper discusses the X-ray background and contributing sources and concludes that traditional ground-based optical spectroscopic observations can very easily be biased against AGN, leading to an under-representation of objects. While this is not a problem for the most luminous AGN (especially those of type 1) for the low luminosity objects and those of type 2 attempting to determine the number density as a function of cosmic epoch then becomes perilously difficult, not only because they are faint, but because their AGN signatures can be missed. Therefore, we must always be cognisant of selection effects that can bias the sample and hence the conclusions, and this is certainly a major cause of concern for AGN studies, especially when discussing evolution.

To conclude this section let me reiterate the point made above; that even identifying AGN activity can be difficult, let alone working out the precise details of their evolutionary tracks. The classification of LINERs (low ionization emission line galaxies) is a case in point; do they or do they not harbour an AGN? According to the strict original definition we find that LINERs are extremely common in the local Universe but they comprise a rat-bag of differing morphological types of galaxy with the spectroscopic classification being the sole unifying theme. The most recent evidence suggests that the situation is even more complex, with a range of activity displayed in LINERS, most, although not all of which appear to possess a true AGN core. On the other hand, hidden AGN are not everywhere; as originally suspected they are absent from the so-called HII-region galaxies.

Given the 'standard model' of orientation and an obscuring torus, one puzzling question has always been where, and what are the type 2 quasars (where we consider only the radio quiet quasar population rather than their radio-loud counterparts where the answer is much clearer)? These must be very luminous objects and should show the same evolution as the quasars. We have already had a hint of this in Sect. 11.4.3, and suspicion is now focusing on the ultra luminous infrared galaxies (ULIRGs) as the type 2 objects. However, much larger and deeper samples will be needed – perhaps from the SIRTF mission to be launched in 2003 – to prove this one way or the other.

11.5 The Central Engine and Accretion Disks

11.5.1 The Accretion Disk Puzzle

Finally, let us look at the central engine itself, namely the issues of fuelling and the accretion disk. At the outset we hinted that all might not be well here, and in fact the situation is just that. In the early days thin accretion disks were the preferred option, mainly because of the simplicity and the potential explanation for what was believed to be the continuum signature in the spectral energy distribution of AGN (mainly the big-blue-bump and a power-law in the X-ray region). However, as astronomers began to get a better handle on the accretion rates and luminosities of AGN, thin disks were found to have a number of problems, including long-term stability in the inner zones. An alternative class of models was a thick 'puffed up' disk, a concept that started life in one particular form as an ion torus in a seminal paper by Rees, Begelman, Blandford and Phinney (1982) – this showed promise for the collimation of material for the production of radio jets – and more recently from studies focussed on stellar sized black holes. This was developed into the Advection Dominated Accretion Flows, or ADAFs, by Narayan and Yi (1994). Since around 1996, this type of accretion disk model has been the flavour of the month.

11.5.2 ADAFs to the Rescue?

The key difference between a thin disk and an ADAF is that much of the radiated energy in the latter is trapped within the flow itself and is carried into the black hole rather than radiating as a hot disk. Therefore, it appears to be a much less radiatively efficient process, explaining the very low luminosities observed. Indeed, it was observations of the Galactic Centre Sgr A* source, which not only has a very low intrinsic luminosity but which is also very low compared to its anticipated accretion rate from the winds of surrounding stars, that provided much of the thrust for ADAFs in AGN in general. ADAFs come in two forms, optically thick and thin. While the former was developed first, it is the latter that seemed to hold most promise, whereby the accreting gas turns out to have such a low density that it is no longer able to cool efficiently by radiation and instead, the viscous energy in the flow is advected to the central black hole along with the gas. This leads to a number of differences from the thin disk; the gas in the ADAF is very hot, almost at the virial temperature, the radial velocity is comparable to the free-fall velocity, the angular velocity is then sub-Keplerian and the stability of the disk is maintained by the strong pressure gradient. This also leads to the characteristic 'puffed-up' shape of a torus rather than a disk. However, further work suggested that under certain conditions close to the Schwarzschild radius the torus is an over-simplification and that the shape is much more complex and probably a more ragged affair with little resemblance to either a disk or torus, but maybe more of a coronal gas.

So the current picture is rather complex. At large radii any accretion disk is likely to be a thin disk while for smaller radii, where the accretion flow rate and cooling become more important, the ADAF comes into its own with the cartoon-like toroidal shape. Indeed, the lower the accretion rate, the larger the domain of the ADAF compared to the disk. As the accretion flow increases there comes a point at which the shrinking ADAF is comparable in output to the disk, and for even higher flows the ADAF solution breaks down and the torus shape is destroyed and turns into a very hot corona (but with a thin disk farther out). Space limitations preclude further detailed discussion but an excellent and readable reference to ADAFs can be found in Narayan (1997). Further developments of accretion disk theories have led to convection dominated accretion flows (CDAFs) and the intriguing sounding adiabatic inflow-outflow solution (ADIOS). For an illuminating treatment of the arguments see Blandford (2001).

11.5.3 The Fuelling Process

Linking the accretion disk to the bulk of the host galaxy is mostly unexplored territory but one of the clues is the orientation of the black hole spin axis (as demonstrated by radio-jets) to one of the principal axes of the host galaxy. Unfortunately, even here the picture is complex, where general alignment of the dust disk and the plane of the accretion disk is not detected; in fact there seems to be no clear alignments for either disk (Seyfert type) or elliptical galaxy hosts. As we saw in the previous section, feeding the central engine is tricky and a good examination of this is given by Schmitt, Pringle, Clarke and Kinney (2002).

However, for AGN there is an apparent fundamental problem in that as Begelman (2003) points out, they are rather "fussy eaters" – the accretion luminosity is much less than the available fuel supply; black holes have problems swallowing the mass presented to them. This leads to the classic problem for accretion disks: how do quasars work? From what we saw earlier it is strongly believed that they have large black hole masses, so how do you get a high accretion rate at high efficiency when the energy transport mechanism seems to break down? ADAFs are almost certainly not the solution to the high luminosity AGN accretion disk. Although it may sound like heresy, one has to question whether ADAFs are the solution at all as a recent and particularly important observation regarding Sgr A* tends to support the idea that even in this case an ADAF solution is not the answer. (The key observation was the unexpected detection of strong polarization from Sgr A* at a wavelength of around 1mm by Aitken et al. (2000), which seems to rule out ADAFs.) For a general critique of the Sgr A* situation and the ADAF solution see, e.g. Agol (2000) and references therein.

11.5.4 They Seek Them Here, They Seek Them There...

So just where do we stand regarding accretion disks for AGN? There is precious little direct evidence for an accretion disk at small radii (although we saw an excellent example for large radii of $\sim 100,000$ Schwarzschild radii in NGC 4258 in Sect. 11.2.2). The X-ray iron line is the only example and there is still no evidence for such a line in the most luminous sources such as quasars (or many sources at all in fact). Maybe all the accretion disk models so far conceived work only for the lowest luminosity sources but not for the luminous Seyfert and quasar categories? Also, the spectral signatures expected from disks of various shapes seems to be missing (but see Lawrence 2003), so we currently face a dilemma in explaining how the accreting material flows towards the black hole and how it radiates in the process. While a thin disk works at large radii, what happens as the disk nears the minimum radius remains unclear, especially for the high luminosity objects. Do we end up with a wind-driven corona and how that couples into the broad-line region is clearly an area of work that is ripe for further study.

On the other hand we do have some clues. Winds and flows are known to be an important feature of the broad-line region due to the observations of what are known as broad-absorption line quasars (BALs). In these objects strong and fast winds are identified from their signature of broad and deep absorption features in the optical spectra. Although this is seen in only a minority of objects, suspicion is again pointed at orientation and obscuration (to provide the absorption lines) effects. However, unlike the type 1 and 2 dusty torus explanation, now we are dealing with a very central phenomenon, something well within the vicinity of the central parsec. The final comment on accretion must be 'watch this space!'

11.6 Conclusions

This review has considered a number of factors for the 'big picture' of AGN, describing the standard model and its successes and limitations. Unification based on orientation and obscuration seems safe and well, but the details are still to be fully worked through. Many questions remain unanswered: what is the precise shape of the molecular torus, how does it link with the central zone, how does the accretion machine work? While we have probed some of these areas, we should note that other areas have been notably omitted. For example, we have conspicuously avoided a general discussion of the geometry and physical interactions of the broad-line region, a topic almost exclusively that of optical and X-ray spectroscopy. It is clear that the picture is complex with primary (?) emission being reflected off clouds of highly ionized gas (the BLR) back onto the putative accretion disk and being reflected/absorbed/re-radiated to a distant observer (see Fig. 11.9). The two new X-ray observatories

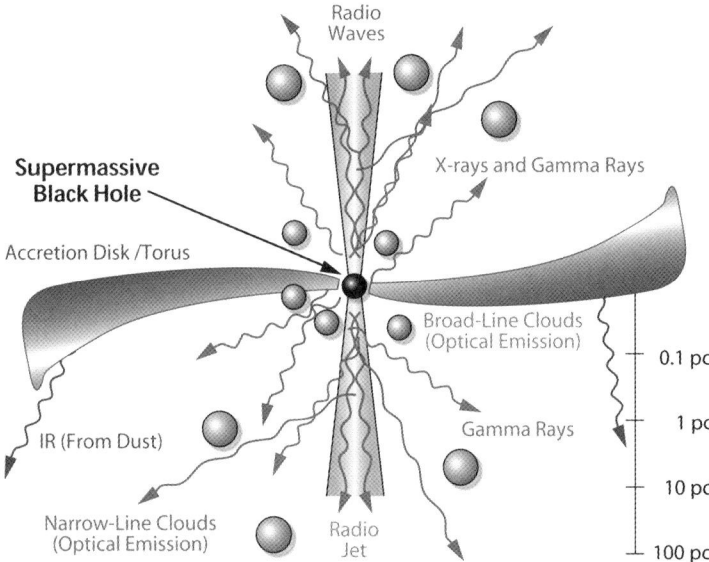

Fig. 11.9. A very recent cartoon of the region surrounding a supermassive black hole accreting material and producing a radio-jet. The torus is no longer shown as the smooth characterisation of Fig. 11.1 and the highly flared disk may provide the necessary obscuration. Note that the scale-length bar is logarithmic in an attempt to bring scale sizes to the picture. (Courtesy of NASA).

(Chandra and XMM-Newton) will provide much needed insight into this complex topic. In fact the first X-ray results are only just emerging and we will leave the central engine for the focus of a future review.

We have also ducked the burning issue of why so few AGN are radio loud: is it to do with the black hole itself (such as spin), accretion rate and/or the surrounding quenching medium? The entire study of relativistic jet physics is fascinating and to understand the pencil-thin beams of radio emission blasting out from the core of a galaxy is mind-blowing; this again will be reserved for future discussion.

So to conclude; while we have made huge progress in understanding the most luminous objects in the Universe, whereby the output of a thousand galaxies of stars is produced by a central engine of dimensions that are minuscule by comparison, many questions remain unanswered. At the 2002 AGN conference in Paris, the final session was devoted to following up a series of the 'top ten questions' posed by Begelman (2003) during his presentation, which considered the current major questions in AGN research. The reader is referred to this paper for the latest in what challenges currently face AGN research as well as the entire conference proceedings as the most up to date bibliography on the subject (e.g. Combes 2003).

References

Agol, E. 2000, Sagittarius A* Polarization: No Advection-dominated Accretion Flow, Low Accretion Rate, and Nonthermal Synchrotron Emission, ApJ., **538**, L121

Aitken, D. et al., 2000, Detection of Polarized Millimeter and Submillimeter Emission from Sagittarius A*, ApJ., **534**, L173

Antonucci, R & Miller, J. 1985, Spectropolarimetry and the nature of NGC 1068, ApJ., **297**, 621

Barger, A.J. et al. 2003, Very High Redshift X-ray Selected AGN in the Chandra Deep Field-North, ApJ., **584**, L61

Begelman, M.C., 2003, All you Wanted to Know about AGN but were Afraid to Ask in "Active Galactic Nuclei: from Central Engine to Host Galaxy" Meudon, July 2002; ASP Conference Series, eds.: S. Collin, F. Combes and I. Shlosman, 290, p23

Blandford, R. 2001, Probing the Physics of AGN: A Summary in "Probing the Physics of Active Galactic Nuclei by Multiwavelength Monitoring" eds. B. M. Peterson, R. W. Pogge & R. S. Polidan. ASP, p499

Cowie, L.L. et al., 2003, The Redshift Evolution of the 2-8 keV X-ray Luminosity Function, ApJ., **584**, L57

Combes, F. 2003, AGN Fueling: The Observational Point of View, in "Active Galactic Nuclei: from Central Engine to Host Galaxy" Meudon, July 2002; ASP Conference Series, eds.: S. Collin, F. Combes and I. Shlosman, 290, p411

Combes, F. 2002, Formation and Evolution of Galactic Black Holes, in "7th cosmology colloquium, High Energy Astrophysics, for and from Space, Paris, 11-15 June" N. Sanchez and H. de Vega, Editors

Della Ceca, R. et al., 2002, An Enshrouded Active Galactic Nucleus in the Merging Starburst System Arp 299 Revealed by BeppoSAX, ApJ., **581**, L9

Fadda, D. et al., 2002, The AGN contribution to mid-infrared surveys. X-ray counterparts of the mid-IR sources in the Lockman Hole and HDF-N, A&A, **383**, 838

Fadda, D. et al., 1998, The Near- and Mid-Infrared Continuum Emission of Seyfert Nuclei: Constraints on the Models of Obscuring Tori, ApJ., **496**, 117

Falcke, H., Wilson, A.S. & Simpson, C. 1998, HST and VLA Observations of Seyfert 2 Galaxies: The Relationship between Radio Ejecta and the Narrow-Line Region, ApJ., **502**, 199

Fan, X. et al., 2001, A Survey of $z > 5.8$ Quasars in the Sloan Digital Sky Survey. I. Discovery of Three New Quasars and the Spatial Density of Luminous Quasars at $z \sim 6$, AJ, **122**, 2833

Fan, X. et al., 2003, A Survey of $z > 5.7$ Quasars in the Sloan Digital Sky Survey II: Discovery of Three Additional Quasars at $z > 6$, A. J. **125**, 1649

Franceschini, A., Hasinger, G., Miyaji, T. & Malquori, D. 1999, On the relationship between galaxy formation and quasar evolution, MNRAS, **310**, L5

Greenhill, L.J. et al., 1995, Detection of a Subparsec Diameter Disk in the Nucleus of NGC 4258, ApJ., **440**, 619

Ivison, R.J. et al., 2002, Deep radio imaging of the SCUBA 8-mJy survey fields: sub-mm source identifications and redshift distribution, MNRAS, **337**, 1

Kormendy, J. & Gebhardt, K. 2002, Supermassive Black Holes in Galactic Nuclei, "20th Texas Symposium on Relativistic Astrophysics", ed. H. Martel & J.C. Wheeler, AIP, in press

Lawrence, A. 2003, The Ghost in the Machine, in "Active Galactic Nuclei: from Central Engine to Host Galaxy" Meudon, July 2002; ASP Conference Series, eds.: S. Collin, F. Combes and I. Shlosman, 290, p65

Magorrian, J. et al. 1998, The Demography of Massive Dark Objects in Galaxy Centers, AJ, **115**, 2285

McLure, R.J. & Dunlop, J.S. 2002, On the black hole-bulge mass relation in active and inactive galaxies, MNRAS, **331**, 795

Miyoshi, M. et al., 1995, Evidence for a Black-Hole from High Rotation Velocities in a Sub-Parsec Region of NGC 4258, Nature, **373**, 127

Moran, E.C., Filippenko, A.V. and Chornock, R. 2002, "Hidden" Seyfert 2 Galaxies and the X-Ray Background, ApJ., **579**, L71

Narayan, R. & Yi, I. 1994, Advection-dominated accretion: A self-similar solution, ApJ., **428**, L13

Narayan, R. 1997, Advective Disks, in "Accretion Phenomena and Related Outflows; IAU Colloquium 163". ASP Conference Series; Vol. 121; eds. D. T. Wickramasinghe; G. V. Bicknell; and L. Ferrario, p.75

Priddey, R. et al. 2003, Quasars as probes of the submm cosmos at $z > 5$, MNRAS, in press

Rees, M., Begelman, M.C., Blandford, R.D. & Phinney, E.S. 1982, Ion-supported tori and the origin of radio jets, Nature, **295**, 17

Reeves, J. 2003, XMM-Newton Observations of AGN; the iron K line and the central engine, in "Active Galactic Nuclei: from Central Engine to Host Galaxy" Meudon, July 2002; ASP Conference Series, eds.: S. Collin, F. Combes and I. Shlosman, 290, p35

Robson, I. et al. 2003, Submillimetre observations of a sample of very high redshift quasars, ibid, p617

Rowan-Robinson, M., 2000, Hyperluminous infrared galaxies, MNRAS, **316**, 885

Scarpa, R., Urry, C.M., Falomo, R., Pesce, J.E. & Treves, A. 2000, The Hubble Space Telescope Survey of BL Lacertae Objects. I. Surface Brightness Profiles, Magnitudes, and Radii of Host Galaxies, ApJ., **532**, 740

Schmitt, H.R., Pringle, J.E., Clarke, C.J. and KinneyA.L. 2002, The Orientation of Jets Relative to Dust Disks in Radio Galaxies, ApJ., **575**, 150

Schödel, R. et al., 2002, A star in a 15.2-year orbit around the supermassive black hole at the centre of the Milky Way, Nature, **419**, 694

Steidel, C.C. et al., 2002, The Population of Faint Optically Selected Active Galactic Nuclei at z $\sim$ 3, ApJ., **576**, 653

Tanaka, Y. al. 1995, Gravitationally Redshifted Emission Implying an Accretion Disk and Massive Black-Hole in the Active Galaxy MCG:-6-30-15, Nature, **375**, 659

Urry, C.M. & Padovani, P 1995, Unified Schemes for Radio-Loud Active Galactic Nuclei, PASP, **107**, 803

Vignati, P. et al., 1999, BeppoSAX unveils the nuclear component in NGC 6240, A&A, **349**, L57

Wang, J-M., Staubert, R. & Ho, L-C. 2002, The Accretion Rates and Spectral Energy Distributions of BL Lacertae Objects, ApJ., **579**, 554

Wilman, R.J., Fabian, A.C., Crawford, C.S. & Cutri, R.M. 2003, XMM-Newton observations of two hyperluminous IRAS galaxies: Compton-thick quasars with obscuring starbursts, MNRAS, **338**, L19

Woo, J-H. and Urry, C.M. 2002, Active Galactic Nucleus Black Hole Masses and Bolometric Luminosities, ApJ., **579**, 530

Yaqoob, T. & Padmanabhan, U. 2003, New Chandra Results on Warm Absorbers and Iron K Lines in AGN, in "Active Galactic Nuclei: from Central Engine to Host Galaxy" Meudon, July 2002; ASP Conference Series, eds.: S. Collin, F. Combes and I. Shlosman, 290, p39

Yu, Q & Tremaine, S. 2002, Observational constraints on growth of massive black holes, MNRAS, **335**, 965

12 The Story of Gamma-Ray Bursts

G. Vedrenne and J.-L. Atteia

12.1 Introduction

Gamma-ray bursts (GRBs) are short flashes of gamma-rays. They were fortuitously discovered by the US military VELA satellites which were used by the USA to monitor possible nuclear explosions within and outside the atmosphere of the Earth. The first event was recorded in 1967 but the existence of gamma-ray flashes coming from the cosmos was not announced to the scientific community until 1973 [56]. GRBs are brief and bright, lasting from less than a second to tens of seconds (sometime hundreds of seconds). It is impossible to predict when and from where they will arrive. During their short lifetime, they are so bright that their flux in gamma-rays may exceed the flux received from all other astronomical sources. This makes them easy to detect with small omni-directional detectors outside the Earth's atmosphere. Gamma-ray bursts were very quickly considered as an exciting new phenomenon by the astronomical community, and the rather easy way to detect them led many groups to explore this new field.

In the first part of this GRB story (Sect. 12.2), we describe the epoch of GRB discovery and present their main properties. We also discuss the first precise positions obtained by triangulation using an interplanetary network of detectors, and the unsuccessful searches for counterparts at other wavelengths. This period lasted nearly 20 years. At the end of this period the most accepted model was based on galactic neutron stars with high magnetic fields accreting enough matter to trigger a thermonuclear explosion. In 1991, NASA launched the Compton Gamma-Ray Observatory (CGRO) with BATSE, a highly sensitive GRB instrument. Section 12.3 discusses the results obtained by BATSE. They demonstrated the isotropy of GRBs on the sky and their radial inhomogeneity, which progressively convinced the scientific community of their extragalactic origin. The definite breakthrough came in 1997 with the Italian Dutch satellite BeppoSAX (Sect. 12.4). This satellite, although not particularly designed for GRB studies, carried two wide-field X-ray cameras and a GRB monitor which together were able to quickly localize GRBs at the level of several arc minutes, and to inform the scientific community within a few hours of the burst detection. This allowed searches for counterparts to commence after only a few hours instead of several weeks as was usual before BeppoSAX. This strategy led to the discovery of GRB afterglows at radio, optical and X-ray wavelengths. The afterglows, unlike the bursts, can be observed with the usual tools of astronomers (large ground- and space-based telescopes) and their study established for the first time the redshift and the distance of several GRB sources, and enabled their host galaxies to be found.

These multiwavelength studies opened a new era in GRB science. But even though we know today that GRBs are huge explosions taking place in very distant galaxies (at typical distances of 10^{10} light years), the real origin of these fireballs is not yet understood. Two main sources are proposed: coalescence of compact objects, or unusual supernovae, which is now the favourite model. Section 12.5 discusses our current understanding of GRBs, their link with supernovae and their place in the 'zoo' of stellar explosions. In this section, we also show how GRBs can be used to probe the remote universe and the properties of young, distant galaxies. After the end of BeppoSAX (in 2002), researchers in this field can now rely on HETE and INTEGRAL to extend the work of BeppoSAX, with even faster localizations. These missions have already brought some surprises. We conclude with Sect. 12.6 which considers some important questions that are still unanswered.

This review intends to show how GRBs have evolved in our view from a mere astronomical curiosity to a complex physical and astronomical phenomenon, which now deserves observation with the most powerful observatories, and which is closely connected with many fundamental issues of modern astrophysics. A significant section of our review is devoted to the discoveries which made possible this evolution. A consequence of our choice to survey a long time period is that we could not mention all of the contributions to the GRB story; we apologize to those who contributed to this story and are not cited here. The research field has grown very quickly in recent years and the present state of research would deserve a full review. So, the reader seeking more information on the history of GRB research or on the present status of the field could consult reviews on GRB observations by Hurley [51], Mazets [67], Vedrenne [112], Fishman & Meegan [29], and Piro [91]. Excellent theoretical reviews have been written by Hartmann [44], Piran [90], and Mészáros [72]. Last but not least, the story of GRBs is presented in full detail in the recent book of J. Katz[1].

12.2 The Early Times

During this period which lasted nearly twenty years, many GRB properties were established. In parallel, the theoretical work on the origin of GRBs led, after many models and scenarios, to the generally accepted idea of a galactic

---

[1] Katz, J. I. 2002, The biggest bangs : the mystery of gamma-ray bursts, the most violent explosions in the universe, by Jonathan I. Katz. Oxford: Oxford University Press, 2002, ISBN 0195145704

origin. The objects responsible for these emissions were thought to be neutron stars accreting matter and emitting GRBs when this matter was burnt during strong thermonuclear flashes at their surface [42, 114]. Although during this period this kind of model was extensively studied, it will not be described here because we know now that it does not explain the GRB population. We will concentrate on the general properties of GRBs discovered during this period and for most of them confirmed later by BATSE.

Morphology and temporal properties of GRBs. Figure 12.1 shows a typical GRB light curve. Very early on it was realized that the light curves of GRBs are quite variable from one burst to the next. The shapes are different as are the time scales which can vary from a few milliseconds to hundreds of seconds. The shortest structures identified in the light curves of long bursts last only a few ms. The analysis of GRB durations shows a bimodal distribution with 2 broad peaks around 0.3 and 30 s (Fig. 12.2). Short duration GRBs are not only shorter, they also have a harder spectrum [25]. Another property was also noticed quite early on: at high energies GRB light curves are sharper with shorter durations. This effect was later quantified with BATSE. Finally, the searches for periods or quasi-periods in the light curves did not give convincing results.

Spectral properties. A typical GRB spectrum is shown in Fig. 12.3 (for GRB 990123), illustrating the non-thermal nature of the emission. The peak

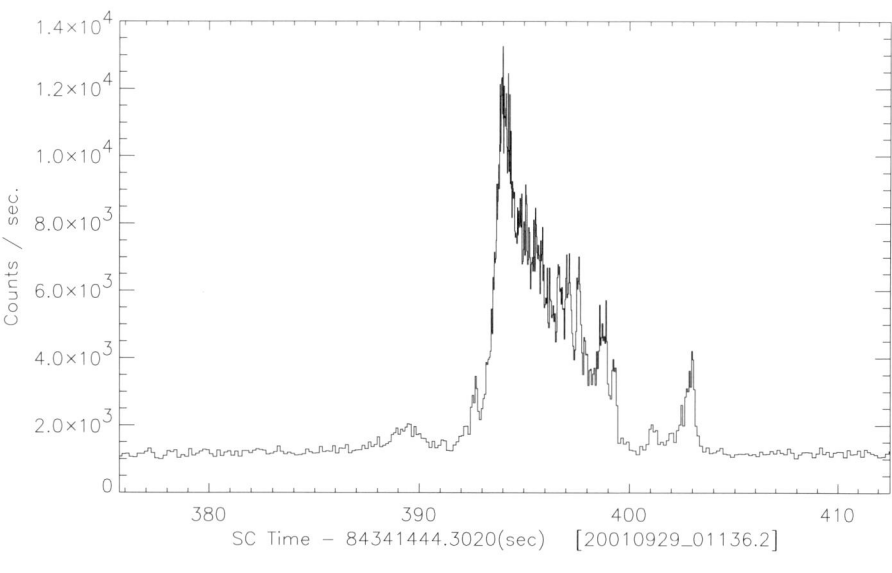

Fig. 12.1. Typical light curve of a gamma-ray burst at gamma-ray energies (from the HETE/FREGATE instrument).

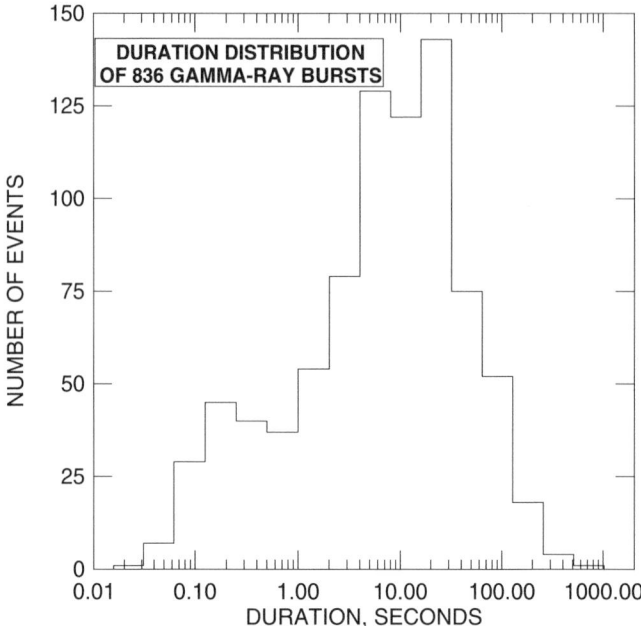

Fig. 12.2. Duration distribution of 836 GRBs (from Hurley 1992 [52]).

of the spectral energy distribution is clearly visible in the lower panel. This peak usually varies from below 100 keV to 1 MeV. Some GRBs are accompanied by X-ray (keV) [78] or high energy gamma-ray (GeV) [53] emissions. Spectral evolution is the rule during GRBs. Due to the lack of spectra integrated over short timescales, the spectral evolution was often studied through the evolution of the hardness of the spectrum (spectral index or hardness ratio). Two very general trends were found: the hardness tends to follow the evolution of the luminosity (hardness-intensity correlation) and the maximum hardness is often observed at the beginning of the peaks (hard-to-soft evolution).

In these early times, the possibility of emission or aborption lines in GRB spectra was extensively studied. Absorption features at a few tens of keV were reported by KONUS [65]. Some doubts were raised as to the reality of these features which were shown to depend crucially on the choice of the continuum [111], but they were often observed in two different detectors increasing the credibility of their detection. In fact their reality was accepted by the international community only after they were confirmed by the GRB detectors of Ginga [77]. Even so, some researchers continued to express the view that these lines could be instrumental effects [80]. These absorption features, seen by Ginga in three GRBs, were interpreted as cyclotron absorption lines in the very strong magnetic fields of neutron stars; this was considered as a strong argument in favour of the production of GRBs by galactic neutron stars. An-

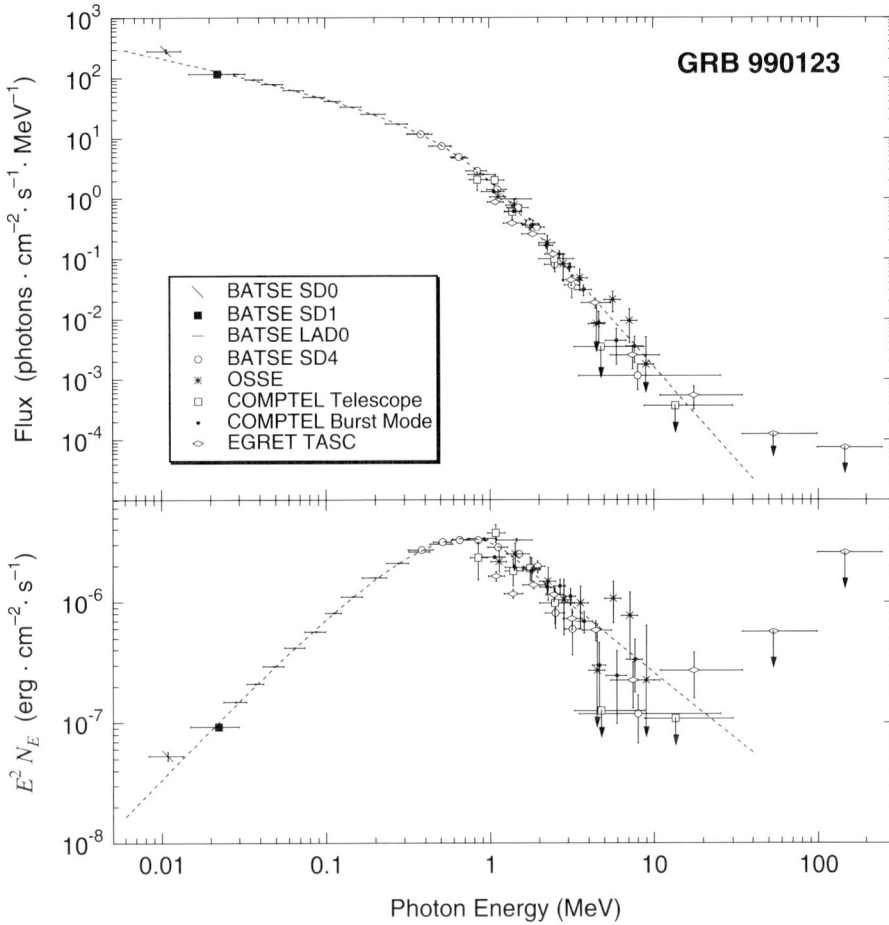

Fig. 12.3. Broadband spectrum of GRB990123 recorded by BATSE, COMPTEL, OSSE, and EGRET on CGRO [17].

nihilation features, possibly gravitationally redshifted, were also reported by KONUS experiments for more than 10 % of the GRBs [65], but here again they were subject to large controversy. In fact we shall see that BATSE [5] and TGRS [86] have found no evidence for absorption or emission features in GRB spectra.

12.2.1 GRB Counterparts

Most scientists were convinced that the mystery of the GRB origin would be solved with the identification of counterparts at longer wavelengths (optical, radio or X-rays). The search for counterparts, however, requires accurate positions and before 1996 the only way to get precise (arcminute) positions was the triangulation method. This method uses the fact that GRBs are transient

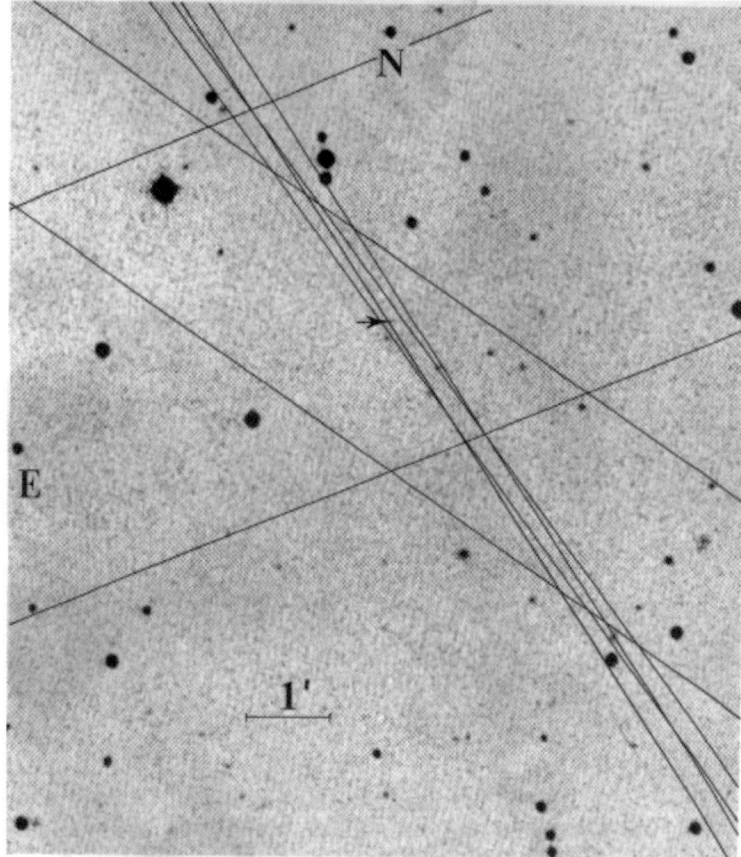

Fig. 12.4. The error box of GRB 790613, illustrating the lack of quiescent optical counterparts in GRB error boxes [6].

signals, and reconstructs their direction on the sky from the measure of their crossing time at various points in the Solar System. The precision of localization is strongly dependant on the distances of the satellites and on the precise knowledge of their positions in space. Space probes going to Mars or Venus have been very useful in improving the precision of localization, sometimes down to about one arc minute. With the Franco-Soviet program SIGNE, and in collaboration with US satellites, a program of precise GRB localizations led by K. Hurley in France, and by T. Cline in the US was soon started; it was called the Interplanetary Network for GRB localization, or IPN. In the 1980s, several GRBs were precisely localized with this method, and significant efforts were devoted to searches for GRB counterparts. They concerned *flaring* (contemporaneous with the GRB), *fading* (which could remain visible in the days or weeks following the burst) or *quiescent* counterparts (always present).

All three types of counterparts were searched for in the optical, while only quiescent counterparts were searched for at radio and X-ray wavelengths.

Flaring counterparts. Two methods were used to search for optical emission simultaneous with GRBs. The first used 'all sky surveys' monitoring a significant fraction of the sky ($> 10\%$) for optical transients brighter than magnitude 5-10. Photographic plates and/or CCD images were recorded every clear night and examined *a posteriori* when a GRB was localized in the part of the sky which was monitored [50]. The requirement of spatial and temporal coincidence was mandatory in these searches, given the high background of optical transients due to meteors, planes, and satellites... The second method used archival plates taken in the direction of well-localized GRBs. The idea behind this technique was to look for past eruptions from GRB sources. The assumption that GRBs were emitted by galactic neutron stars led to the conclusion that they had to repeat to explain an observed GRB rate of a few 10^2 yr^{-1}, with only one neutron star created every thirty years or so. This work, which required hundreds or thousands of archival plates of the GRB field, used plate collections available at the Harvard, Sonneberg or Ondrejov observatories [39, 106]. Many groups contributed to this work which revealed some interesting possible counterparts. Nevertheless after several years of deep studies the general feeling was that these searches had not provided convincing counterparts and that the few interesting possible cases could be explained by background optical transients, possibly due to flare stars, meteors, planes or satellite flashes. We now understand better all these negative results because the all-sky surveys were not sensitive enough to detect the prompt optical emission from most GRBs (see Sect. 12.4), while archival plate searches were doomed to fail because GRB sources are not galactic and do not repeat.

Fading counterparts. Searches for fading counterparts were plagued by the long delay necessary to localize GRBs by triangulation, typically several weeks, due to the need to exchange magnetic tapes between researchers in the US, in France and in the Soviet Union. The identification of GRB sources, in 1997, was finally the result of the discovery of GRB fading counterparts. In the 1980s these searches started too late. We know now that the fading optical emission from GRBs (the afterglow) decreases quickly and is only detectable above magnitude 21-22 during the few hours following the burst (up to 2-3 days after the GRB in the most favourable cases).

Quiescent counterparts. This is certainly the field which demanded the most effort and attention: quiescent counterparts to GRBs were searched for at optical, X-ray and radio wavelengths. These searches were based not only on deep observations within some small error boxes (see Fig. 12.4), but also on the statistical comparison of GRB positions with those of other astronomical objects (e.g. pulsars, quasars, X-ray binaries...). These searches revealed a few interesting objects in some of the error boxes, in particular a radio source [48] and an X-ray source [16] in the error box of GRB 781119. However after much work, no consistent picture had emerged from these studies, leading to the general consensus that GRBs have no quiescent counterparts. X-ray observations played an important rôle in these early studies. At the beginning of the 1980s several classes of galactic neutron stars were known, such as the radio pulsars, the X-ray pulsars and low mass X-ray binaries (bursting or not). X-ray observations were very powerful for detecting accreting neutron stars in the Galaxy. In this context many observers were convinced that X-ray observations could be the missing link between gamma-ray bursters and their quiescent counterparts. HEAO and EXOSAT observed a few well-localized GRBs but did not find the expected X-ray emission [15, 94]. The lack of X-ray counterparts had two important consequences for galactic GRB sources: it showed that they should have low accretion rates (of the order of 10^{-15} $M_\odot$ yr^{-1}) and that they should not be too close to the Sun (hundreds of parsecs).

12.2.2 The Angular and Intensity Distributions

Angular distribution. In the absence of any identification with known celestial sources, measuring the distribution of GRBs on the sky was a good way to approach their origin. Galactic sources in particular should be concentrated in the galactic plane. Omnidirectional gamma-ray detectors have generally been used for GRB detection. While a single detector is unable to give the arrival direction of the burst by itself, a crude direction (at the level of a few degrees) can be obtained by comparing the count rates received by detectors with different orientations on the same spacecraft (because the response of a detector to a parallel photon flux varies approximately as the cosine of the angle of the photon beam to the detector normal). Mazets and collaborators [66] were the first to use this method with the KONUS experiments. This technique would also be used later by BATSE on CGRO. KONUS and the IPN, with different methods and for differents GRBs, obtained maps of nearly a hundred GRB positions on the sky. As an example, the analysis of the distribution of 88 GRBs localized by the IPN showed that this distribution was isotropic within the statistical limits [3]. This analysis also provided an opportunity to estimate a lower limit for the repetition time scale of GRBs, which could vary from 0.5 to 10 years depending on whether the bursts were assumed to have an extended luminosity function, or to be monoluminous.

Intensity distribution. For galactic sources, the isotropy was puzzling. In fact all other known populations of galactic neutron stars were more or less strongly concentrated in the galactic plane. The most accepted interpretation of isotropy was that GRB sources were nearby galactic objects and that the instruments were not sensitive enough to see them beyond their scale height in the disk of the Galaxy. It was commonly accepted that more sensitive detectors (seeing farther sources) would be needed to disclose the GRB galactic structure. Since the IPN and KONUS localized only half the GRBs they detected, the brightest ones, the work based on localized GRBs did not allow a check on the distribution of the full sample of detected GRBs. A method to check the spatial homogeneity of non-localized sources is to measure their intensity distribution. For local, spatially homogeneous objects the number of visible sources increases as the cube of the distance, while the brightness of the sources decreases as the square of the distance. Consequently for an homogeneous distribution, the number of visible sources must grow like the intensity at the power -1.5. This is how the intensity distribution provides insight into the source radial distribution.

Several measures of burst intensity have been used, each having specific biases and selection effects. Early work used the total fluence (the fluence is the GRB energy flux integrated during the whole duration of the burst) or the peak photon flux. Because GRB detectors trigger when they detect an excess of photons above the background in fixed time and energy intervals, their threshold in fluence varies strongly from burst to burst and the fluence distribution is not appropriate to test the spatial homogeneity of GRBs. The peak photon flux is a much better quantity for this purpose, especially if it is measured on a timescale and in an energy range matching those used for the trigger. While the peak photon flux is acceptable when only one instrument is considered, the difficulties return when the results of different detectors are to be compared. Finally, after many years of controversy, M. Schmidt [107] proposed to measure GRB intensities (at least for the purpose of testing their spatial homogeneity) by C_{max}/C_{min}, the ratio of the maximum count rate measured during the GRB to the minimum count rate needed to trigger the instrument. This measure of intensity directly leads to the test $\langle V/V_{max} \rangle$, which measures for each GRB its position within the volume accessible to the instrument. For homogeneous sources V/V_{max} is uniformly distributed between 0 and 1 with an average value of 0.5. The agreement on the use of $\langle V/V_{max} \rangle$ resulted, in 1990-1992, in the publication of the values measured by several GRB detectors (summarized in [26, 112]), simultaneously with the first results of BATSE. As expected, smaller detectors have $\langle V/V_{max} \rangle$ closer to 0.5, with e.g. $\langle V/V_{max} \rangle = 0.46 \pm 0.02$ for PVO with an effective area of 11 cm^2; $\langle V/V_{max} \rangle = 0.43 \pm 0.03$ for KONUS and SIGNE with effective areas of 50-60 cm^2; $\langle V/V_{max} \rangle = 0.37 \pm 0.03$ for Phebus with an effective area of 90 cm^2 and $\langle V/V_{max} \rangle = 0.32 \pm 0.02$ for BATSE with an effective area of 2000 cm^2. Another interesting property discovered at that time is that detectors working at high energies have lower $\langle V/V_{max} \rangle$ than detectors working at lower energies [26]. This was explained later by the fact that intrinsically bright GRBs have harder spectra, allowing GRB detectors which operate above 100 keV to see them at larger distances. This again illustrates how the intrinsic properties of GRBs (luminosity, shape of the light curve, spectral hardness) make difficult the interpretation of their intensity distribution.

To conclude this first part, many GRB properties were established during the period of 20 years following the discovery of these strange sources. The temporal and spectral characteristics of GRBs were however unable to limit the large number of theories proposed to explain them. Cyclotron lines observed by Ginga were nevertheless suggestive of high magnetic fields at the surface of neutron stars and therefore of a galactic origin for the bursts. Strong constraints on galactic models were however provided by the lack of quiescent X-ray counterparts, by the absence of repetitions and by the apparent spatial isotropy of the sources [43, 85]. In the late 1980s, there was a strong need for more sensitive detectors which could measure the spatial distribution and definitely state on their galactic origin.

12.3 CGRO: A New Step in Understanding the Origin of GRBs

BATSE (for Burst And Transient Source Experiment) on CGRO is certainly the most ambitious experiment especially designed to study GRBs. BATSE consisted of 8 pairs of NaI detectors. Each pair included a Large Area Detector and a Spectroscopy Detector. The surface area of a Large Area Detector was 2000 cm^2. With a sensitivity 5 to 10 times better than previous GRB detectors, BATSE detected an average of 1 GRB per day during the 9-year long life of CGRO. These detectors were oriented in different directions to cover all the sky, and to localize GRBs crudely, with the same techniques already used by Mazets with the KONUS experiments. Due to the large dimensions of the detectors, the spectrum and the time profile of each burst were analyzed with unprecedented statistics. This mission confirmed the diversity of GRB lightcurves. Moreover the large number of events registered allowed a lot of statistical studies which revealed some new spectro-temporal properties of GRBs (see for instance [61, 95, 103].

Among the important advances achieved by BATSE we note:

– The isotropy of the celestial distribution of GRBs which was established after just 6 months of observation [68], and was confirmed with a high statistical significance after 9 years. A detailed assessment of the level of isotropy of the GRB population as a whole and of selected subsets can be found in Briggs et al. (1996). Figure 12.5 displays the distribution on the sky of 1637 GRBs from the 4th BATSE Catalog [83].
– The inhomogeneity of the GRB spatial distribution is measured with the test $\langle V/V_{max} \rangle$. After 6 months of operation Meegan et al. [68] found $\langle V/V_{max} \rangle = 0.33 \pm 0.02$. After 9 years and nearly 1000 GRBs used for this analysis, the value was essentially the same but with a smaller error bar. So what was suspected before BATSE is now well established: GRBs are not homogeneously distributed in space. More detailed studies of the intensity distribution showed that the departure from homogeneity was due to a deficit of faint bursts. The important finding of BATSE is that this deficit is not accompanied by any kind of anisotropy: faint bursts are more rare but just as isotropic as bright GRBs.

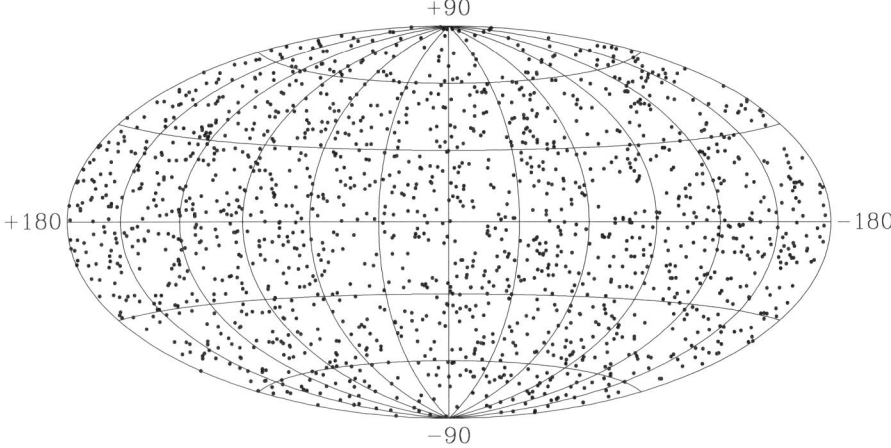

Fig. 12.5. Distribution on the sky of 1637 GRBs from the 4$^{\rm th}$ BATSE Catalog [83].

- The clear confirmation of a bimodality in the duration distribution of GRBs: the duration histogram shows 2 peaks around 0.6 sec and 31 sec [58]. The boundary between the two classes is around 2 sec, and short GRBs are 'harder', containing a larger proportion of photons above 100 keV.
- The detailed analysis of a large number of GRB spectra, which permitted characterization of the continuum of the burst as a broken power law, with a smooth break [4]. The spectral evolution during GRBs was confirmed while the presence of sharp spectral features (lines) was not: BATSE showed no absorption features at several tens of keV which would have confirmed the 'cyclotron lines' [5] and no emission lines near 400 keV. This is very important since the absorption lines were considered as one of the strongest argument in favour of galactic neutron star models.

BATSE enabled the discovery of other important properties of GRBs, such as the presence of precursors to some GRBs [57], the existence of very long GRBs lasting more than 20 minutes [110], and a correlation of the GRB spectral hardness with their apparent brightness [27, 64]. An interesting result was established by Fenimore et al. [28] who quantified the stretching of GRB light curves at low energies. More details on the results of BATSE have been reported in a very complete paper by G. Fishman and C. Meegan [29]. The beautiful results of BATSE definitely excluded for the first time a galactic disk population of GRBs, leaving open two possibilities for their origin:

- GRBs confined within an extended galactic halo. The isotropy of observed GRBs requires a size significantly greater than the distance between the Sun and the Galactic Center. In 1994, after three years of BATSE oper-

ation, Hakkila et al. [40] showed that BATSE data were compatible with a corona extending beyond 125 kpc but smaller than 400 kpc.
- GRBs at cosmological distances. They are naturally compatible with the observed isotropy, while inhomogeneity is explained by cosmological effects and source evolution. Of course in this case the energy content in GRBs is enormous.

Of the two models remaining after the BATSE results we can say without anticipating the last part of this story that the extended halo appeared as an *ad hoc* model which was strongly connected with the idea that galactic neutron stars were implicated in GRBs. Major questions were raised by this model, among them: How to populate such a large halo with neutron stars? Why do neutron stars in the halo produce GRBs while neutron stars in the galactic disk do not?

Concerning the cosmological scenario it always had its strong defenders, an eminent one among them being B. Paczyński [84]. During the BATSE period many searches were undertaken to settle such a model, by looking for cosmological effects in GRBs.

- Time dilation. Norris et al. [81] found evidence that dim bursts last about 2 times longer than the bright one. But Mitrofanov et al. [76] found no evidence for such time dilation.
- Spectral redshift. As mentioned previously, faint GRBs are clearly softer than bright bursts [27, 64]. While this trend is compatible with cosmological sources, various authors suggested that it was rather the consequence of an intrinsic hardness-*luminosity* correlation which was not blurred by the distance [62]. Again the hardness-intensity correlation could not be used to measure the distance to GRBs.

The main problem in the search for distance estimators with the gamma-ray data alone was the great diversity and variability of the light curves and the lack of lines in the spectra. Recent studies using the growing sample of GRBs with measured redshifts (see next section) suggest however that this task is not impossible and good correlations have already been found between the intrinsic GRB luminosity and their spectral lags [82] (the delay of low energy photons with respect to higher energy photons) and between the intrinsic GRB luminosity and the variability of the light curve [100]. To conclude, BATSE on CGRO had provided decisive information on the GRB phenomenon but it had not yielded the distance of the sources. This was to be the role of BeppoSAX.

12.3.1 GCN: The Gamma-Ray Burst Coordinate Network

BATSE localizations were too large to permit efficient searches for quiescent or fading counterparts, and the data arrived too late to search for flaring counterparts. The early failure of the tape recorders on-board CGRO led to the decision to transmit the data to the ground in real-time via the TDRSS communications satellites. This reduced to a few seconds the delay to get the data to the ground and allowed searches for flaring counterparts of long GRBs. S. Barthelmy set up a very efficient network for the fast distribution of BATSE positions to observers around the world (BACODINE, for BAtse COordinate DIstribution NEtwork [8]). Later renamed GCN (GRB Coordinate Network), this network would also be used to distribute GRB positions measured by BeppoSAX, HETE-2, and INTEGRAL. It is the GCN which permitted the detection of the first flaring counterpart of a GRB in January 1999, with ROTSE, a wide field optical camera [1].

12.4 BeppoSAX: Its Decisive Role in Understanding GRBs

The BeppoSax satellite (SAX for Satellite italiano per Astronomia X, and Beppo from the nickname of the famous Italian physicist Giuseppe Occhialini) launched from Cape Canaveral on April 30, 1996 was put in an almost equatorial orbit at an altitude of 600 km. This orbit is well protected from high-energy cosmic rays by the Earth's magnetic field; so the background of X- and gamma-ray detectors due to these high-energy particles is minimized. The GRB detection on BeppoSAX was based on the GRB Monitor (GRBM), composed of 4 identical CsI(Na) scintillator slabs, about 1000 cm^2 each. This system with a nearly 4π field of view worked between 40 and 700 keV. The localization was provided by two Wide Field Cameras (WFC) which were position sensitive proportional counters with a coded mask in front of their collimated field of view. The energy range was 2-26 keV and the field of view $40° \times 40°$ at zero response. After the detection of a GRB by the GRBM, the scientific team looked for the burst in the WFCs, and if the burst was visible in a WFC it could be localized with a precision of 3 arc minutes. This capability to obtain arcminute positions in a few hours was the major advantage of BeppoSAX over previous missions and the key to its success.

In addition to the fast localization capability provided by BeppoSAX, it was possible to quickly re-orient the spacecraft to observe GRBs with the narrow field instruments (NFI), sensitive telescopes with a field of view of $1.5°$. These new pointings could be acquired within 5 to 8 hours of the GRB. After this typical time it was possible to look for GRB emission in the X-ray domain with the NFI, composed of one low energy concentrator (0.1 - 10 keV), three medium energy concentrators (2 - 10 keV), and one High Pressure Gas Scintillation Proportional Counter (4 - 120 keV). This very complete set of instruments [14] on the same satellite and the possibility to react quickly (within a few hours) have been enormous advantages over previous missions. Since the positions obtained with the WFC were immediately communicated to the scientific community, this allowed the activation of space- and ground-telescopes working at other wavelengths a few hours after GRB detections. This strategy led to the discovery of the X-ray afterglows of many GRBs, and the detection of these afterglows in visible, radio, and infrared domains. As we shall see with a few examples this resulted in the measure of GRB redshifts and in the identification of their host galaxies. We will also understand why in the past, even when precise locations were obtained by the IPN, it had been impossible to find the afterglows of GRBs. In most cases the precise locations were obtained a too long time after the GRB preventing the detection of a fast fading object in the error boxes. We illustrate below the contribution of BeppoSAX to GRB science with the presentation of a few 'milestone' GRBs. A more detailed discussion of GRB afterglows is given in Sect. 12.5.

GRB 970228. This GRB was detected by the GRBM on February 28, 1997 at 2h58 UT. This was a moderately intense, multipeak, classical GRB [20]. One of the two WFCs localized this GRB with a precision of 3 arcmin. The decision to look at this burst with the narrow field instruments was quickly taken and the observation started 8 hours after the trigger. The first observation, which lasted about 14000 s, identified a previously unknown X-ray source with a flux of 3×10^{-12} erg cm^{-2} s^{-1} in the 2-10 keV energy range. During the observation it appeared the source was fading. A new observation (lasting 16000 s) was decided for March 3. Figure 12.6 shows the two images of the GRB region. Clearly the X ray source has a very fast decreasing intensity. The decrease followed a power law, $t^{-1.3\pm 0.1}$. This was the very first detection of an X-ray afterglow following a GRB. One week

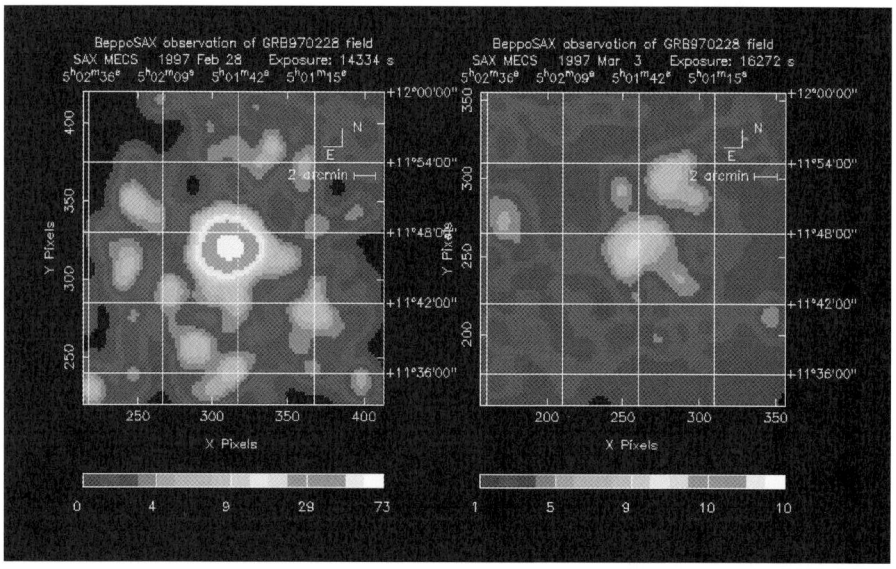

Fig. 12.6. X-ray afterglow of GRB 970228 [20].

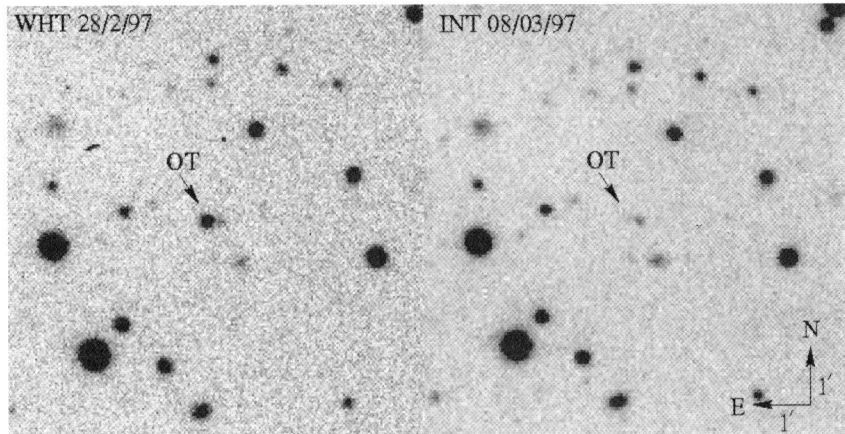

Fig. 12.7. Optical afterglow of GRB 970228 [88].

later Asca detected the source [117] and confirmed that the afterglow was decaying as $t^{-1.3}$. The X-ray satellite ROSAT also detected this afterglow at a flux level compatible with the extrapolation of the decay law established by BeppoSAX, and provided an improved position for the X-ray source [33]. The BeppoSAX localization was also confirmed by the IPN which included the Ulysses spacecraft far from the Earth [54].

In parallel with the X-ray observations, an observational campaign started using optical telescopes. An optical transient was discovered [88] thanks to the Isaac Newton Telescope and the William Herschel Telescope (see Fig. 12.7). The observations took place the night of the GRB appearance and one week later. Between these two detections the flux had decreased dramatically by a factor of 300. This was the first true detection of an optical afterglow following a GRB. During this period the excitement was at its maximum and two pointings of the HST (Hubble Space Telescope) were decided. A nebulosity around the source was detected [34, 104] and interpreted as the host galaxy of the GRB.

GRB 970508. This is the first GRB whose distance has been measured. After the localization of the burst by one WFC on May 8, 1997, a follow-up observation was quickly decided; it started 5.7 hours after the trigger. The X-ray and optical afterglows were identified but their light curves were more complex than in the case of GRB 970228. More important for this burst was the brightness of the optical counterpart, which permitted measurement of its spectrum with the Keck Telescope [69]. This spectrum showed absorption lines of FeII, MgI and MgII at cosmological redshifts z=0.767 and z = 0.835 due to galaxies in the line of sight to the GRB. This showed that GRB 970508 was at a distance of several gigaparsecs, and marked the end of a 24-year long debate on the distance scale of GRBs and on the amount of energy released in these events. This burst also gave the opportunity to identify the first radio afterglow with the VLA, which proved the relativistic expansion of the fireball [31] (see Sect. 12.5.1). Finally the host galaxy was identified with the Keck Telescope [11] and the HST [35] when the afterglow had faded below detection, several months after the GRB. The afterglow was found to be exactly coincident with the centre of its host galaxy [35], leading to the conclusion that the GRB was within the galaxy and not behind it. The spectrum of the galaxy exhibited emission lines which were identified with lines from OII and NeIII at a redshift z=0.835 [11].

GRB 990123. In order to take advantage of the rapid dissemination of BATSE localizations permitted by the GCN (Sect. 12.3.1), Akerlof and his collaborators [55] built a wide-field optical camera (16° x 16°) able to pinpoint GRB positions in less than 3 seconds. This instrument, called ROTSE (Robotic Optical Transient Search Experiment) was installed in New Mexico in 1998 and looked for every BATSE burst which was visible at night from this site. GRB 990123 was detected by BATSE at 9h47 UT on January 23, 1999 while New Mexico was still in the night, and localized with an error of 10°. ROTSE started to take pictures of the region of the burst only 22 seconds after the BATSE trigger. In the meantime this burst was detected by the GRBM on BeppoSAX and localized by the WFC within an error region of 5', eleven degrees off the center of the BATSE position [92]. On the morning of January 24, Akerlof and collaborators looked at the ROTSE data in the error box of the WFC, and they discovered a new star which reached mag 9 and disappeared after 15 minutes. Later the spectroscopy of the afterglow of GRB 990123 showed that this burst exploded at a redshift z=1.6. This was the first observation of a contemporaneous optical emission from a GRB. Astronomers would have to wait almost four years before they could achieve a comparable level of success, with the observation of GRB 021211 with RAPTOR, Super-LOTIS and KAIT, following a localization sent by HETE 22 seconds after the trigger.

These few examples illustrate the fantastic progression of the field after the launch of BeppoSAX. We now understand that GRBs are the witnesses of fantastic explosions which release in just a few seconds as much energy as the Sun during its entire life. Six years after the discovery of the afterglow of GRB 970228, the follow-up of several tens of GRBs localized with BeppoSAX, the IPN, HETE or INTEGRAL, has given us a much more complete and complex view of this phenomenon which is briefly discussed in the next section.

12.5 A New Window on Stellar Explosions and on the Early Universe

The study of GRB afterglows is connected with many interesting astrophysical issues which are briefly highlighted below. The reader interested in more details should read the excellent review on afterglow observations and theory written by van Paradijs, Kouveliotou & Wijers [89].

– Beaming of the relativistic jet.
  This question was first raised by Rhoads in 1997 [101]. He demonstrated that the afterglows of beamed GRBs must have a break in their light curve; this distinctive feature has been called a beaming break. Such a beaming break was clearly detected two years later in GRB 990510 with a beaming angle estimated to be 4° [41]. Recently Frail et al. [32] showed that the 'apparently' more energetic GRBs are also the most highly beamed. When the size of the jet is used to correct the apparent energy release, GRBs seem to have a nearly constant energy output peaked around 5×10^{50} erg. An interesting consequence of beamed GRBs is the likely existence of 'orphan' afterglows (afterglows not associated with a visible GRB), observable when our line of sight is close to the jet axis but does not intersect the GRB jet.
– GRB environment and progenitors.
  The synchrotron shock model of the afterglow (see Sect. 12.5.1 below) permits probing of the environment of GRBs when their broad-band energy spectrum can be measured. This method was first used by Galama et al. on GRB 970508 [36]. They showed that its ejecta was expanding in a medium of density n=0.03 cm^{-3}. Figure 12.8 displays the broad-band spectrum which was used for this analysis, and the fit with a synchrotron shock model. This method was later applied to several GRBs by Panaitescu &

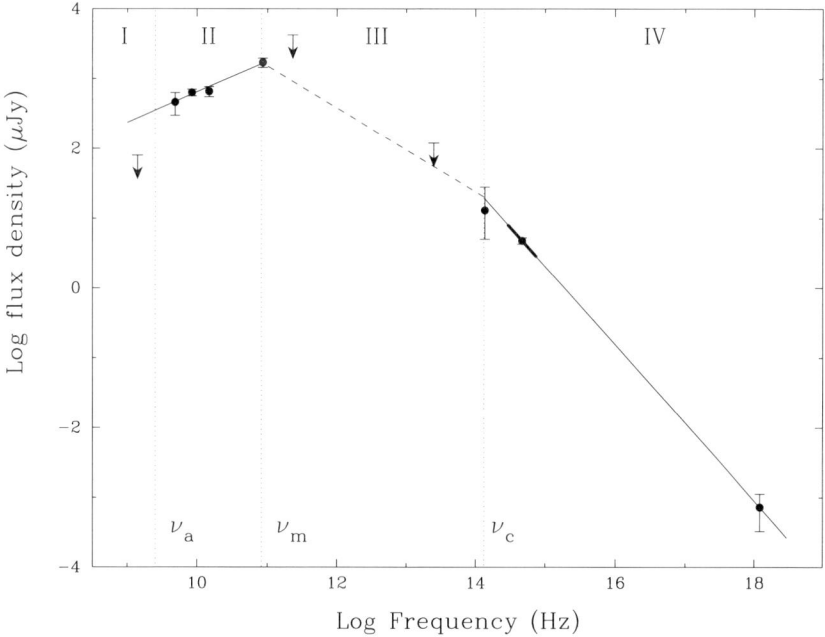

Fig. 12.8. Radio to X-ray energy spectrum of GRB 970508, measured on May 21, 1997 [36].

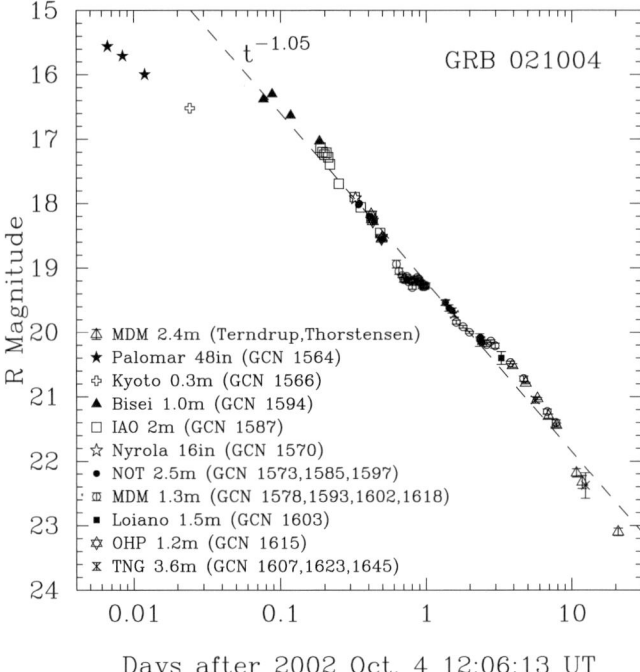

Fig. 12.9. Light curve of the optical afterglow of GRB 021004 from 10 minutes to 10 days after the GRB (from http://www.astro.columbia.edu /~jules/021004.ps).

Kumar [87]. Another way to probe the immediate environment of GRBs is the analysis of the short term variability of the afterglows (Fig. 12.9) as discussed in a series of recent papers on GRB 021004 ([30] and refs. therein).

X-ray afterglows also provide useful data on the GRB environment. Emission lines, possibly from iron, have been detected in four GRB afterglows [93] and a transient absorption edge has been seen in the prompt emission of GRB 990705 [2]. Recent observations of the afterglow of GRB 011211 with XMM, eleven hours after the GRB, led to the detection of emission lines of magnesium, silicon, sulfur, argon, calcium, and possibly nickel [99]. These X-ray lines show that some GRBs explode in regions which are enriched with metals. While Fe lines seem primarily due to photoionisation or reflection of the X-rays close to the GRB source, the lines observed by XMM are more likely due to thermal emission from a hot plasma [99].

Beyond the immediate environment of the source, high resolution spectroscopy in the optical can be used to study the line of sight to the GRB, and the composition of its host galaxy (e.g. [75]). This technique, which requires bright afterglows, was made possible thanks to the capability of HETE to provide arcminute localizations in a few minutes, instead of hours for BeppoSAX.

Regarding the nature of the progenitor, Bloom et al. did a very interesting study on the position of 20 bursters inside their host galaxies [12]. They find that the burster distribution is similar to the distribution of the UV light, suggesting a strong connection of GRB activity with regions of star formation (at least for long GRBs since no counterpart of a short GRB has ever been found).

– GRB-supernova link[2]

The best evidence for a connection between GRBs and supernovae comes from the temporal and spatial coincidence of GRB 980425 with supernova SN 1998 bw [37]. More recently, GRB 020903, a very soft GRB detected by HETE was also found to coincide in time and direction with a supernova at z=0.25 [108]. More circumstantial evidence comes from red bumps in the light curves of some GRB afterglows after 10-20 days (e.g. [13]). These bumps, however, might have other causes, like the reactivation of the shock by relativistic protons coming from the desintegration of the free neutrons within the fireball [9, 24]. Finally, the detection of X-ray emission lines from elements like magnesium, silicon, or sulfur, usually produced by supernovae, in the spectrum of the afterglow of GRB 011211 (see above) and GRB 020813 [19] is further evidence of a close link between GRBs and supernovae.

– Redshift distribution of GRBs.

Today more than 30 GRB distances have been measured ranging from z=0.25 to z=4.5 (with the exception of GRB 980425, probably associated with a supernova at z=0.0085 [37]). The realization that GRBs could be visible at high redshifts came very soon after the discovery of the first afterglows. Wijers et al. [113] mentioned that if the GRB rate is proportional to the star formation rate, GRBs should be 10 to 20 times more frequent at z=1 than in our local Universe. In these conditions the flattening of the intensity distribution observed by BATSE indicates that it detects GRBs much beyond z=1. Lamb & Reichart [60] showed that the prompt GRB emission must be easily visible out to z=10, and that the afterglows might also be seen at large redshift if searched for in the infrared. In fact, GRB afterglows cannot be found in the optical beyond z=5, due to the Lyman α absorption, and they must be searched for in the infrared. These searches are just beginning now, raising hopes for the detection of GRBs at redshifts z $= 5 - 10$.

With their cosmological distances and their high energy content, GRBs appear today as promising sources to explore the young Universe, and the formation of the first stars and galaxies. There is no doubt that those who,

---

[2] On 29 March 2003, HETE-2 detected a bright nearby GRB ($z = 0.168$). After a few days, spectroscopic observations of the afterglow revealed the appearance of a very energetic supernova. This is the first direct evidence linking GRBs to stellar collapse [49, 109].

in the 1980s, considered the resolution of the GRB mystery as a major astrophysical question were right.

12.5.1 The Physics of Gamma-Ray Bursts

Until now we have not considered the physical processes at work in GRBs. The measure of GRB distances evidenced the enormous amount of energy released in a short duration, and revived the fireball model which describes the consequences of a huge energy release in a very small volume. In its original form, and for the physical conditions expected in GRBs, the fireball model predicts an optically thick highly super Eddington pair fireball emitting blackbody radiation when it becomes transparent. In this simple form it cannot explain the non-thermal spectra observed in GRBs and their time scales. An even more severe problem is that in the presence of a small amount of matter, most of the fireball energy is transferred to baryons and only a very small fraction is emitted as radiation. The solution to these problems came with a series of papers by Rees & Meszaros [71, 96, 97] considering the crucial role of shocks in converting into radiation the energy stored by the baryons.

The model works as follows: In the first instance the radiation energy is converted into kinetic energy. A relativistic wind is produced with Lorentz factors of 10^2 to 10^3. The GRB emission originates either from internal shocks in the relativistic wind or from the external shock which forms when the relativistic layer decelerates and is energized by sweeping up matter from the surrounding medium. Internal shocks which might be the best way to explain the complex temporal structure of GRBs are due to inhomogeneities within the relativistic outflow with faster shells catching with slower ones: these encounters give sub-relativistic shocks producing bursts of gamma-rays

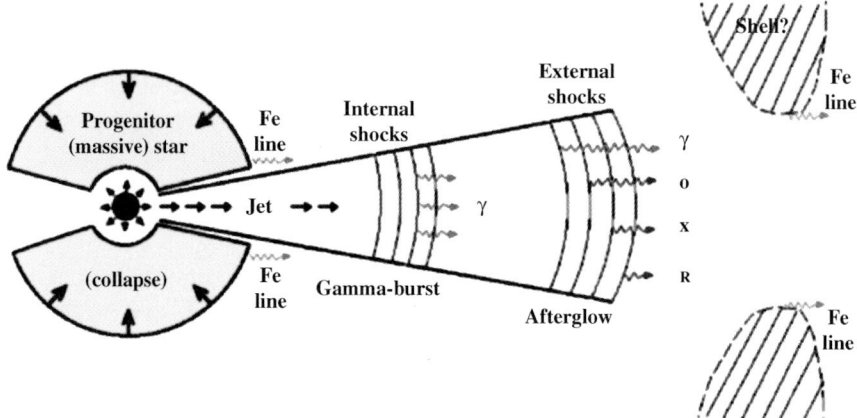

Fig. 12.10. Schematic view of internal shocks and external shocks giving rise to the GRB and its afterglow [72].

[21]. External shocks can lead to longer afterglows observed at lower energies. A third ingredient of this picture is a reverse shock which could explain X-ray or optical emission contemporaneous with the burst [22, 105]. This model is schematically illustrated in Fig. 12.10. In order to produce GRBs, it is essential that the expanding wind reaches Lorentz factors greater than 10^2 [90]. The acceleration, and collimation of the fireball by the central engine are issues which are poorly understood and merit further theoretical work. The physics of GRBs encompasses other important issues which require additional observations such as the amount of energy released in GRBs (including non-electromagnetic emissions of high energy neutrinos and cosmic rays, and gravitational waves), the radiation processes at work during the GRB, or the importance of the photospheric emission of the fireball when it becomes transparent.

While it predicts the main features of GRB emission, the fireball model gives no indication of the nature of the sources. In fact, it can be reconciled with two popular scenarios: collapsars and mergers of compact objects.

- A collapsar is a massive and rapidly spinning star whose central core collapses to a black hole [63, 115].
- Mergers of compact objects [73, 79]. Binaries with two neutron stars in orbit emitting gravitational waves have been observed in our Galaxy. Other binary systems including a neutron star orbiting a black hole should also exist, even if they have not been observed yet. The compact objects in these systems must coalesce after several 10^8 years following the loss of angular momentum due to the emission of gravitational waves.

It is expected that these two models lead to the formation of a very fast spinning massive black hole with a large ring of matter ([98] and refs. therein). Such systems are able to liberate the energy needed to explain GRBs, over the correct timescales, via accretion of disk material or extraction of the rotational energy of the hole itself [10].

While collapsars are short lived stars and produce GRBs in star forming regions, mergers of compact objets have much longer lifetimes (of the order of 10^8 years), and are expected to result in coalescences occurring at the periphery of the GRB host galaxy. This is why collapsars are preferred to explain long duration GRBs which appear to be associated with star formation regions in their host galaxies (Sect. 12.5). Short duration GRBs (lasting less than 2 seconds) remain a mystery. Only a few of them had decent follow-up (their localization is difficult because they yield far fewer photons than long duration GRBs), and no afterglow of a short/hard GRB has been found yet. Some authors have suggested that short duration GRBs might be the result of the coalescence of compact objects. Additional clues to the nature of the progenitors could come from the detection of high energy neutrinos, high energy cosmic-rays, or gravitational waves. New generations of instruments in these areas might well make a contribution to this field very soon.

12.6 The Story Continues...

After 30 years of observational and theoretical efforts, the flaring, fading and quiescent counterparts of GRBs have finally been found, and we are beginning to understand the nature of these events. While BeppoSAX ceased operations in 2002, HETE-2 [102], INTEGRAL [46], and the IPN are now operating, and they are complemented by an ever growing set of telescopes devoted to the fast follow-up of GRBs. Some of the questions mentioned in the previous section have already been addressed with this new generation of instruments. However, despite the considerable advances of recent years, stimulating mysteries remain:

- The origin of short/hard GRBs. They could be due to mergers of compact objects. In the absence of detected afterglows, however, we have no clue on their distances, space density, possible host galaxies, and potential relation to the long duration GRB population. The solution might come from fast localizations with HETE and from early X-ray observations with SWIFT (see below).
- Dark GRBs. One discovery of BeppoSAX is the existence of long duration GRBs with no optical afterglow, which have been called dark GRBs. The quick localizations sent by HETE-2 in the minutes following the bursts, permitted the identification of rapidly fading afterglows from GRBs which would have been considered as 'dark GRBs' with slower instruments. It is interesting to note that the three GRBs observed to date in the optical during the minutes following the burst (GRB 990123, GRB 021004, and GRB 021211) had optical afterglows reaching magnitude 14-15 at these early times. While high redshifts or dust extinction might still explain some 'dark GRBs', it is now clear that a large fraction of them are GRBs with fast fading afterglows.
- X-ray flashes. The X-ray emission of GRBs has been known for a long time, but BeppoSAX showed that some 'GRBs' emit their energy *only* in X-rays below 50 keV. These events have been called X-ray flashes. The broad-band coverage of HETE-2 allows the detection of both X-ray flashes and GRBs, showing that X-ray flashes do not constitute a distinct class from the classical GRBs, but rather an extension of the GRB population towards events which are intrinsically faint and soft [7]. These soft events have been renamed X-ray rich GRBs to take into account their close link with GRBs. Theoretical explanations on the origin of these X-ray rich GRBs are still debated [74, 116, 118].
- High redshift GRBs. While the farthest GRB detected to date is at a redshift $z=4.5$, the most distant GRBs have probably not been found yet. The detection of afterglows from GRBs at large redshifts (beyond $z=5$) is only possible in the infrared and represents a great challenge for the coming years.

- GRB progenitors. X-ray and optical spectroscopy of the afterglow provide critical information on the immediate surroundings of the source. X-ray spectroscopy has provided good evidence for supernovae elements near GRB 011211 and GRB 020813 [19, 99]. High-resolution optical spectroscopy requires bright afterglows to be caught early after the burst. It has been used for the first time for GRB 020813 and GRB 021004 [75], allowing detailed studies of the gas very close to the GRB.
- The physics of GRBs. The physics at work during the first seconds of a GRB is far from being fully understood. The main theoretical issues concern the nature of the central source and the generation of the relativistic outflow, the efficiency of internal shocks, the emission processes at work during GRBs, and the origin of the early optical emission and of the GeV emission. On the observational side, one important challenge is to obtain multi-wavelength observations of the *prompt emission*. This requires large field of view instruments with broad-band energy coverage which will become available in the near future.

Answers to many of these questions will benefit from future missions called SWIFT, AGILE, GLAST, ECLAIRs, REM, and ARAGO. SWIFT[3] will come first (at the end of 2003) and provide accurate multi-wavelength observations of the early afterglow. AGILE[4], GLAST[5] and ECLAIRs[6] will approach the mysterious physics of the explosion. REM[7] and ARAGO[8] are members of the next generation of robotic telescopes which will allow longer and deeper observations of the afterglows over a broader energy range.

There is no doubt that GRBs will continue to fascinate astronomers in the coming years, while they try to elucidate their many remaining mysteries with these powerful instruments.

Acknowledgements

The authors gratefully acknowledge the review of this manuscript by R. Mochkovitch.

References

[1] Akerlof, C., Balsano, R., Barthelmy, S. et al. 1999, Nature, **398**, 400
[2] Amati, L., Frontera, F., Vietri, M., et al. 2000, Science, **290**, 953
[3] Atteia, J.-L., Barat, C., Hurley, K. et al. 1987, ApJ Suppl., **64**, 305

---

[3] http://swift.gsfc.nasa.gov
[4] http://agile.mi.iasf.cnr.it/Homepage/index.shtml
[5] http://glast.gsfc.nasa.gov
[6] http://www.cesr.fr/~barret/ECLAIRs.html
[7] http://golem.merate.mi.astro.it/projects/rem/
[8] http://arago.cesr.fr/description.htm

[4] Band, D., Matteson, J., Ford, L., et al. 1993, ApJ, **413**, 281
[5] Band, D., Ford, L., Matteson, J., et al. 1997, ApJ, **485**, 747
[6] Barat, C., Hurley, K., Niel, M., et al. 1984, ApJ, **280**, 150
[7] Barraud, C., Olive, J-F., Lestrade, J.P., et al. 2003, A&A **400**, 1021
[8] Barthelmy, S., Butterworth, P., Cline, T., et al. 1995, Ap&SS, **231**, 235
[9] Beloborodov, A., 2003, ApJ Lett., **585**, L19
[10] Blandford, R., & Znajek, 1977 MNRAS, **179**, 433
[11] Bloom, J., Djorgovski, S., Kulkarni, S., Frail, D., 1998, ApJ Lett., **507**, L25
[12] Bloom, J., Kulkarni, S., Price, P., 2002, ApJ Lett., **572**, L45
[13] Bloom, J., Kulkarni, S., Djorgovski, S., 2002, A. J., **123**, 1111
[14] Boella, G., Butler, R., Perola, G., et al. 1997a, A&AS, **122**, 299
[15] Boer, M., Atteia, J-L., Gottardi, M., et al. 1988, A&A, **202**, 117
[16] Boer, M., Motch, C., Greiner, J., et al. 1997, ApJ Lett., **481**, L39
[17] Briggs, M. S. , Paciesas, W.S., Pendleton, G.N., et al. 1996, ApJ, **459**, 40
[18] Briggs, M., Band, D., Kippen, R., et al. 1999, ApJ, **524**, 82
[19] Butler, N., Marshall, H., Ricker, G., et al. 2003, ApJ submitted, astro-ph/0303539
[20] Costa, E., Frontera, F., Heise, J., et al. 1997, Nature, **387**, 783
[21] Daigne, F. & Mochkovitch, R., 1998, MNRAS, **296**, 275
[22] Daigne, F. & Mochkovitch, R., 1999, A&AS, **138**, 523
[23] Daigne, F. & Mochkovitch, R., 2002, MNRAS, **336**, 1271
[24] Derishev, E., Kocharovsky, V., & Kocharovsky, Vl. 1999, ApJ, **521**, 640
[25] Dezalay, J.-P., Barat, C., Talon, R., et al. 1992, AIP Conf. Proc., **265**, 304
[26] Dezalay, J.-P., Atteia, J.-L., Barat, C., et al. 1994, A&A, **286**, 103
[27] Dezalay, J.-P., Atteia, J.-L., Barat, C., et al. 1997, ApJ Lett., **490**, L17
[28] Fenimore, E., In't Zand, J., Norris, J. et al. 1995, ApJ Lett., **448**, L101
[29] Fishman, G. J. & Meegan, C. A. 1995, ARA& A , **33**, 415
[30] Fox, D., Yost, S., Kulkarni, S., et al. 2003, Nature, **422**, 284
[31] Frail, D., Kulkarni, S., Nicastro, S. 1997, Nature, **389**, L261
[32] Frail, D., Kulkarni, S., Sari, R. 2001, ApJ Lett., **562**, L55
[33] Frontera, F., Greiner, J., Antonelli, L., et al. 1998, A&A, **334**, 69
[34] Fruchter, A., Livio, M., Macchetto, L., et al. 1997, IAUC, 6747
[35] Fruchter, A., Pian, E., Gibbons, R, et al. 2000, ApJ, **545**, 664
[36] Galama, T., Wijers, R., Bremer, M., et al. 1998, ApJ, **500**, 101
[37] Galama, T., Vreeswijk, W., van Paradijs, J., et al. 1998, Nature, **395**, 670
[38] Galama, T., Wijers, R., Vreeswijk, P., et al. 1999, A&A Suppl., **138**, 451
[39] Greiner, J., Flohrer, J., Wenzel, W., & Lehmann, T. 1987, Ap&SS, **138**, 155
[40] Hakkila, J., Meegan, C., Pendleton, G., et al. 1994, ApJ, **422**, 659
[41] Harrison, F., Bloom, J., Frail, D., et al. 1999, ApJ Lett., **523**, L121
[42] Hameury, J-M., Bonazzola, S., Heyvaerts, J., & Ventura, J., 1982, A&A, **111**, 242
[43] Hartmann, D., Woosley, S. E., & Epstein, R. I. 1990, ApJ, **348**, 625
[44] Hartmann, D., 1991, Annals of The New York Academy of Sciences, **647**, 575
[45] Heise, J., in't Zand, J., Kippen, R., & Woods, P., 2001, *Proc. of "Gamma-Ray Bursts in the afterglow era", 2nd Workshop*, Ed.E. Costa, F. Frontera, & J. Hjorth. Berlin Heidelberg: Springer, 2001, p. 16.

[46] Hermsen, W. & Winkler, C. 2002, *Proceedings of the XXII Moriond Astrophysics Meeting "The Gamma-Ray Universe"*, Ed.A. Goldwurm, D.N. Neumann, J. Trân Thanh Vân, Th'e Giói, Vietnam, 2002, p. 393
[47] Higdon, J. C. & Schmidt, M. 1990, ApJ, **355**, 13
[48] Hjellming, R. M. & Ewald, S. P. 1981, ApJ Letters, **246**, L137
[49] Hjorth, J., Sollermann, J., Møller, P., et al. 2003, Nature, **423**, 847
[50] Hudec, R., Borovicka, J., Wenzel, W., et al. 1987, A&A, **175**, 71
[51] Hurley, K. 1986, AIP Conf. Proc., **141**, 1
[52] Hurley, K. 1992, AIP Conf. Proc., **265**, 3
[53] Hurley, K. et al., 1994, Nature, **372**, 652
[54] Hurley, K., Costa, E., Feroci, M., et al. 1997, ApJ Lett., **485**, L1
[55] Kehoe, R., Akerlof, C., Belsano, R., et al. 2001 In: Supernovae and Gamma-Ray Bursts: The Greatest Explosions since the Big Bang, Eds. M. Livio, N. Panagia and K. Sahu, Cambridge, Cambridge Univ. Press, p. 47
[56] Klebesadel, R., Strong, I., & Olson, R. 1973, ApJ Lett., **182** L85.
[57] Koshut, T., Kouveliotou, C., Paciesas, W. et al. 1995 ApJ, **452**, 145
[58] Kouveliotou, C., Meegan, C., Fishman, G., et al. 1993, ApJ Lett., **413**, L101
[59] Kulkarni, S., Frail, D., Wieringa, M., et al. 1998, Nature, **395**, 663
[60] Lamb, D. Q. & Reichart, D. E. 2000, ApJ, **536**, 1
[61] Liang, E. & Kargatis, V. 1996, Nature, **381**, 49
[62] Lloyd, N., Petrosian, V., & Mallozzi, R. 2000, ApJ, **534**, 227
[63] MacFadyen, A. & Woosley, S.E., 1999, ApJ, **524**, 262
[64] Mallozzi, R., Paciesas, W., Pendleton, G. et al. 1995, ApJ, **454**, 597
[65] Mazets, E., Golenetskii, S., Aptekar, R., et al., 1981 Nature, **290**, 378
[66] Mazets, E. P., Golenetskii, S. V., Ilinski, V. N., et al. 1981, Ap & SS, **80**, 3
[67] Mazets, E. P. 1988, Advances in Space Research, **8**, 669
[68] Meegan, C. A., Fishman, G. J., Wilson, R. B., et al. 1992, Nature, **355**, 143
[69] Metzger, M., Djorgovski, S., Kulkarni, S., et al. 1997, Nature, **387**, 878
[70] Mészáros, P. & Rees, M. J. 1992, MNRAS, **257**, 29
[71] Mészáros, P. & Rees, M. J. 1993, ApJ, **405**, 278
[72] Mészáros, P. 2002, ARA&A, **40**, 137
[73] Mochkovitch, R., Hernanz, M., Isern, J., & Martin, X., 1993, Nature, **361**, 236
[74] Mochkovitch, R., Daigne, F., Barraud, C., & Atteia, J-L., 2003, Proc. of "Gamma-ray bursts in the afterglow era", 3rd workshop, Rome 2002
[75] Möller, P., Fynbo, J., Hjorth, J., et al. 2002, A&A, **396**, L21
[76] Mitrofanov, I., Chernenko, A., Pozanenko, A., et al. 1996, ApJ, **459**, 570
[77] Murakami, T., Fujii, M., Hayashida, K., et al., 1988, Nature, **335**, 234
[78] Murakami, T., Inoue, H., Nishimura, J., et al. 1991, Nature, **350**, 592
[79] Narayan, R., Paczyński, B., & Piran, T., 1992, ApJ Lett., **395**, L83
[80] Niel, M., Jourdain, E., & Roques, J. P. 1990, A&A, **228**, L1
[81] Norris, J., Bonnell, J., Nemiroff, R., et al. 1995, ApJ, **439**, 542
[82] Norris, J., Marani, G., & Bonnell, J., 2000, ApJ, **534**, 248
[83] Paciesas, W., Meegan, C., Pendleton, G., et al. 1999, ApJ Suppl., **122**, 465
[84] Paczyński, B., 1986, ApJ, **308**, L43
[85] Paczyński, B., 1990, ApJ, **348**, 485
[86] Palmer, D., Seifert, H., Teegarden, B. et al. 1996, AIP Conf. Proc., **384**, 218
[87] Panaitescu, A., & Kumar, P., 2001, ApJ Lett., **560**, L49

[88] van Paradijs, J., Groot, P., Galama, T., et al. 1997 Nature, **386**, 686
[89] van Paradijs, J., Kouveliotou, C., Wijers, R., 2000, ARA&A
[90] Piran, T. 1999, Physics Reports, **314**, 575
[91] Piro, L., 1999, Procs. of "The neutron star-BH connection, NATO/Kluwer, **567**, 431, eds Kouveliotou, Ventura, van den Heuvel
[92] Piro, L., et al. 1999, GCN 199
[93] Piro, L., 2002, AIP Conf. Proc., 662, in press
[94] Pizzichini, G. et al. 1986, ApJ Letters, **301**, 641
[95] Quilligan, F., McBreen, B., Hanlon, L., et al. 2002, A&A, **385**, 377
[96] Rees, M. J. & Meszaros, P. 1992, MNRAS, **258**, 41P
[97] Rees, M. J. & Meszaros, P. 1994, ApJ Lett., **430**, L93
[98] Rees, M. J., 1999, A&AS, **138**, 491
[99] Reeves, J., Watson, D., Osborne, J., et al. 2002, Nature, **416**, 512
[100] Reichart, D., Lamb, D., Fenimore, E., et al. 2001, ApJ, **552**, 57
[101] Rhoads, J., 1997, ApJ Lett., **487**, L1
[102] Ricker, G., et al. 2003, AIP Conf. Proc., 662
[103] Ryde, F. & Svensson, R. 2000, ApJ Lett., **529**, L13
[104] Sahu, K., Livio, M., Petro, L., et al. 1997, Nature, **387**, 476
[105] Sari, R., & Piran, T., 1999, ApJ, **520**, 641
[106] Schaefer, B. E. 1981, Nature, **294**, 722
[107] Schmidt, M., Higdon, J. C., & Hueter, G. 1988, ApJ Lett., **329**, L85
[108] Soderberg, A., Price, P., Fox, D., et al. 2002, GCN, 1554
[109] Stanek, K. Z., Matheson, T., Garnavich, P. M., et al. 2003, ApJ Lett., **591**, L17
[110] Tikhomirova, Ya. & Stern, B. 2002, *Proc. 37^e Rencontres de Moriond*, p.235
[111] Vedrenne, G., & Jourdain, E., 1991, AIP Conf. Proc., **232**, 317
[112] Vedrenne, G., 1991, Annals of The New York Academy of Sciences, **647**, 556
[113] Wijers, R., Bloom, J., Bagla, J. & Natarajan, P., 1998, MNRAS, **294**, 13
[114] Woosley, S.E., & Wallace, R.K., 1982 ApJ, **258**, 716
[115] Woosley, S.E., 1993 ApJ, **405**, 273
[116] Yamazaki, R., Ioka, K., & Nakamura, T., 2002 ApJ Lett. **571**, L31
[117] Yoshida, A., Kawai, N., Otani, C., et al. 1997, IAUC 6593
[118] Zhang, B., & Mészáros, P., 2002, ApJ, **581**, 1236

13 Update on Gravitational Wave Research

L.P. Grishchuk

13.1 Introduction

The concept of gravitational radiation has been with us for quite a long time. Thinking of the relativistic gravitational field in parallel with the familiar case of the electromagnetic field, it was natural to expect that there should exist waves of the gravitational field similar to the waves of the electromagnetic field. As Einstein [1] put it in 1913: "The conviction had to come that Newton's law of gravitation is as incapable of describing all gravitational phenomena as Coulomb's laws of electrostatics and magnetostatics are of electromagnetic phenomena". Decades of hard work have followed. In the beginning, the research was purely theoretical. Through doubts and controversies, the conceptual and mathematical issues of gravitational radiation have been clarified. Then, in the 1960s, the experimental work began with the pioneering effort of J. Weber [2]. In the 1990s, gravitational waves have been observed indirectly via the measurement of secular changes in the orbital parameters of the binary system of neutron stars that includes the pulsar PSR 1913+16 [3]. These days, gravitational waves are routinely taken into account in theoretical and observational studies ranging from orbital evolution of close pairs of compact stars to the cosmology of the early Universe. The experimental progress has also been very impressive.

We are now at a special and decisive point. First, the relevance and importance of gravitational-wave research is fully recognised by the communities of physicists and astronomers. The worldwide network of scientific collaborations has been established, reflecting the necessity of coincident observations at various instruments and the need for joint analysis of the data [4], [5]. Second, the scientific runs have begun at the recently assembled sensitive laser-interferometric observatories - American LIGO [6] and British-German GEO600 [7]. The French-Italian VIRGO [8] will be operational soon. Meanwhile, the Japanese TAMA300 [9], along with the international network of bar detectors [10], continue to collect data at their level of sensitivity. The design sensitivity of the three LIGO interferometers plus VIRGO and GEO600 meets the realistic astrophysical predictions, so the direct detection of powerful (even if rare) sources, such as coalescing stellar-mass black holes, becomes possible. Third, the European Space Agency (ESA) and US Space Agency (NASA) have agreed to share the costs of the space-based Laser Interferometer Space Antenna (LISA) [11]. LISA is planned to be launched around the year 2011, preceded by a technology demonstration mission. Plans for advanced ground-based detectors, such as LIGO II, the Japanese cryogenic laser interferometer LCGT, and, possibly, the European EURO, are also maturing very quickly. These instruments of the next generation will detect a host of well anticipated sources, such as compact binary stars, but really fundamental discoveries are also expected. Fourth, there exists a strong competition on the side of purely astronomical means for indirect detection of gravitational waves. In fact, the anisotropy and polarisation measurements of the Cosmic Microwave Background radiation (CMB) are likely to bring us decisive information on the fundamentally important relic gravitational waves much earlier than will be achieved by direct methods. By all counts, gravitational-wave research is now one of the most exciting and promising areas of physical science.

The weakness of gravity as a physical interaction is the strength of gravitational waves as a tool for scientific research. It is difficult to detect gravitational waves because they carry their energy practically without scattering and absorption. But it is exactly because of this difficulty that we have any chance to learn about what was happening at the beginning of the Universe, or in the depth of an exploding supernova, or in the vicinity of, what we think are, merging black holes. It is also likely that the differing properties of gravitational waves will be a discriminating signature of different fundamental physical theories and modified gravities.

In this review, we will start, in Sect. 13.2, with an elementary theory of gravitational waves. Then, in Sect. 13.3, we will discuss the current status of gravitational-wave experiments. In Sect. 13.4 we will focus on astrophysical sources of gravitational waves and the new physics that will be learned from their observation. Section 13.5 is devoted to gravitational waves and cosmology. In particular, some recent results from the WMAP satellite will be discussed in the context of gravitational-wave research, including the 'implications for inflation'. Finally, we briefly summarise this update in Sect. 13.6. The background material on gravitational-wave research can be found in [12–14], and some of the previous reviews in [15–19].

13.2 Elementary Theory of Gravitational Waves

There are some similarities in the mathematical description of the gravitational field and the electromagnetic field. The electromagnetic field can be described by the components of the 4-vector potential A^μ. The quantities A^μ are functions of time and spatial coordinates x^α, $x^\alpha = (ct, x, y, z)$, and they obey the wave-like dynamical equations - the Maxwell equations. Other quantities, such as electric and magnetic components **E**, **H**, and the energy-momentum tensor of the electromagnetic field, are calculable from A^μ. The quantities A^μ allow the gauge freedom. This means that some seemingly different solutions of the field equations are in fact equivalent solutions in the sense of their physical manifestations. Analogously, the gravitational field can be described by 10 components of the 4×4 symmetric tensor $h^{\mu\nu}$ as functions of x^α. The quantities $h^{\mu\nu}$ obey the nonlinear wave-like dynamical equations - the Einstein equations. The energy-momentum tensor of the gravitational field $t^{\mu\nu}$ is calculable from the tensor $h^{\mu\nu}$. The quantities $h^{\mu\nu}$ allow the gauge freedom, so that some seemingly different solutions of the gravitational field equations describe in fact the physically equivalent configurations.

In contrast to the electromagnetic field, the gravitational field $h^{\mu\nu}$ is a nonlinear field. This means that a sum of two solutions of the field equations is not a new solution, and the gravitational field, along with matter fields, is a source for itself. However, in a number of situations one can neglect the nonlinearity of the gravitational field. In this approximation we come to the notion of weak gravitational fields and linearised gravitational waves.

Many properties of linearised gravitational waves resemble those of electromagnetic waves. Gravitational waves propagate with the velocity of light c and have two independent transverse polarization states. In its action on free masses, a gravitational wave (g.w.) exhibits some analogs of the electric and magnetic contributions of an electromagnetic wave (em.w.) acting on free electric charges. The gravitational wave field is dimensionless and its strength can be characterized by a dimensionless amplitude h. The amplitude h decreases in the course of propagation from a localized source in inverse proportion to the travelled distance: $h \propto 1/r$. Gravitational waves carry away from a radiating system its energy, angular momentum and linear momentum.

The linearised g.w. satisfies the wave equation

$$h^{\mu\nu,\alpha}{}_{,\alpha} + \eta^{\mu\nu}h^{\alpha\beta}{}_{,\alpha,\beta} - h^{\nu\alpha,\mu}{}_{,\alpha} - h^{\mu\alpha,\nu}{}_{,\alpha} = 0, \qquad (13.1)$$

where the ordinary derivative is denoted by a comma and $\eta^{\mu\nu}$ is the metric tensor of the Minkowski space-time:

$$\mathrm{d}\sigma^2 = \eta_{\mu\nu}\mathrm{d}x^\mu \mathrm{d}x^\nu = c^2\mathrm{d}t^2 - \mathrm{d}x^2 - \mathrm{d}y^2 - \mathrm{d}z^2. \qquad (13.2)$$

The first term in eqn. (13.1) is the familiar d'Alembert (wave) operator. A plane-wave solution to eqn. (13.1) is given by

$$h^{\mu\nu} = a^{\mu\nu} e^{ik_\alpha x^\alpha}, \qquad (13.3)$$

where $k_\alpha k^\alpha = 0$, reflecting the fact that a g.w. propagates with the velocity of light. Because of this condition, the field equations (13.1) require the 10 components of the constant matrix $a^{\mu\nu}$ to satisfy 4 constraints: $a^{\mu\nu}k_\nu = 0$. Then, the quantities a^{00} and a^{0i} can be expressed in terms of 6 components of the matrix a^{ij}. This matrix itself includes the part $\tilde{a}^{ij}$ satisfying the further 4 constraints: $\tilde{a}^{ij}k_j = 0$, $\tilde{a}^{ij}\eta_{ij} = 0$. These remaining 2 degrees of freedom (sometimes called the TT-components) fully determine the observational manifestations of the plane wave and its energy-momentum characteristics. Indeed, it is easy to show that the gravitational energy-momentum tensor

$$t_{\mu\nu} = \frac{c^4}{32\pi G} \left[h^{\alpha\beta}{}_{,\mu} h_{\alpha\beta,\nu} - \frac{1}{2} h_{,\mu} h_{,\nu} \right] \qquad (13.4)$$

depends only on the TT-components of the field:

$$t_{\mu\nu} = \frac{c^4}{32\pi G} k_\mu k_\nu \left[2\tilde{a}^{ij} \tilde{a}^*_{ij} \right], \qquad (13.5)$$

where we have dropped (as we normally do in the case of electromagnetic waves) the purely oscillatory terms.

How does a gravitational wave affect the free (i.e. not subject to any other forces or constraints) masses? In the case of an em.w., we could have noticed the displacement of a charged particle with respect to a neutral particle placed initially at the same point. In the case of a gravitational wave, there are no particles neutral to the gravitational interaction, so we need to study the relative displacement of particles separated initially. This brings us to the analysis of tidal effects of a gravitational wave, similar to the tidal effects of a Newtonian gravitational field. Let one of the free masses define the origin of our coordinate system, $x^i_{(1)} = 0$, whereas the second mass is placed initially at $x^i_{(2)} = l^i$ and its (small) displacement caused by the gravitational wave is denoted by ξ^i. Then the equations of motion of the second mass are:

$$\frac{d^2 \xi^j}{dt^2} = \frac{1}{2} \omega^2 \tilde{a}^j_k l^k e^{i\omega t}. \qquad (13.6)$$

These equations depend only on the TT-components of the wave.

For a wave propagating, say, in z-direction it is convenient to write the TT-components explicitely:

$$h_{xx} = -h_{yy} = h_+ \sin(\omega t - kz + \psi), \quad h_{xy} = h_\times \cos(\omega t - kz + \psi). \qquad (13.7)$$

Then, the relevant solution to eqn. (13.6) reads

$$x = l_1 - \frac{1}{2} l_1 h_+ \sin(\omega t + \psi) - \frac{1}{2} l_2 h_\times \cos(\omega t + \psi),$$
$$y = l_2 - \frac{1}{2} l_1 h_\times \cos(\omega t + \psi) + \frac{1}{2} l_2 h_+ \sin(\omega t + \psi),$$
$$z = l_3. \qquad (13.8)$$

The masses that lie (on average) on the ring $(l_1)^2 + (l_2)^2 = l^2, l_3 = 0$, oscillate around their ring positions. In Fig. 13.1 we show a quarter of the cycle caused by a wave with the + polarization state (i.e. when $h_+ \neq 0$, $h_\times = 0$). The pattern of oscillations enforced by the × polarization state (i.e. when $h_\times \neq 0$, $h_+ = 0$) can be obtained from Fig. 13.1 by its rotation by the

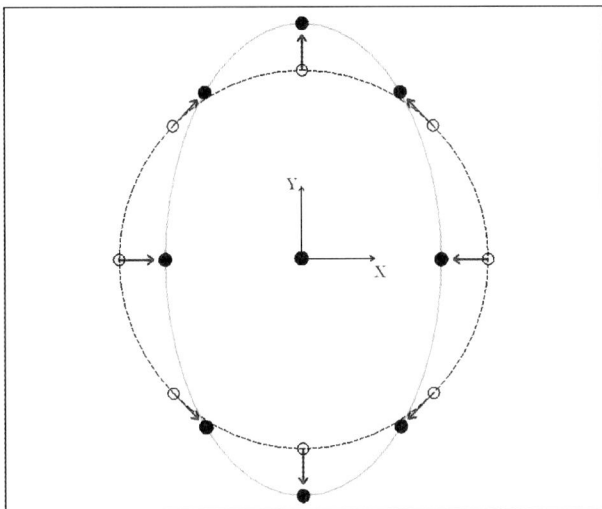

Fig. 13.1. Motion of free particles in the field of a linearly-polarised gravitational wave.

angle 45° in the (x, y) plane. The general motion of a particle is a linear superposition of these oscillations, as seen in eqn. (13.8). The oscillatory deformation of a sphere of masses, surrounding the central mass at the origin, is described by a linear combination of spherical harmonics $Y_{lm}(\theta, \phi)$ with $l = 2$ and $m = \pm 2$, and where the polar axis is taken along the z direction.

Equations (13.8) suggest that the motion of free particles is confined strictly to the planes $z = const$. This conclusion comes about only because we have so far neglected the smaller terms, containing the products of h_+, $h_\times$ with the extra small factor l/λ, where l is the separation between masses and $\lambda = 2\pi c/\omega$ is the gravitational wavelength. When these terms are taken into account, the perturbed positions x, y and, most importantly, z receive further contributions. The gravitational wave drives the masses not only in the plane of the wave-front, but also, to a smaller extent, back and forth in the propagation direction [16]. This extra component of motion is similar to the one caused by the magnetic field and the Lorentz force of the em.w. acting on a charged particle. In Fig. 13.2 we show typical displacements of the masses forming (on average) the plane $z = 0$, when the 'magnetic' contribution to their motion is also taken into account. It is seen from this figure that the entire plane of particles is being bent in an oscillatory fashion. This 'magnetic' component of motion will play a certain role in the analysis of observations with laser interferometers [20].

The amplitudes $h_+, h_\times$ are determined by the source of the gravitational waves. To find the amplitudes, we should replace the zero in the right hand side of eqn. (13.1) by the source term $(16\pi G/c^4)T^{\mu\nu}$, where $T^{\mu\nu}$ is the energy-momentum tensor of the source, and seek the retarded solutions to the wave

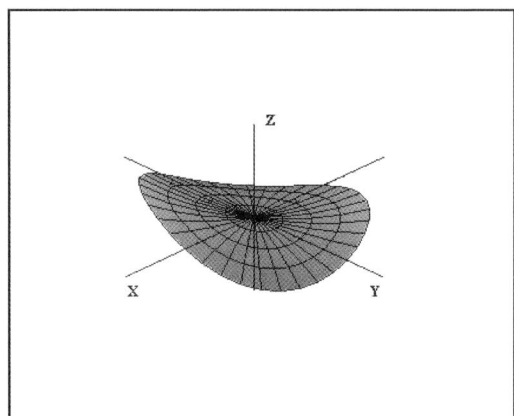

Fig. 13.2. Motion of free particles with the "magnetic" component taken into account.

equation. Assuming that the distance R_0 to the source is much larger than the wavelength, one can write

$$h_+(k_\alpha) = \frac{1}{R_0}\frac{2G}{c^4}[\hat{T}^{11}(k_\alpha) - \hat{T}^{22}(k_\alpha)], \quad h_\times(k_\alpha) = \frac{1}{R_0}\frac{4G}{c^4}\hat{T}^{12}(k_\alpha), \quad (13.9)$$

where the Fourier components $\hat{T}^{\mu\nu}(k_\alpha)$ are defined by

$$\hat{T}^{\mu\nu}(k_\alpha) = \int \left(\int T^{\mu\nu}(t_r, r_0) d^3 r_0\right) e^{-ik_\alpha x^\alpha} d^4 x$$

and t_r is retarded time. These formulas allow one to evaluate the typical amplitude h from a given source:

$$h \sim \frac{1}{R_0}\frac{GM}{c^2}\left(\frac{v}{c}\right)^2, \quad (13.10)$$

where M is the total mass of the source and v is the characteristic non-spherical velocity of the matter bulk motion. For a bound system like a binary star, $(v/c)^2 \sim GM/c^2 a$, where a is the size of the system. An accurate calculation for a binary in a circular orbit, consisting of masses M_1, M_2 separated by the distance a, and after averaging over the orbital period and orientation of the orbital plane, gives

$$h = \left(\langle h_+^2 \rangle + \langle h_\times^2 \rangle\right)^{1/2} = \left(\frac{32}{5}\right)^{1/2} \frac{1}{R_0}\frac{G^{5/3}}{c^4}\frac{M_1 M_2}{(M_1+M_2)^{1/3}}(\pi f)^{2/3}, \quad (13.11)$$

where the emitted g.w. frequency f (in Hz) is

$$f = \frac{1}{\pi}\left[\frac{G(M_1+M_2)}{a^3}\right]^{1/2}. \quad (13.12)$$

Note that the amplitude h depends on a particular combination of masses:

$$\frac{M_1 M_2}{(M_1 + M_2)^{1/3}} = \mathcal{M}^{5/3}, \quad \text{where} \quad \mathcal{M} = \frac{(M_1 M_2)^{3/5}}{(M_1 + M_2)^{1/5}}, \quad (13.13)$$

so that $h \propto (1/R_0)\mathcal{M}^{5/3} f^{2/3}$ and $\mathcal{M}$ is sometimes called a chirp mass.

13.3 Current Status of Gravitational-Wave Detectors

The possible methods of detecting gravitational waves, as well as the associated difficulties, can be seen from the discussion above. Figure 13.1 is helpful for understanding the principles of mechanically coupled detectors (bar detectors) and electromagnetically coupled detectors (laser interferometers). A bar detector is essentially a mechanical oscillator consisting of two masses connected by a spring. As an illustration, one can think of two elastically connected masses lying, say, at the opposite ends of the x axis in Fig. 13.1. The size of bar detectors is normally small, a couple of metres or so. A bar detector is a relatively narrow-band instrument. It is mostly sensitive to g.w. frequencies in the vicinity of the main eigen-frequency of the bar. In the presently operating instruments the resonant frequency is around $\sim 1 kHz$. There are some advantages in using spherically shaped mechanical detectors. The construction of such detectors is currently taking place (see, for example, [21]). While the operating bar detectors continue to collect useful information [22–24], we will concentrate on laser interferometers.

The laser interferometer technique is based on free masses-mirrors whose relative distances are monitored by light beams travelling between the mirrors. A ground-based interferometer is normally an L-shaped configuration. The light beamsplitter is in the corner of L, while the reflecting mirrors are at the ends of the configuration and near the beam splitter. The central mass in Fig 13.1 is an illustration of the beamsplitter and corner mirrors, whereas the end mirrors are located, say, in the positive directions of x and y. Obviously, mirrors in real interferometers are not free masses, they are suspended like pendulums. But they behave essentially as free masses in their motion along the corresponding arm. The length of the arms of an interferometer can be large. For instance, it ranges from 600 meters in GEO600 to $4 km$ in the largest of the LIGO interferometers. In contrast to bar detectors, laser interferometers are relatively broad-band instruments. They are sensitive to frequencies in the interval $\approx (30 - 10^4) Hz$. Figure 13.1 shows the simplest case of a single monochromatic wave with one polarisation state arriving from the orthogonal direction to the detector's plane, but the response of the detector to the general case of the incoming wave is also calculable.

It is clear from eqn. (13.8) that the relative displacement of free masses is proportional to the incoming wave amplitude h: $\delta l/l \approx h$. Which numerical values of h can be expected from astrophysical sources in the surrounding

Universe? Consider one of the most powerful and efficient emitters - a pair of compact stars orbiting each other in a tight orbit. Specifically, consider two neutron stars with masses $M_1 = M_2 = 1.5 M_\odot$ at the late stage of their inward spiral. Orbiting each other at separation $a = 100 km$, they emit gravitational waves at frequency $f = 200 Hz$. The emitted intensity of radiation is very high by astronomical standards: $3 \times 10^{52} erg/sec$. Let the distance to the source be $R_0 = 100 Mpc$. We cannot make this distance much shorter than $100 Mpc$, because the expected number of such events per year in a smaller volume of the surrounding Universe would be less than 1. Then, eqn. (13.11) says that the amplitude at Earth is $h \approx 10^{-22}$. This is an incredibly small number. It enters any conceivable method of detection of gravitational waves and explains why it is so difficult to observe them. In a $4km$ long interferometer, we need to measure the mirror's displacements at the level of $4 \times 10^{-17} cm$.

The interferometer monitors the time-dependent difference of the relative distance variations in the two arms. Using the terminology of solid state physics, this quantity is called the (dimensionless) strain. Regardless of the presence or absence of the useful g.w. signal, there will always be some strain noise $n(t)$ in the output of the detector. The mean-square value of the noise can be expressed as an integral over the noise power spectral density $S_n(f)$:

$$\overline{n^2(t)} = 2 \int_0^\infty S_n(f) df. \qquad (13.14)$$

The square root of $S_n(f)$ is the noise amplitude $\sqrt{S_n(f)}$. This quantity has the dimensionality of $Hz^{-1/2}$, and it is this quantity that is usually plotted on the sensitivity graphs. In Fig. 13.3 (taken from [6]) we show the recent status of the noise amplitudes in the $4km$ interferometer of the LIGO Livingston site. One can see the great progress that has been made towards reaching the design goal of sensitivity, shown by the solid line. The lower frequency part of the sensitivity curve is dominated by seismic perturbations, the central part by thermal noise in the mirror's suspensions, and the higher frequency part by the shot noise of the laser light. The advanced LIGO sensitivity curve will be lower than the solid line in Fig. 13.3 by a factor of 10 across all the frequencies. This upgrade of LIGO is planned to take place in about the year 2007. The advanced instruments will also allow optical configurations in which the sensitivity can be increased within a certain narrow frequency band at the expense of lowering the sensitivity outside the chosen band.

To compare qualitatively the astrophysical signal S, represented by the dimensionless amplitude h, with the detector's r.m.s. noise N, we should calculate, as eqn. (13.14) suggests, the product $\sqrt{S_n(f)}\sqrt{\Delta f}$, where Δf is the appropriate bandwidth in the noise spectrum. Taking the initial LIGO design sensitivity $3 \times 10^{-23} Hz^{-1/2}$ at $f = 200 Hz$ and $\Delta f = f$, we would get $\sqrt{S_n(f)}\sqrt{\Delta f} = N = 5 \times 10^{-22}$. This N is a factor of 5 higher than the signal $S = 10^{-22}$ from the coalescing neutron stars considered above. The ratio $S/N = 1/5$ would suggest that, by a significant margin, the signal is

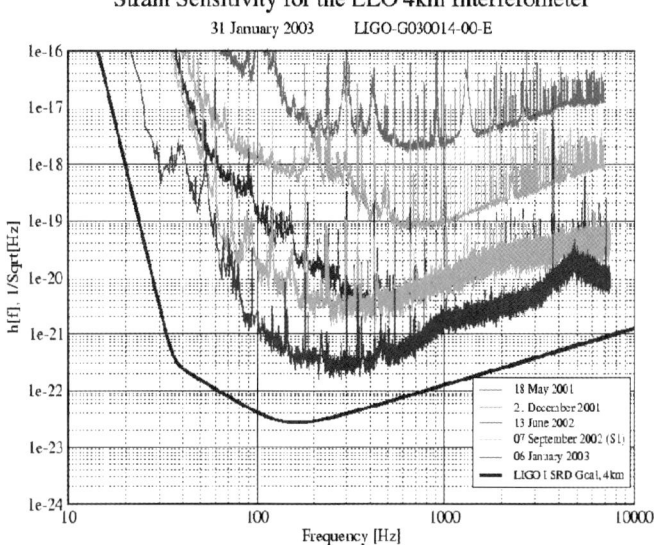

Fig. 13.3. The status of sensitivity of one of the LIGO interferometers.

not measurable. The reality, however, is somewhat better than this estimate. One should take into account the *a priori* knowledge of the expected waveform. If the waveform is known, one can use the well developed technique of matched filtering. This method recovers the signal by, effectively, reducing the appropriate Δf and the relevant amount of the detector's noise. For a quasi-periodic signal, this is achieved by using a long observation time T: $1/T \sim \Delta f \ll f$. In the case of a coalescing binary, the dominant g.w. frequency is increasing with time, but the binary still executes almost $n = 200$ cycles before changing its frequency by a factor of 2. This points to a possible increase of the estimated S/N by a factor $\sqrt{n}$. The importance of accurate modelling of the expected waveforms is well appreciated. At present, this is an important challenge for theorists (see, for example, [25]).

Figure 13.4 (taken from [18]) shows the accurately calculated S/N ratios for the initial laser interferometers. As expected, the S/N grows with the total mass M of the coalescing binary, because the signal gets larger. However, binaries with a total mass greater than about $80 M_\odot$ radiate at too low frequencies before merging. At these low frequencies the detector noise is high, as seen in Fig. 13.3. Therefore, the S/N decreases again. Certainly, the equal mass binaries with a total mass greater than $M \sim 10 M_\odot$ can only be pairs of black holes, according to our present understanding of stars and stellar evolution.

The LIGO-GEO collaboration has reported [26] the first upper limits on coalescing binary systems, as well as other possible g.w. sources. These limits were derived from observations at the currently achieved sensitivity. They are

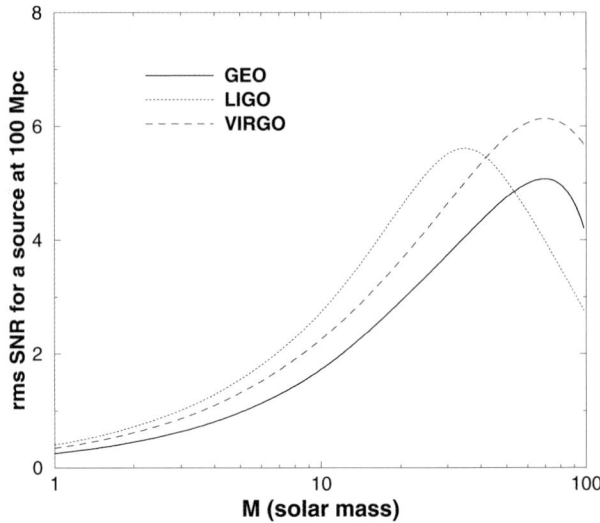

Fig. 13.4. Signal to noise ratio in initial interferometers as a function of total mass for inspiral signal from binaries of equal masses and averaged over source inclination.

not yet significant from the astrophysical point of view, but, outside the frequency interval probed by the bar detectors, they are tighter than other upper limits experimentally established so far. The necessary steps for increasing the sensitivity in the initial and advanced ground-based interferometers are well recognised and are being taken up.

The space-based interferometer LISA will be sensitive to g.w. in the interval $10^{-4} - 10^0 Hz$, that is, to lower g.w. frequencies as compared with ground-based instruments. Some new types of astronomical sources will be accessible to this detector. LISA will consist of three spacecraft, forming an equilateral triangle of side 5 million km, in a heliocentric orbit, lagging behind the Earth by $20°$. The phase shifts of the laser light travelling along all three sides of the triangle will monitor the light-travel distance between the small drag-free test masses inside the spacecraft. The nominal lifetime of the mission is 5 years. Figure 13.5 shows the LISA design sensitivity curve and some interesting sources of gravitational waves. This single figure attempts to show the noise curve of the entire assembly of spacecraft together with the strengths of g.w. sources of different types - from quasi-periodic to stochastic - and therefore the figure should be treated with some care. The instrumental noise level is shown here in bins of $3 \times 10^{-8} Hz$, which are appropriate for a 1 year integration time. In other words, the r.m.s. instrumental noise $\sqrt{S_n(f)}$ in units of $(Hz)^{-1/2}$ (see eqn. (13.14)) is multiplied with $\sqrt{3 \times 10^{-8} Hz}$ at each frequency, thus producing a dimensionless spectral quantity. (For the latest amendments to the LISA noise curve see, for example, [27].) This is done mainly in order to emphasize that during this observational time the

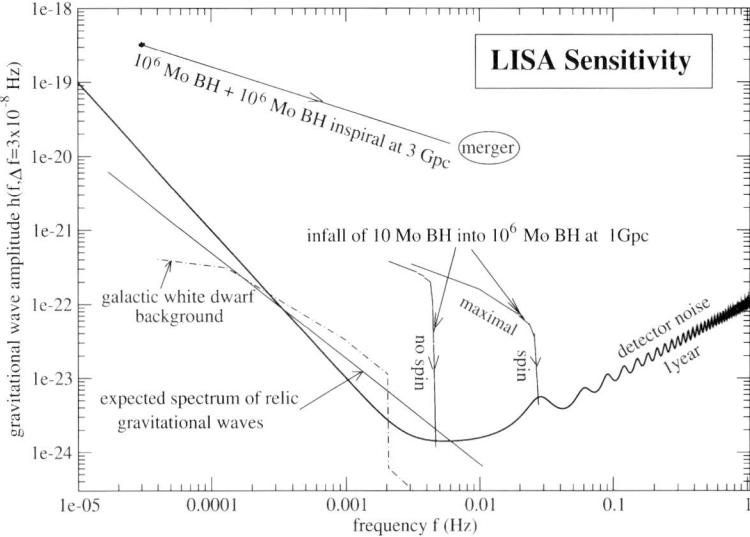

Fig. 13.5. Some interesting sources in comparison with LISA sensitivity.

g.w. signals from many of the galactic white-dwarf binaries can be resolved and removed from the data. This means that their g.w. noise will not prevent us from seeing something much more interesting - a stochastic background of relic gravitational waves. The sharp drop in signal from the galactic binary white dwarfs, shown in Fig. 13.5 at $f = 2 \times 10^{-3} Hz$, illustrates this assumed operation with the data [18, 28, 29], and not the total lack of galactic binaries radiating at frequencies higher than $2 \times 10^{-3} Hz$. The continuation of this curve at $f > 2 \times 10^{-3} Hz$ shows the much smaller g.w. noise from unresolved extragalactic white dwarf binaries. In accordance with this way of describing the dimensionless instrumental noise, the dimensionless g.w. signal amplitudes are also calculated in bins of $3 \times 10^{-8} Hz$ around any given frequency f. In the next section, we will present more details on astrophysical sources for ground-based and space-based instruments.

13.4 Gravitational Waves and Astrophysics

It is common to divide the sources into groups, depending upon whether they are accessible to ground-based or space-based instruments. It is also common to call them, respectively, the high-frequency and low-frequency sources.

13.4.1 High-Frequency Sources

We will focus on the sources that are likely to be the first ones detected by the ground-based instruments. It is clear from eqn. (13.10) that a powerful source of gravitational waves should involve large masses and relativistic velocities. A pair of neutron stars spiraling inwards towards each other under the influence of the g.w. radiation reaction force is a primary example. In the last few seconds of their inward spiral, before the neutron stars 'touch' each other and merge, the frequency of gravitational radiation increases from $200Hz$ to about $1200Hz$. The orbital velocity increases from $v/c \approx 0.2$ to $v/c \approx 0.4$. In the last few seconds, the binary emits $1.1 \times 10^{53} ergs$ of energy in the form of gravitational waves. This is 2% of its total rest-mass energy Mc^2. What can be more powerful and efficient a source of gravitational radiation than this one? Only a pair of even more massive and even more compact stars. Astronomers call them stellar-mass black holes.

If the coalescence of a binary neutron star (NS+NS) or a binary black hole (BH+BH) happened in our Galaxy, it would be easily detectable by the already operating instruments. The problem is that these catastrophic events are expected to take place only once in a very long time. In order to have a reasonable chance of seeing, say, three events per year, we have to survey a large volume of space, which includes many galaxies. This means that the instrument's sensitivity should be so high that the events could be seen from the edges of this large volume. This consideration explains the importance of theoretical evaluations of the event rates in a typical galaxy.

The event rate of coalescing NS+NS systems is partly constrained by pulsar observations in our Galaxy. Three NS+NS binaries are known, each involving a pulsar, whose coalescence time is less than the Hubble time and is, on average, 3×10^8 years. Starting with these three binaries, one would evaluate the NS+NS event rate as 1 per 100 million years. However, we observe only about 1% of the galactic volume, so the coalescence rate can easily be raised to $10^{-6} yr^{-1}$ [30]. Very likely, the event rate for NS+NS systems is significantly higher than this estimate, if only because of the fact that not all NS+NS systems include a currently observable pulsar. The observational situation with black holes is more uncertain. There are a dozen BH candidates in X-ray binary systems, but they are all in pairs with non-degenerate companions. So far, there is no observational evidence for NS+BH or BH+BH binaries. Nevertheless, one can make some evaluations on the basis of the star formation rates.

It is believed that the neutron star progenitors have masses greater than $10 M_\odot$, whereas the black hole progenitors have masses greater than $80 M_\odot$. The Salpeter function for the star formation rate is

$$\frac{dN}{dt d(M/M_\odot)} \simeq \left(\frac{M}{M_\odot}\right)^{-2.35} yr^{-1}. \qquad (13.15)$$

Integrating this function over M and using the lower limit of integration, one finds the ratio of the expected numbers of progenitors:

$$\frac{N(M > 80 M_\odot)}{N(M > 10 M_\odot)} = \left(\frac{80 M_\odot}{10 M_\odot}\right)^{-1.35} \simeq 0.06. \qquad (13.16)$$

It is reasonable to think that, despite all the complexities and differences in binary evolution, the ratio of coalescence rates will also be given by approximately the same quantity,

$$\frac{\mathcal{R}_{BH}}{\mathcal{R}_{NS}} = \left(\frac{80 M_\odot}{10 M_\odot}\right)^{-1.35} \simeq 0.06. \tag{13.17}$$

This expectation turns out to be in rough agreement with the results of detailed numerical population synthesis calculations.

Numerical calculations take into account all the available observational information. Their advantage is in that one can follow not only the channels leading to the NS+NS, NS+BH, BH+BH systems, most interesting for gravitational-wave astronomy, but also other evolutionary outcomes, which allow comparison with observations in their own right. The population synthesis results cannot be less reliable than purely 'observational' estimates, since they are controlled by the same available observational data. Discrepancies in final results exist because of uncertainties in astrophysics, not because one of the methods is inherently less reliable than another. This is especially true with regard to the NS+BH and BH+BH binaries, where the current purely 'observational' evaluations would have to begin with zero.

The results of conservative population synthesis calculations [18, 31] have been checked on their consistency with other evolutionary outcomes. The calculations show that the NS+NS rate is expected to be at the level $\mathcal{R}_{NS} = 3 \times 10^{-5} yr^{-1}$, whereas the BH+BH rate is at least one order of magnitude lower. For further estimates we will take it at the level $\mathcal{R}_{BH} = 0.06\, \mathcal{R}_{NS} = 2 \times 10^{-6} yr^{-1}$. These rates for a typical galaxy, $\mathcal{R}_G$, determine the rates for a given cosmological volume, $\mathcal{R}_V$, which includes many galaxies. When deriving $\mathcal{R}_V$, it is convenient to use a conservative estimate for the baryon content of the Universe. This brings us to the relationship

$$\mathcal{R}_V \approx 0.1 \mathcal{R}_G \left(\frac{r}{1\ Mpc}\right)^3. \tag{13.18}$$

Thus, it is expected that within a volume of radius $r = 100\ Mpc$ and during 1 year, there will be 3 of the NS+NS events and only 0.2 of the NS+BH or BH+BH events. An increase of the radius to $r = 200\ Mpc$ increases the volume and the event rates by a factor of 8. It is interesting to note that, despite all the diversity of approaches in the literature, there exists some tendency to convergence in the final results [32–36].

The derived event rates are the basis for the calculation of the expected detection rates. The important fact is that the mass of a typical neutron star is 1.4 $M_\odot$, whereas the mass of a typical black hole is $(10-15)\ M_\odot$. The averaged mass of the observed black hole candidates in the X-ray binaries is $M_{BH} \simeq 8.5 M_\odot$. A pair of black holes is a more powerful source of gravitational waves than a pair of neutron stars, and therefore black hole binaries can be seen by a given instrument from much larger distances. When it comes to the detection rates, the lower event rate of BH+BH sources is more than compensated for by their larger masses. Indeed, the optimal signal to noise ratio is [15, 18, 37]: $S/N \propto \mathcal{M}^{5/6}/r$. At a fixed S/N, the detection volume is proportional to r^3 and therefore to $\mathcal{M}^{5/2}$. The detection rate $\mathcal{D}$ for binaries of a given class is the product of their event rate $\mathcal{R}_V$ and the detector's registration volume $\propto \mathcal{M}^{5/2}$ for these binaries. Therefore, one obtains

$$\frac{\mathcal{D}_{BH}}{\mathcal{D}_{NS}} = \frac{\mathcal{R}_{BH}}{\mathcal{R}_{NS}} \left(\frac{\mathcal{M}_{BH}}{\mathcal{M}_{NS}}\right)^{5/2} = 0.06 \left(\frac{8.5 M_\odot}{1.4 M_\odot}\right)^{5/2} \simeq 5.5. \quad (13.19)$$

This remarkable result is consistent with the more accurate numerical calculation of S/N displayed in Fig. 13.4.

It is seen from Fig. 13.4 that g.w. signals from NS+NS at $r = 100 Mpc$ cannot be regarded detectable by initial interferometers. The situation is much better for the heavier pairs of NS+BH and BH+BH. If the total mass of a BH+BH binary is near $(20-30)M_\odot$, then $S/N \approx 2$ even if the binary is placed at $r = 200 Mpc$. It follows from the event rate $\mathcal{R}_{BH}$ discussed above, that in this larger volume one expects a couple of BH+BH events per year. Simultaneous observations on two or three instruments will significantly diminish the probability of false alarms to such events. This is why it is argued in [18] that coalescing black holes will probably be the first sources detected by the initial ground-based interferometers, when they reach their planned sensitivity. Of course, the discussed estimates are statistical by the very nature of things, and they have significant systematic uncertainties. It will not be very surprising if the reality is somewhat better or somewhat worse than what the mean values suggest. It is important, however, that even the most pessimistic evaluations of the NS+NS and BH+BH rates indicate that there should be many detections per year in the advanced interferometers.

If coalescing black holes are detected first, it is likely that at the beginning we will only have proof that the objects are black holes in an astronomical sense - an inward spiralling pair of heavy compact masses. The next step will be to try to understand their real physical nature. The general-relativistic black holes possess event horizons, which are supposed to merge into the resulting black hole, which will then emit a damped train of ringdown waves at specific frequencies, and so on [37, 38]. This fascinating physics will be testable when the good quality data are available. This analysis will also require a continuation of the intense effort on the side of analytical and numerical calculations (for a recent review of numerical relativity, see [39]).

Direct detection of the first sources will be not the end, but only the beginning of observational g.w. astronomy. In the long run, the aim of g.w. science is to explore a great variety of sources, many of which can hardly be seen in electromagnetic radiation. Needless to say that there is also a great chance of discovering new and totally unexpected sources. In addition to coalescing binary stars, many other sources will eventually be detected. For instance, the long-recognised importance of the core collapse of massive stars

[40, 41] as g.w. sources has been reinforced by the mounting evidence of asymmetries during supernova explosions, by the likelihood of forming and quick collapse of very massive early stars, by the association of supernovae events with gamma-ray bursts, etc. (For a recent review and extensive list of references, see for example [42].) The tidal disruption of a neutron star by its companion, the various sorts of stellar instabilities [43], the slightly deformed spinning neutron stars, young pulsars, and low-mass X-ray binaries [44], [45] - are also the astrophysically important and interesting g.w. sources that will be studied, very likely, by advanced detectors.

13.4.2 Low-Frequency Sources

It is common to call the low-frequency g.w. sources as 'sources for LISA'. Some of them are displayed in Fig. 13.5. As explained in the previous section, the dashed line shows the g.w. confusion noise from the binary white dwarfs, WD+WD, mostly concentrated in the disk of our Galaxy. LISA will not only detect thousands of WD+WD systems radiating at $f > 2 \times 10^{-3} Hz$, but is capable of doing this so accurately that their contributions can be removed individually from the data. Some known binaries consisting of degenerate and normal stars will also be detectable. Their angular coordinates and distance can be measured with high precision [46]. In fact, the well identified galactic binaries serve as guaranteed sources for LISA, and they will help in testing LISA's performance.

There is growing evidence for the existence of binary supermassive black holes (SMBH, $M \geq 10^6 M_\odot$) in the centres of merging galaxies. (For a recent review of SMBH formation see [47]). Certainly, a coalescing pair of SMBH is an extremely bright g.w. emitter. These sources are located at cosmological distances, so that the redshift of the incoming gravitational waves becomes important. But if the total mass of the pair is not significantly larger than $10^7 M_\odot$, the coalescing pair can still be visible with LISA at frequencies up to $f \approx 10^{-4} Hz$, even if the source is located at the redshift $z = 3$. The upper line in Fig. 13.5 shows the effective gravitational wave amplitude, relative to the plotted LISA instrumental noise, for a source consisting of two SMBH. The dot shows the signal when the inward spiralling pair was only 1 year prior to merger. The presence of spins of SMBH changes the LISA signal [48]. The trouble with these super-powerful sources is that their event rate is very uncertain. One of the difficult issues is whether a SMBH forms by direct collapse of gas in deep galactic potential wells or by hierarchical build-up of pre-galactic structures. Nevertheless, it is estimated that LISA may see 0.1-1 events per year, or maybe a factor of 10 more [49].

It seems certain that, from time to time, a SMBH will be closely approached by nearby compact stars - white dwarfs, neutron stars and stellar-mass black holes. A compact star orbiting a SMBH is a powerful g.w. source. However its orbit, and the emitted waveform, depend strongly on the spin of

SMBH. Fig. 13.5 shows two curves (taken from [19]) describing the performance of a $10 M_\odot$ BH completing its inward spiral toward a $10^6 M_\odot$ SMBH. The left curve corresponds to a non-spinning SMBH and a circular orbit of a stellar-mass BH. The right curve describes the signal from a prograde, circular, equatorial orbit around a nearly maximally spinning Kerr SMBH. Both curves begin at frequencies when the time to merger is 1 year. A serious theoretical problem is the construction of reliable templates for these sources. One should expect that a typical orbit will be eccentric, non-equatorial, and subject to radiation-reaction force corrections. Correspondingly, the waveforms could be extremely complicated. Some progress in this area is reported in [50].

Finally, Fig. 13.5 shows the expected level of relic gravitational waves. This is a fundamentally important signal from the very early Universe. Its explanation requires cosmological notions and some elements of quantum physics. Since the relic gravitational waves and primordial density perturbations are, at present, one of the most fascinating and active areas of research, we devote a separate section to it.

13.5 Gravitational Waves and Cosmology

13.5.1 Generation of Relic Gravitational Waves and Primordial Density Perturbations

In many situations one can neglect the non-linearity of the gravitational field, that is, the interaction of gravitational waves with other gravitational fields and with themselves. However, this is not always the case. The most dramatic example is the interaction of gravitational waves with the strong variable gravitational field of the very early Universe. A gravitational wave can be thought of as a harmonic oscillator, while the smooth variable gravitational field of the surrounding Universe may be thought of as a gravitational pump field. The g.w. oscillator is parametrically coupled to the gravitational pump field. This specific coupling is a consequence of the non-linear nature of the Einstein equations. The coupling provides a mechanism for the superadiabatic (parametric) amplification of classical waves and the quantum-mechanical generation of waves from their zero-point quantum oscillations [51]. The word 'superadiabatic' emphasizes the fact that this effect takes place over and above whatever happens to the wave during very slow (adiabatic) changes of the pump field. That is, we are interested in the increase of occupation numbers, rather than in the gradual shift of energy levels. The word 'parametric' emphasizes the mathematical structure of the wave equation. It is a change of a parameter of the oscillator caused by the pump field, namely, a sufficiently rapid variation of its frequency, that is responsible for the considerable increase of energy of that oscillator.

It is common to write the perturbed gravitational field of a homogeneous isotropic universe in the form:

$$ds^2 = a^2(\eta)[-d\eta^2 + (\delta_{ij} + h_{ij})dx^i dx^j]. \tag{13.20}$$

The gravitational field perturbations $h_{ij}(\eta, \mathbf{x})$ can be expanded over spatial Fourier harmonics $e^{\pm i \mathbf{n} \cdot \mathbf{x}}$, where $\mathbf{n}$ is a constant (time-independent) wave-vector,

$$h_{ij}(\eta, \mathbf{x}) = \frac{\mathcal{C}}{(2\pi)^{3/2}} \int_{-\infty}^{\infty} d^3\mathbf{n} \sum_{s=1,2} \overset{s}{p}_{ij}(\mathbf{n}) \frac{1}{\sqrt{2n}} \left[\overset{s}{h}_n(\eta) e^{i\mathbf{n}\cdot\mathbf{x}} \overset{s}{c}_\mathbf{n} + \overset{s}{h}_n^*(\eta) e^{-i\mathbf{n}\cdot\mathbf{x}} \overset{s\dagger}{c}_\mathbf{n} \right]. \tag{13.21}$$

The polarisation tensors $\overset{s}{p}_{ij}(\mathbf{n})$, $s = 1, 2$ have different forms, depending on whether the h_{ij} represent gravitational waves or density perturbations. In the case of gravitational waves, the $\overset{s}{p}_{ij}$ describe the two familiar 'plus' and 'cross' polarisations introduced in Sect. 13.2. In the case of density perturbations, the polarisation tensors are:

$$\overset{1}{p}_{ij}(\mathbf{n}) = \sqrt{\frac{2}{3}} \delta_{ij}, \quad \overset{2}{p}_{ij}(\mathbf{n}) = -\sqrt{3}\frac{n_i n_j}{n^2} + \frac{1}{\sqrt{3}} \delta_{ij}. \tag{13.22}$$

The Einstein equations for the gravitational field perturbations h_{ij} with the polarisation tensors (13.22) can only be satisfied if the h_{ij} are accompanied by perturbations in the density of matter. This is why this class of perturbations is called density perturbations. The difference between $\overset{s}{p}_{ij}(\mathbf{n})$ for, respectively, gravitational waves and density perturbations is responsible for the difference in polarisation patterns of the CMB radiation, caused by these two classes of gravitational perturbations (see Sect. 13.5.2 below).

For a classical field h_{ij}, the quantities $\overset{s}{c}_\mathbf{n}$, $\overset{s\dagger}{c}_\mathbf{n}$ are complex numbers. For a quantized field, they are annihilation and creation operators satisfying the conditions

$$[\overset{s'}{c}_\mathbf{n}, \overset{s\dagger}{c}_\mathbf{m}] = \delta_{s's}\delta^3(\mathbf{n}-\mathbf{m}), \quad \overset{s}{c}_\mathbf{n}|0\rangle = 0,$$

where $|0\rangle$ (for each mode $\mathbf{n}$ and s) is the initial vacuum state defined at some η_0 in the very distant past, long before the superadiabatic regime for the given mode has started. The normalization constant $\mathcal{C}$ is determined by the requirement that initially each mode contained only the zero-point energy $\frac{1}{2}\hbar\omega$. Then, $\mathcal{C} = \sqrt{16\pi} l_{Pl}$ for gravitational waves and $\mathcal{C} = \sqrt{24\pi} l_{Pl}$ for density perturbations, where $l_{Pl} = (G\hbar/c^3)^{1/2}$ is the Planck length. Obviously, the initial vacuum amplitude, and the entire field, should vanish, if the Planck constant $\hbar$ is formally sent to zero.

The calculation of quantum-mechanical expectation values and correlation functions provides the link between quantum mechanics and macroscopic physics. Using the representation (13.21) and definitions above, one finds the variance of the gravitational field perturbations:

$$\langle 0|h_{ij}(\eta,\mathbf{x})h^{ij}(\eta,\mathbf{x})|0\rangle = \frac{C^2}{2\pi^2}\int_0^\infty n^2 \sum_{s=1,2} |\overset{s}{h}_n(\eta)|^2 \frac{dn}{n}. \qquad (13.23)$$

The quantity

$$h^2(n,\eta) = \frac{C^2}{2\pi^2} n^2 \sum_{s=1,2} |\overset{s}{h}_n(\eta)|^2 \qquad (13.24)$$

gives the mean-square value of the metric (gravitational field) perturbations in a logarithmic interval of n and is called the (dimensionless) power spectrum. The power spectrum of metric perturbations is a quantity of great observational importance. It defines the temporal structure and amplitudes of the g.w. signal in the frequency bands of direct experimental searches. It is also crucial for calculations of anisotropy and polarisation induced in CMB by relic gravitational waves and by other gravitational field perturations.

To find the power spectrum at any given moment of time (for instance, today or at the moment of decoupling of CMB from the rest of matter) we need to know the mode functions $\overset{s}{h}_n(\eta)$ at those moments of time. In the case of gravitational waves, the mode functions $\overset{s}{\mu}_n(\eta)$ (where $\overset{s}{\mu}_n(\eta) \equiv a(\eta)\overset{s}{h}_n(\eta)$) are governed by the equation for the parametrically disturbed oscillator [51]:

$$\overset{s}{\mu}_n'' + \overset{s}{\mu}_n \left[n^2 - \frac{a''}{a}\right] = 0. \qquad (13.25)$$

The equation describing density perturbations in the very early Universe can also be reduced to the equation very similar to eqn. (13.25), and the generating mechanism will work without change. As soon as the pump field (represented by the cosmological scale factor $a(\eta)$) is known, and since the initial conditions are fully determined, the mode functions $\overset{s}{h}_n(\eta)$, as well as other properties of the generated fields, are unambiguously calculable. The important (and, strictly speaking, unknown) era of cosmological evolution is the stage preceeding the radiation-dominated era. We call it an initial (i) era and characterize it, for simplicity of calculation, by a set of power-law scale factors: $a(\eta) = l_o |\eta|^{1+\beta}$, where l_o and β are constants.

The main properties of cosmological perturbations generated as a result of superadiabatic (parametric) amplification of their zero-point quantum oscillations are as follows (for more details see [18, 52] and references there):

1. The initial vacuum state is described by a Gaussian wavefunction. As a result of quantum-mechanical Schrödinger evolution, the vacuum state transforms into a multi-particle state known as a squeezed vacuum state. The distributions of amplitudes and phases acquire strongly unequal variances. In the general expression for the gravitational field mode,

$$h_\mathbf{n} = A_1 \sin(n\eta + \phi_1)\cos \mathbf{n}\cdot\mathbf{x} + A_2 \sin(n\eta + \phi_2)\sin \mathbf{n}\cdot\mathbf{x}, \qquad (13.26)$$

the amplitudes A_1 and A_2 are drawn from a broad Gaussian distribution, whereas the phases ϕ_1 and ϕ_2 are practically fixed and equal up to $\pm\pi$. The field is a stochastic collection of standing waves and is characterised by a strongly modulated power spectrum. This complicated statistical picture of generated cosmological perturbations is often replaced in the literature by a single word: 'Gaussian'.

2. The generated gravitational field perturbations act on all sorts of matter together. There is no reason why the inhomogeneities in different sorts of matter (if more than one component of matter was dynamically important in the very early Universe) should be displaced and move with respect to each other, which would constitute the so-called isocurvature, or entropy, perturbations. This is why the generated density perturbations are often called 'adiabatic'.

3. The primordial spectrum (i.e. the spectrum before processing at the radiation-dominated and matter-dominated stages) as a function of the wavenumber n is fully determined by the variable pump field as a function of time η. Every interval of spectrum that can be meaningfully approximated as a power-law function of n, was generated by an interval of a power-law evolution $a(\eta) \propto |\eta|^{1+\beta}$. Explicitly, for the spectrum of eqn. (13.24) one finds: $h^2(n) \propto (l_{Pl}/l_o)^2 n^{2(\beta+2)}$. Specifically for density perturbations, one often uses the spectral index n, related to β by n $= 2\beta + 5$. If the very early Universe was goverened by a scalar field (the central assumption of inflationary scenarios), then, at every power-law interval of evolution and, hence, at every power-law interval of the generated spectrum, there must be $\beta \leq -2$ and n ≤ 1. The 'red' spectra $\beta < -2$ (n < 1) possess a serious theoretical difficulty: the mean-square value of the field becomes power-law divergent in the limit of very long waves (lower limit of integration in eqn. (13.23)). The case $\beta = -2$ (n $= 1$) is called the flat, or Harrison-Zeldovich-Peebles, or 'scale-invariant', spectrum.

4. Gravitational waves are generated inevitably, wheras the generation of density perturbations requires additional assumptions about the coupling of matter fields to gravity. The primordial (unprocessed) amplitudes of density perturbations can be as large as g.w. amplitudes, but never much larger. The observed CMB anisotropy in lower multipoles (caused by primordial gravitational field perturbations) may have nothing to do with quantum mechanics, but if it does, the contribution of relic gravitational waves to lower multipoles must be substantial.

5. The parametric mechanism is universal, and the generation of primordial cosmological perturbations takes place regardless of whether the Universe 13 billion years later will appear to astronomers as spatially-flat, or not. In any case, if Ω_{total} is not identically 1, the present-day Ω_{total} is usually regulated 'by hand' through the duration of the initial era. This parameter of

duration does not affect seriously the amplitudes and spectral slopes of primordial perturbations.

13.5.2 Detection of Relic Gravitational Waves

The direct detection of relic gravitational waves will be fundamentally important for uncovering the physics of the very early Universe. In Fig. 13.6 we show the expected spectrum of today's r.m.s. amplitudes $h(\nu)$ (square root of eqn. (13.24)) as a function of frequency ν in Hz. The graph shows the piecewise envelope of the spectrum and ignores its oscillations. Almost everything in this graph is the processed spectrum; the primordial part survives only at frequencies $\nu < \nu_H$, where $\nu_H = c/\lambda_H = H \approx 2 \times 10^{-18} Hz$ is the Hubble frequency. This particular spectrum was derived under the assumption that a significant fraction of the observed large-scale CMB anisotropy is caused by relic gravitational waves and that the primordial spectral index is $\beta = -1.9$ (n = 1.2). This value of n follows from the COBE data [53] and it has been recently reinforced, albeit with broad error bars, by more sophisticated analysis [54]. The evaluations of n based on larger data sets usually lead to smaller n's, varying in a narrow interval around n = 1. However, this precision appears to be artificial (caused by the excessively rigid *a priori* assumptions about the tested models) and the story does not seem to be over. In any case, we use Fig. 13.6 for the analysis of detectability of relic gravitational waves.

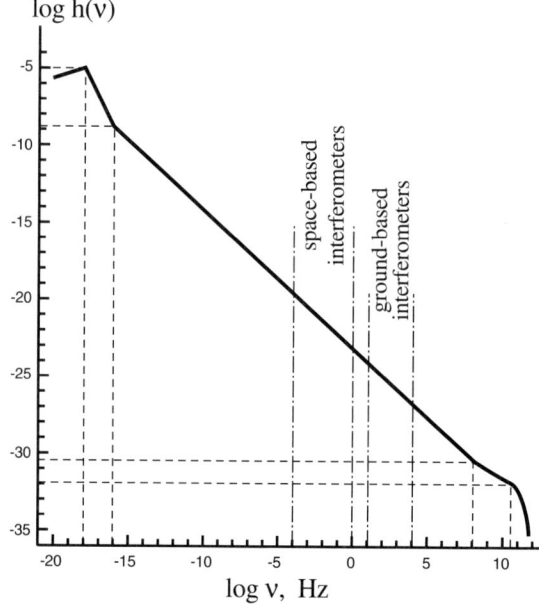

Fig. 13.6. Expected envelope of the spectrum $h(\nu)$ for the case $\beta = -1.9$ (n = 1.2).

The r.m.s. values $h(\nu)$ are directly entering the detectability evaluation, but they also determine the $\Omega_{gw}(\nu)$-parameter, which is useful for comparison of the g.w. background with other energy components. This parameter can be calculated according to the formula

$$\Omega_{gw}(\nu) = \frac{\pi^2}{3}h^2(\nu)\left(\frac{\nu}{\nu_H}\right)^2. \tag{13.27}$$

For example, one has $h(\nu) \approx 10^{-20.5}$, $\Omega(\nu) \approx 10^{-11}$ at $\nu = 10^{-3} Hz$, and $h(\nu) \approx 10^{-25}$, $\Omega(\nu) \approx 10^{-10}$ at $\nu = 10^2 Hz$. It is seen from Fig. 13.5 that the part of the spectrum accessible to LISA is higher than the instrumental noise and can be measured. The ground-based advanced interferometers are also promising. The target value of the LIGO-II at $\nu = 10^2 Hz$ is $h_{ex} = 10^{-23}$. The gap in two orders of magnitude can be covered by cross-correlation of the outputs of two or more detectors. The S/N will be better than 1 if the common integration time exceeds $10^6 sec$. This does not seem to be a hopeless task.

It is possible that relic gravitational waves will be first observed indirectly, with the help of polarisation measurements of CMB. The problem is to distinguish the polarisation pattern caused by gravitational waves from that caused by density perturbations. The polarisation arises as a result of the Thompson scattering of CMB photons on free electrons [55]. To produce a net polarisation, the electrons should be illuminated by CMBR having a non-zero quadrupole anisotropy. The two polarisation patterns are distinguishable if the quadrupole anisotropies are distinguishable. For our purposes of illustration, it is sufficient to consider the emission (rather than scattering) of electromagnetic waves by free electrons. The electrons are set in motion by, respectively, gravitational waves and density perturbations.

The motion of free particles in the field of a linearly-polarised gravitational wave is shown in Fig. 13.1. Imagine that the moving particles are free electrons in the early Universe, rather than free mirrors of an interferometer in laboratory, that we were discussing in Sect. 13.3. Then, the directions of the induced oscillations of the electrons, indicated by arrows in Fig. 13.1, are, at the same time, the directions of the electric fields of the electromagnetic waves emitted by these oscillating electrons. The pattern of arrows seen on this figure is the pattern of polarisation components that will be seen on the sky.

The motion of free electrons induced by the gravitational field of a density perturbation is different. Two polarisation tensors of a density perturbation are given by eqn. (13.22). The first one describes a trivial purely spherically-symmetric deformation of a sphere of free particles. The second one, the subject of our interest, describes the deformation which has the angular dependence of the spherical harmonic $Y_{lm}(\theta, \phi)$ with $l = 2$ and $m = 0$, in contrast to the combination of $Y_{lm}(\theta, \phi)$ with $l = 2$ and $m = \pm 2$, attributable to a gravitational wave. In both cases, the polar axis z is taken along the

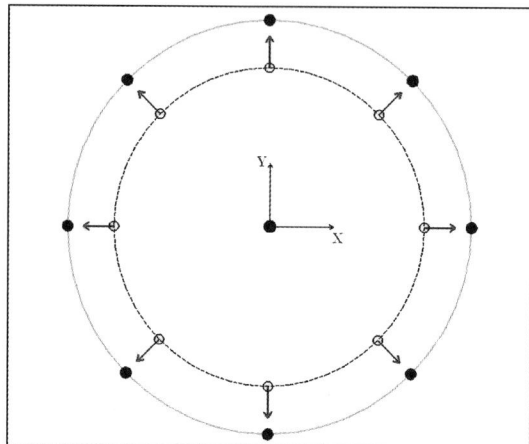

Fig. 13.7. Motion of free particles in the gravitational field of a density perturbation.

wave-vector **n** of the perturbation. Figure 13.7 shows a ring of electrons set in motion by the gravitational field of a density mode. The pattern of these arrows is, at the same time, the pattern of polarisation components generated by a density perturbation. Comparing Figs. 13.1 and 13.7 one concludes that the polarisation patterns are distinguishable, if one can study the distribution of polarisation Stokes parameters over a sufficiently large portion of the sky. In real conditions, the free electrons will be influenced by the superposition of many perturbation modes with arbitrary wave-vectors, but the net difference between these two sorts of perturbations should survive.

The gap between the expected g.w. signal and the polarisation detection capabilities is relatively small. Taking into account the current impressive activity in this area, one can hope that some decisive observational information about relic gravitational waves will be obtained fairly soon.

13.5.3 Gravitational Waves and Inflation

The theoretical and experimental studies of relic gravitational waves are endangered by absurd claims that are prevalent in inflationary literature. Inflationists claim that the amount of relic gravitational waves should be zero if the primordial spectrum is flat, that is, if n = 1. There seems to be no need to bother about relic gravitational waves, as the observations indicate that n may indeed be close to 1. Since the 'pillars of inflation' are popular in a part of astrophysics community, and sometimes are said to be 'confirmed', it is important to put matters straight.

The inflationary scenario operates with a scalar field φ and the scalar field potential $V(\varphi)$. Having accepted the general concept of parametrically amplified quantum fluctuations, inflationary theorists are performing their own calculations. The quantum-mechanical content of these calculations is usually limited to vague words, such as that 'inflation amplified quantum fluctuations onto macroscopic scales'. Being unsure why and where the Planck constant $\hbar$ should enter the calculations, inflationists never write it explicitly; and when it is written implicitly, in the form of the Planck mass, $M_{Pl} = (\hbar c/G)^{1/2}$, it always stands in the wrong place, in the denominator of the final expression instead of the nominator. With this sort of 'quantisation', inflationary theorists derived their contribution to the subject of cosmological perturbations – the 'standard inflationary result'. The 'standard inflationary result' predicts the infinitely large amplitudes of today's density perturbations in the limit of the flat spectrum n = 1. Indeed, one will always be able to recognise in inflationary papers the evaluations relating the final (f) amplitudes of the perturbations to the initial (i) values of φ and other quantities: $(\delta\rho/\rho)_f \sim (h_S)_f \sim (\zeta)_f \approx (\zeta)_i \sim (H^2/\dot\varphi)_i \sim (V^{3/2}(\varphi)/V'(\varphi))_i \sim H_i/\sqrt{1-n}$. The denominator of the last expression is zero for n = 1. The nominator is the Hubble parameter H_i at the initial stage. H_i is much larger than the Hubble parameter at the subsequent radiation-dominated stage, and, in any case, H_i is not zero. Therefore, the predicted final amplitudes go to infinity, if the spectral index n = 1. Inflationists are hiding this absurd prediction of infinitely large density perturbations by composing the ratio of the gravitational wave amplitude h_T to the predicted divergent amplitude of the scalar metric perturbations h_S (the so called 'tensor-to-scalar ratio' or 'consistency relation': $h_T/h_S \approx \sqrt{1-n}$) and declaring that it is the amount of gravitational waves that should be zero, or almost zero, at cosmological scales and, hence, down to laboratory scales.

Certainly, the 'standard inflationary result' is in full disagreement not only with theoretical quantum mechanics, but with available observations too, as long as the error-boxes of the observationally derived spectral index n are centered at n $\approx$ 1 and include n = 1. To be consistent with inflationary predictions, the data should not allow 'blue' spectra n > 1, as the scalar field cannot produce them, and the density amplitudes should go to infinity, when one processes the data assuming that n = 1. This spectacular failure of inflationary calculations is systematically painted by inflationists and their followers as a great success. The most recent example is the analysis of WMAP satellite data [56]. The authors praise and follow inflationary derivations, and conclude that the 'tensor/scalar ratio r' is 'consistent with zero'. The 'standard inflationary result' is written in that paper (their formula (17)) as:

$$\Delta_{\mathcal{R}}^2 = \frac{V/M_{Pl}^4}{24\pi^2 \epsilon_V}, \qquad (13.28)$$

and the 'tensor/scalar ratio r' (their formula (18)) as:

$$r = 16\epsilon_V. \qquad (13.29)$$

Combining the second formula with the first one, one can easily see that if the WMAP data demonstrate that 'r is consistent with zero', then the

WMAP data should also be consistent with an infinite numerical value of the density amplitudes $\Delta_\mathcal{R}^2$ and, hence with an infinite numerical value of the induced CMB anisotropies. If the WMAP data are not consistent with such an infinite numerical value of density perturbations, then the only 'implication for inflation' that follows from the WMAP observations is that the single concrete formula derived by inflationists – their 'standard inflationary result' – is shown to be wrong.

It seems to the author that overenthusiasm for inflation has reached unscientific, even ecclesiastical proportions. For example, this is how inflation is characterized in the educational programme 'Astronomy' of the Smithsonian Institution [57]: "...the inflationary scenario is the best current theory of the Universe... It has met four critical observational tests...". It is unclear which four tests the inflationary school credits to itself, rather than to direct consequences of quantum mechanics and general relativity, but one can recall that even general relativity was characterized until quite recently as a theory that has met only three critical observational tests (1 of which, gravitational redshift, is not a test of specifically general relativity). As for professional papers, one reads almost every day claims that the CMB and galaxy surveys are "in spectacular agreement with an inflationary Λ-dominated cold dark matter cosmology" (compare, for example, with [58] and [59]). Somewhere in the text, authors usually admit that, say, the observed quadrupole anisotropy is way out of the predicted value, and that the probability of finding such a result within the 'standard' model is 1.5×10^{-3} [58, 60]. Surely a theory which is admitted to have only a one in a thousand chance of being consistent with one of its crucial observational tests is not in 'spectacular agreement' with the cosmos we are trying to understand and should not be a subject given to self-congratulation.

13.5.4 Gravitational Waves and Quadrupole Anisotropy

The accurate measurements of CMB by WMAP reiterate the issue of the gravitational wave contribution to the lower order multipoles. The best strategy is to rely on conclusions of general physics and to use the minimum number of extra hypotheses. If the general considerations suggest (see Sect. 13.5.1) that the contributions of gravitational waves and density perturbations should be of the same order of magnitude, it is this conclusion that should be tested most thoroughly.

Figure 13.8 shows [61] the WMAP observational points (triangles) and the best fit curve (solid line) of the Λ-dominated cosmology without gravitational waves. The predicted quadrupole ($l = 2$) anisotropy is a factor of 8 higher than the actual observed value. The thin solid line (reproduced from [52], fig.12) shows the contribution of density perturbations alone in a model with $z_{dec} = 1000$, $z_{eq} = 5000$, n = 1 that fits the position and value of the peak at $l = 220$. This line is surrounded by the 1σ uncertainty belt (shown by two dashed lines) arising due to the lack of ergodicity on a

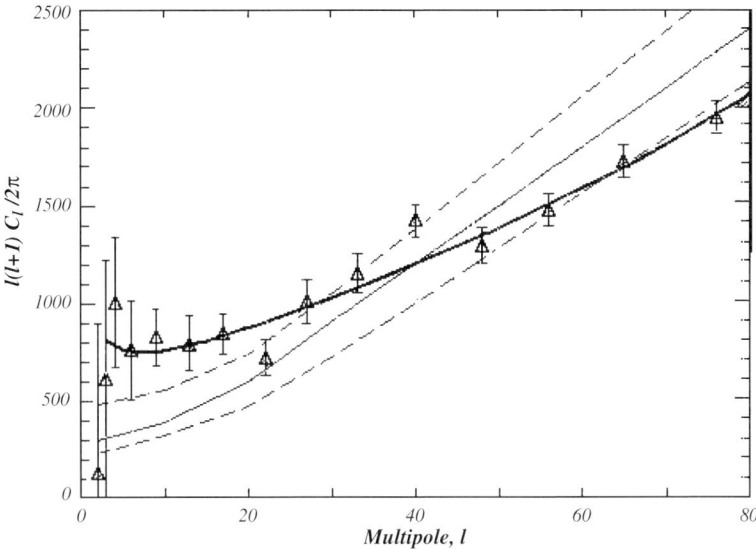

Fig. 13.8. The WMAP data and some theoretical models

2-sphere. At sufficiently large l's the belt is approximately symmetric and its size is $\Delta C_l \approx \sqrt{2/(2l+1)}C_l$. But the asymmetry grows towards small l's, and specifically at $l = 2$ the size of the deviation down is only 0.4 part of the deviation up [62]. [The accurate displaying of this fundamental uncertainty makes the actual quadrupole even further out of the uncertainty belt of the Λ-dominated cosmology, than what is implied by the usually plotted symmetric 'cosmic variance'.] Comparing the thin line with the data points, it is difficult to avoid the conclusion (advocated in [52]) that in fact there exists an excess, rather than a deficit, of power at small multipoles, and the most natural explanation of this excess is the anticipated contribution of gravitational waves. The increase of the spectral index up to n = 1.2 makes the agreement with the observed quadrupole even better and implies a somewhat larger amount of gravitational waves [52]. One should remember, however, that all the experimental C_l data, together with all the cosmological parameters, is a finite set of numbers. At the same time, in our hands is the (strictly speaking, unknown) shape of the primordial spectrum, i.e. a continuous function and an infinite set of numbers. A perfect agreement with any observed C_l's and practically any set of cosmological parameters, not only $\Lambda = 0$, can be achieved at the expense of the properly chosen primordial spectrum. Unfortunately, the era of 'precision cosmology' is still at some distance from us.

The nature of the observed quadrupole anisotropy deserves special attention. Most likely, it is caused by superposition of very long gravitational and density waves. It is known [63] that a gravitational wave produces the $Y_{2,2}$ and $Y_{2,-2}$ CMB anisotropy, whereas a density perturbation produces the $Y_{2,0}$ CMB anisotropy. We have discussed this distinction in Sect. 13.5.2 in connection with the CMB polarisation. The actual quadrupole distribution over the sky was measured by COBE [64]. In Galactic coordinates,

$$Q(\theta,\phi) = Q_1(3\cos^2\theta - 1)/2 + Q_2 \sin 2\theta \cos\phi + Q_3 \sin 2\theta \sin\phi$$
$$+ Q_4 \sin^2\theta \cos 2\phi + Q_5 \sin^2\theta \sin 2\phi, \qquad (13.30)$$

where the measured (least noisy) components are [65]:

$$Q_1 = 19.0 \pm 7.4, \quad Q_2 = 2.1 \pm 2.5, \quad Q_3 = 8.9 \pm 2.0,$$
$$Q_4 = -10.4 \pm 8.0, \quad Q_5 = 11.7 \pm 7.3. \qquad (13.31)$$

Even if this $Q(\theta, \phi)$ is produced by a single gravitational wave or a single density perturbation, it is only in a special coordinate system that it can be reduced to the combination of $Y_{2,2}$ and $Y_{2,-2}$ or to $Y_{2,0}$, respectively. To find out what we are dealing with, we have to build invariants, that is, quantities independent of the rotation of the coordinate system. One of invariants is

$$Q_{rms}^2 = (4/15)[(3/4)Q_1^2 + Q_2^2 + Q_3^2 + Q_4^2 + Q_5^2], \qquad (13.32)$$

another one (see, for example, [66]) is

$$D = (4/5^{3/2})[(1/4)Q_1(Q_1^2 + Q_2^2/2 + Q_3^2/2 - Q_4^2 - Q_5^2)$$
$$+ 2Q_2 Q_3 Q_5 + Q_4(Q_2^2 - Q_3^2)]. \qquad (13.33)$$

Q_{rms} is always positive, wheras D can be negative, but it always satisfies the condition $|D| \leq Q_{rms}^3$. For a pure density perturbation, $|D| = Q_{rms}^3$; and for a pure gravitational wave, $D = 0$.

Calculating the invariants and the formal errors from the data set (eqn. (13.31)) we find

$$Q_{rms} = (12.6 \pm 3.4)\mu K, \quad -D^{1/3} = (6.9 \pm 12.9)\mu K \qquad (13.34)$$

As expected, the available noisy data do not allow one to prefer one of hypotheses over another. But, for sure, there is no indications whatsoever that the quadrupole anisotropy is produced by a density perturbation alone. If anything, the central values of Q_{rms} and D indicate that the contribution of gravitational waves should be substantial. Hopefully, the WMAP quadrupole data will be more accurate, and then this analysis should be repeated.

13.6 Summary

Gravitational-wave physics is a mature and at the same time a very young science. In a sense, the relativistic gravity (general relativity) itself is still a young science. The enormous progress in technical developments and observational verifications of the theory is accompanied by difficult issues of its adequate description and interpretation, the necessity of bringing it to a closer contact with other branches of physics.

It is rumoured that Nobel recognition eluded the great astronomer Edwin Hubble because of his reluctance to say about his discovery what the establishment wanted him to say. Apparently, the scientific integrity of Hubble allowed him to say what he believed he discovered – the nonstationarity of the system of nearby galaxies – whereas he was required to admit that he discovered the 'expansion of space'. But the 'expansion of space' is still alive and well. [For example, 'As bizarre as it may seem, space itself is expanding – specifically, the vast regions of space between galaxies' [57].] It is regularly proposed to be measured. The logic seems to be impecable. If 'space' expands by a factor of 2 in 10 billion years, why would not the Earth or an atom expand by 10% in 1 billion years? The gravitational-wave research is plagued in a similar fashion. It is often stated that 'gravitational waves are oscillations of space-time itself'. The next phrase seems to be logically unavoidable: 'gravitational waves act tidally, stretching and squeezing any object that they pass through'. If this phrase were correct, we would never be able to notice gravitational waves. The device measuring, say, the displacements of free mirrors in an interferometer would be 'stretched and squeezed' as well. In this situation, we can probably find comfort in the wise observation [67]: 'I agree that much of what one reads in the literature is absurd. Often it is a result of bad writing, rather than bad physics. I often find that people who say silly things actually do correct calculations, but are careless in what they say about them.'

It seems to the author that, in the long perspective, the value of gravitational wave research will be in its influence on fundamental physics. Meanwhile, let us hope that the next gravitational-wave update will be devoted to the fascinating nature of concrete astrophysical sources of gravitational waves detected by the existing instruments.

13.7 Acknowledgements

I am grateful to S. Babak, D. Baskaran, and B. Sathyaprakash for help and useful discussions.

References

[1] A. Einstein. Zum gegenwärtigen Stande des Gravitationsproblems, Phys. Z. **XIV**, 1249-1266 (1913)
[2] J. Weber. Phys. Rev. **117**, 306 (1960); *General Relativity and Gravitational Waves*, (New York: Interscience Publ. 1961)
[3] J. H. Taylor. Rev. Mod. Phys. **66**, 711 (1994)
[4] http://igec.lnl.infn.it

[5] http://www.ligo.org
[6] http://www.ligo.caltech.edu
[7] http:/www.geo600.uni-hannover.de
[8] http://www.virgo.infn.it
[9] http://tamajo.mtk.nao.ac.jp
[10] P. Astone et al. arXiv:astro-ph/0302482
[11] http://www.lisa.jpl.nasa.gov
[12] L. D. Landau and E. M. Lifshitz. *Classical Theory of Fields*, (Oxford: Pergamon Press, 1975)
[13] C. W. Misner, K. S. Thorne and J. A. Wheeler. *Gravitation*, (San Francisco: W.H.Freeman and Co., 1975)
[14] S. Weinberg. *Gravitation and Cosmology*, (New York: J. Wiley and Sons, 1972)
[15] K. S. Thorne. In *Three Hundred Years of Gravitation*, Eds. S. W. Hawking and W. Israel, (Cambridge:CUP, 1987) p. 330
[16] L. P. Grishchuk. Usp. Fiz. Nauk **121**, 629 (1977) [Sov. Phys. Usp. **20**, 319 (1977)]
[17] B. F. Schutz. Class. Quant. Grav. **16**, A131 (1999)
[18] L. P. Grishchuk, V. M. Lipunov, K. A. Postnov, M. E. Prokhorov and B. S. Sathyaprakash. Usp. Fiz. Nauk **171**, 3 (2001) [Physics-Uspekhi **44**, 1 (2001); arXive: astro-ph/0008481]
[19] C. Cutler and K. S. Thorne. In: *General Relativity and Gravitation*, Eds. N.T.Bishop and S.D.Maharaj, (World Scientific, 2002) p. 72
[20] D. Baskaran and L. P. Grishchuk. Gravitational Lorentz force in the field of gravitational wave, Class. Quant. Grav. (to be submitted)
[21] http://www.minigrail.nl; E. Coccia et al. Phys. Rev. D **57**, 2051 (1998)
[22] P. Astone et al. Class. Quant. Grav. **19**, 5449 (2002)
[23] L. S. Finn. arXive: gr-qc/0301092
[24] P. Astone et al. arXive: gr-qc/0304004
[25] L. Blanchet. In: *General Relativity and Gravitation*, Eds. N.T.Bishop and S.D.Maharaj, (World Scientific, 2002) p. 54
[26] B. Barish. AAAS meeting, http://php.aaas.org/meetings
[27] T. A. Prince, M. Tinto, S. L. Larson, and J. W. Armstrong. Phys. Rev. D**66**, 122002 (2002)
[28] N. J. Cornish and S. L. Larson. arXive: astro-ph/0301548
[29] A. Krolak and M. Tinto. arXive: astro-ph/0302013
[30] E. S. Phinney. Astroph. J. Lett. **380**, L17 (1991)
[31] V. M. Lipunov, K. A. Postnov and M. E. Prokhorov. Pis'ma Astron. Zh. **23**, 563 (1997) [Astron. Lett. **23**, 492 (1997)]
[32] V. Kalogera, R. Narayan, D. N. Spergel, and J. H. Taylor. Astroph. J. **556**, 340 (2001)
[33] C. Kim, V. Kalogera, and D. R. Lorimer. arXive: astro-ph/0207408
[34] M. S. Sipior and S. Sigurdsson. Astroph. J. **572**, 962 (2002)
[35] F. A. Rasio. arXive: astro-ph/0212211
[36] S. Sigurdsson. Black Holes and Pulsar Binaries, arXive: astro-ph/0303312
[37] E. E. Flanagan and S. A. Hughes. Phys. Rev. D**57**, 4535 (1998)
[38] A. Buonanno and T. Damour. Phys. Rev. D**62**, 064015 (2000)
[39] L. Lehner. Class. Quant. Grav. **18**, R25 (2001)

[40] M. Rees, R. Ruffini, and J. A. Wheeler. In: *Black Holes, Gravitational Waves, and Cosmology* (Gordon and Breach, New York, 1974)
[41] D. M. Eardley. In: *Gravitational Radiation*, eds. N.Deruelle and T.Piran (North Holland, Amsterdam, 1983) p.257
[42] K. C. B. New. *Gravitational Waves from Gravitational Collapse*, arXive: gr-qc/0206041
[43] N. Andersson. *Gravitational waves from instabilities in relativistic stars*, arXive: astro-ph/0211057
[44] M. A. Papa, B. F. Schutz, and A. M. Sintes. arXive: gr-qc/0011034
[45] P. R. Brady, T. Creighton, C. Cutler, and B. F. Schutz. Phys. Rev. D**57**, 2101 (1998)
[46] C. Cutler and A. Vecchio. In: *Proceedings of the 2nd LISA Symosium*, ed. W.M.Folkner, (Am. Inst. Phys., 1998) p.95
[47] S. Shapiro. arXive: astro-ph/0304202
[48] A. Vecchio. *LISA observations of rapidly spinning massive black hole binary systems*, arXive: astro-ph/0304051
[49] G. Kauffmann and M. G. Haehnelt. MNRAS **311**, 576 (2000)
[50] B. S. Sathyaprakash and B. F. Schutz. *Templates for stellar mass black holes falling into supermassive black holes*, arXive: gr-qc/0301049
[51] L. P. Grishchuk. Zh. Eks. Teor. Fiz. **67**, 825 (1974) [Sov. Phys. JETP **40**, 409 (1975)]; Ann. NY Acad. Sci. **302**, 439 (1977)
[52] D. Dimitropoulos and L. P. Grishchuk. Int. J. Mod. Phys. D**11**, 259 (2002); S. Bose and L. P. Grishchuk. Phys. Rev. D **66**, 043529 (2002); L. P. Grishchuk. In: *2001: A Relativistic Spacetime Odyssey*, Eds. I.Ciufolini, D. Dominici, L. Lusanna. (World Scientific, 2003) p. 223 arXiv:gr-qc/0202072
[53] G. F. Smoot et al. Astroph. J. Lett. **396**, L1 (1992); C. L. Bennet et al. Astroph. J. Lett. **464**, L1 (1996); K. M. Gorski et al. Astroph. J. **464**, L1 (1996).
[54] D. Maino, A. J. Banday, C. Baccigalupi, F. Perrotta, K. M. Gorski. *Astrophysical components separation of COBE-DMR 4yr data with FastICA*, arXive:astro-ph/0303657
[55] M. J. Rees. Astroph. J**153**, L1 (1968); A. G. Polnarev. Sov. Astron. **29**, 607 (1985); M. Zaldarriaga and U. Seljak. Phys. Rev. D. **55**, 1830 (1997); M. Kamionkowski, A. Kosowsky, and A. Stebbins. Phys. Rev. D. **55**, 7368 (1997)
[56] H. V. Peiris et al. *First year WMAP observations: implications for inflation*, arXive:astro-ph/0302225
[57] http://www.si.edu/scienceandtechnology/astronomy
[58] G. Efstathiou. *Is the low CMB quadrupole a signature of spatial curvature?*, arXive:astro-ph/0303127
[59] S. L. Bridle, O. Lahav, J. P. Ostriker, and P. J. Steinhardt. *Precision cosmology ? Not just yet...*, arXive:astro-ph/0303180
[60] D. N. Spergel et al. *First year WMAP observations: determination of cosmological parameters*, arXive:astro-ph/0302207
[61] I am grateful to P. Mauskopf and D. Baskaran for composing Fig. 13.8.
[62] L. P. Grishchuk Phys. Rev. D**53**, 6784 (1996); L. P. Grishchuk and J. Martin Phys. Rev. D**56**, 1924 (1997)
[63] L. P. Grishchuk and Ya. B. Zeldovich. Asron. Zhurn. **55**, 209 (1978) [Sov. Astron. **22**, 125 (1978)]

[64] A. Kogut, G. Hinshaw, A. J. Banday, C. L. Bennet, K. Gorski, G. F. Smoot, and E. L. Wright. arXive astro-ph/9601060
[65] I am grateful to G. Smoot for the clarifying correspondence.
[66] J. C. R. Magueijo. Cosmic confusion, arXive: astro-ph/9412096
[67] I am grateful to S. Weinberg for the permission to quote his e-mail message of 25 Feb 2003

Index

2dF Galaxy Redshift Survey   181–182, 211–227
- distribution of redshifts   217–218
- fibre positioning   212
- large-scale distribution of galaxies   218–219
- power spectrum   224–225
- sky coverage   214–215
- spectroscopy   215–216
- target galaxies   213

Active Galactic Nuclei, accretion disc puzzle   248, 250–251
- and starbursts   245–247
- central black hole mass   235–236, 240–242
- classification 'zoo'   231–232
- discovery   231
- evidence for supermassive black holes   235–240
- evolution   244–247
- fuelling of central engine   245, 249–250
- key observations   236–238
- luminosities   241–242
- mass accretion rates   242–243
- obscuring torus   232–233
- selection effects in studies   247–248
- standard model   233–235
Advection Dominated Accretion Flows   248–250
astrometry missions, future   80
Atacama Large Millimetre Array   165

BeppoSax satellite   267–270
binary star statistics   6–7
BL Lac objects   242–243
black holes,
- in galactic nuclei, relation between stellar mass and bulge mass   149
- in luminous galaxies, origin   147–150
brown dwarf desert   11, 16, 31–33
brown dwarfs   5–6
Burst And Transient Source Experiment (BATSE)   264–266

cataclysmic variables, 'period gap'   51
- alternation between high and low transfer rates   53–54
- mass-transfer rates   51–53
- maximum period   50
- minimum period   54–56
- orbital-period distribution   48–49, 54–56, 57–58
- range in mass-transfer rates   50–51, 56–57
- range in secondary star masses   56–57
Cepheid variables, as distance indicator   65–67
- period-luminosity relationship   66
Chandra X-ray observatory   162–163
close binary stars, evolution   47–58
collapsars, as model for GRB   275
Compton Gamma-Ray Observatory (CGRO)   264–266
Convection Dominated Accretion Flows   249
cosmic distance scale   61–81
Cosmic Microwave Background   89–107
- angular scale of fluctuations   98
- anisotropy in temperature   90, 91, 115
- balloon-borne experiments   92, 102
- blackbody temperature   89

312 Index

- BOOMARANG data   104
- correlation of temperature and polarization   105
- data analysis   103
- fluctuations   90, 99–100
- ground-based experiments   92, 102–103
- MAXIMA data   104
- power spectrum   93–94, 99
- satellite experiments   91
- WMAP data   104–105, 106, 142

cosmological 'distance ladder'   61–62, 63–73
Cosmological Constant, the   227
cosmological parameters   227
- power spectra   97

'Dark Ages'   139
dark energy   113–114, 133
- candidates   125–127
- nature   123
- non-baryonic component   125, 133
- non-supersymmetric candidates   130–132
deceleration parameter 112
distance determination, problems   61
Doppler wobble, see Radial velocity measurements

early Universe, adiabatic perturbations   95
- epoch of recombination   139–140
- epoch of reionisation   140–144
- isocurvature perturbations   95
- physical processes   94–102
eclipsing binaries, as distance indicator   74, 81
exoplanets, see extrasolar planets
extrasolar planet, transiting system   9
extrasolar planets   8–13, 21–42
- 51 Pegasi   9, 21
- 51 Pegasi class   29
- 51 Pegasi class, metallicity of parent stars   36–37
- atmospheres   37–38
- detection by astrometric measurement   24
- detection by radial velocity measurement   23–24

- discovery   22–27
- future searches   42
- gravitational microlensing   26
- host stars   12–14
- mass distribution   10, 11
- mass function   33–34
- masses   30
- metallicity of parent stars   13–14, 34–37
- orbital eccentricities   39–40
- orbital evolution   15–16, 39–40
- orbital radii   30
- parent stars   34–37
- period distribution   40–42
- properties   21–42
- transits   24–26

fibre-coupled integral-field systems   196–197
Friedmann-Lemaitre-Robertson-Walker model (FLRW model)   111–112
fundamental plane, as distance indicator   69

G dwarf spectroscopic binaries, mass function   11
Galactic Centre   238–240
galaxies, distant Lyman-α 167–170
- distant, optical selection   166–170
- dusty submillimetre   164–166
- formation   155
- haloes   158–159
- identification of clusters   221–222
- large-scale distribution in 2dF survey   218–219
-- in Sloan Survey   218–220
- maximum redshift   158, 160
- mergers   155–156
- power spectra in spatial distribution   223–226
- properties and spatial distribution   219–221
- very distant, Lyman-α emission line   170–175
- with $z \geq 5$ 175–176
- X-ray   162–164
Gamma-Ray Burst Coordinate Network   266–267

Gamma-Ray Bursts (GRB) 166, 255–277
- angular distribution 262
- astrophysical aspects 270–274
- cosmological effects 266
- detection of afterglows 267–270
- discovery 255
- duration distribution 258
- intensity distribution 262–264
- light curves 257
- physical processes 274–275
- remaining mysteries 276–277
- searches for counterparts 259–262
- spectral properties 257–259
globular clusters, black holes 148
gravitational lenses, as distance indicator 75–76, 79–80
- time delay measurements 79–80
Gravitational radiation, concept 281
Gravitational wave detectors, current status 287–291
Gravitational waves 281–307
- and inflation 302–304
- and quadrupole anisotropy 304–306
- and WMAP data 304–306
- high-frequency sources 291–295
- low frequency sources 295–296
- relic, detection 300–302
- relic, generation 296–299
- theory 282–287
GRB 9702298, optical afterglow 268–269
GRB 970508, distance measurement 269–270
GRB 990123, distance measurement 270
Greisen Zatsepin Kuzmin limit 133, 134
Gunn-Peterson effect 141

'hot Jupiters' 13

image-slicing 196
inflation, predictions 90
inflationary theory, problems 302–304
initial mass function 1, 3–4
integral-field spectroscopy 195–198

James Webb Space Telescope 142–143, 146, 150

L dwarfs 7
Lambda Cold Dark Matter cosmology 158–159
Large Magellanic Cloud, distance 76–79
last scattering 90
- surface 96, 98
LIGO interferometers 287–289
LIMPs 130–131
LISA interferometer 290–291
luminosity function 1–2
Lyman-α line, blue side cut-off 173–175

M dwarfs 7
magnetic braking, in close binary system 47–49
Main Sequence fitting, as distance indicator 64
mass density, of Universe 222
mass transfer, in close binary system 47
megamasers, as distance indicator 73–74, 80–81
mergers of compact objects, as model for GRB 275
microwave radiation, astrophysical sources 100–101
Mira variables, as distance indicator 65
multi-fibre spectroscopy 185, 189–190
multi-object spectroscopy, development 184–186
- multi-actuator devices 194
- pick-place multi-fibre systems 191–194
- with robots 190–195
multi-slit spectroscopy 186–188

neutralino 125, 126–130
- detection 127–130
neutrinos, three-fold mixing 132

optical spectroscopy, contribution to contemporary astrophysics 182–183
- future challenges 202

- high-resolution spectroscopy 201–202
- improvements in, gratings 199–200
- ingredients for progress 181–182
- optical coatings 200
- sky subtraction 200–201
- spectrograph design 198
- with extremely large telescopes 202–203

particle horizon 90
Planck Surveyor Satellite 106
planet formation 15–16
planet, definition 22
planetary mass companions, alternative hypothesis 9
planetary nebula luminosity function, as distance indicator 70–71
present-day mass function 1, 3–4
primordial density perturbations 296–299
protoplanetary discs 15

quintessence 119–120

radial velocity measurements, for detection of extrasolar planets 23–24
radio galaxies, at high redshift 161–162
redshift space distortion 222
redshift-distance relation, as distance indicator 72–73
Roche lobe, in close binary system 47
RR Lyrae variables, as distance indicator 65

Sagittarius A* 238–240
scale factor of the Universe, time evolution 112
Sloan Digital Sky Survey 141, 160, 211–227
- fibre positioning 212
- large-scale distribution of galaxies 218, 220

- power spectrum 225–226
- sky coverage 214–215
- spectroscopy 215–216
- target galaxies 213
Sloan photometric camera 213–214
SN1987A light echo, as distance indicator 73
SN1997ff 117–118
SNAP satellite 119, 122
sound horizon 97–98
'standard candles' 67, 113
Standard Model of cosmology 111
stars, first generation 144–147
substellar dwarfs in clusters and the field 4–5
Sunyaev-Zel'dovich effect 101, 106
- as distance indicator 74–75, 81
supernovae, as distance indicator 71–72
- Type Ia, high-redshift 114–115
supersymmetry 125–126
surface brightness fluctuations, as distance indicator 69–70, 81

T dwarfs 7
tip of red giant branch, as distance indicator 67–68
trigonometrical parallax, as distance indicator 63–64
Tully-Fisher method, as distance indicator 68–69

Universe, early, see early Universe
- energy densities 115–116
- mass density of 222

Wilkinson Microwave Anisotropy Probe (WMAP) 89
- results 93–94, 104–105, 142
WIMPs 124, 130

XMM-Newton 162
X-ray galaxies 162–164

Printing: Mercedes-Druck, Berlin
Binding: Stein+Lehmann, Berlin